Psychopharmacology Series 4

Transduction Mechanisms of Drug Stimuli

Editors

F. C. Colpaert R. L. Balster

With 57 Figures

Springer-Verlag Berlin Heidelberg New York
London Paris Tokyo

F. C. Colpaert, Ph. D.
FONDAX-Groupe de Recherches Servier
7 rue Ampère
F-92800 Puteaux
France

R. L. Balster, Ph. D.
Department of Pharmacology and Toxicology
Medical College of Virginia
Virginia Commonwealth University
Richmond, VA 23298
USA

The diagram on the cover shows a model for olfactory stimulus transduction. (See Fig. 3, p. 9)

ISBN-13: 978-3-642-73225-6 e-ISBN-13: 978-3-642-73223-2
DOI: 10.1007/978-3-642-73223-2

Softcover reprint of the hardcover 1st edition 1988

Typesetting, printing and binding: Brühlsche Universitätsdruckerei, Giessen
2125/3130-543210

Preface

The following papers were collected on the occasion of the Third International Meeting on Drug Discrimination and State Dependency, which was held in Antwerp and Beerse, Belgium, on 5 and 7 July 1986. The meeting was a satellite of the inaugural congress of the European Behavioural Pharmacology Society, which was founded on 6 July 1986 in Antwerp. This Third International Meeting was also the occasion on which the community of researchers on Drug Discrimination and State Dependency honored Drs. H. Barry III, D. Overton, and J. Rosecrans for their outstanding contributions in this field.

Authors were asked to discuss available data within the theme of Transduction Mechanisms of Drug Stimuli. This theme recognizes that drugs can act as stimuli, much in the same manner as external events do. It also focuses on the mechanisms by which these stimuli are transduced at different levels of analysis, e.g., behavioral, pharmacological, and molecular. Some of the chapters in this volume discuss the mechanisms of transduction of the discriminative effects of several important classes of drugs. Other chapters deal with the methods and research strategies by which these mechanisms can be analyzed. Collectively, the papers in this volume thus reflect the current status of knowledge in an area of behavioral pharmacology that has made relatively rapid progress in recent years.

The following companies have generously supported the Third International Meeting on Drug Discrimination and State Dependency: Abbott Laboratories, Bristol-Myers Company, Burroughs Wellcome Company, CIBA-GEIGY Corporation, Coulbourn Instruments Inc., E. I. du Pont de Nemours & Company, Pfizer Inc., Schering Corporation, Searle Research and Development, State Systems Inc., and the Upjohn Company. J. Willetts assisted with the preparation of the subject index.

F. C. Colpaert
R. L. Balster

Contents

Transduction

Molecular Mechanisms

Author Index

You will find the addresses at the beginning of the respective contribution

Transduction

Drugs as Chemical Stimuli

R. L. BALSTER

Contents

Abstract

Analogies can be drawn at many levels between psychoactive drug effects as discriminative stimuli and stimulus control by traditional exteroceptive stimuli. The most useful analogy may be between drug stimuli and the chemical senses. At the behavioral level, drug discrimination studies have already drawn heavily from behavioral science research on stimulus control. Training of a drug discrimination is operationally identical to training discriminations based on exteroceptive stimuli. Concepts from generalization studies and research with compound stimuli have also been used in drug discrimination research. At the cellular level, a parallel can also be drawn between the study of mechanisms for drug stimulus effects and research on olfactory transduction. One theory of olfactory transduction proposes receptor-mediated actions of odorants, with subsequent biochemical events involving the cyclic neucleotide cascade, that is very similar to theories of receptor-mediated drug action. There is a broad basis for cross-fertilization between the study of drug stimuli and behavioral and cellular studies in classical sensory systems.

1 Introduction

One of the major conceptual advances in behavioral pharmacology research was the appreciation that drug effects could subserve stimulus functions in stimulus-response learning theory. A major landmark in the development of this conception was the publication of the proceedings of a conference on The Stimulus Properties of Drugs (Thompson and Pickens 1971). Drugs have been demonstrated to

Department of Pharmacology and Toxicology, Medical College of Virginia, Virginia Commonwealth University, Richmond, VA 23298, USA

Psychopharmacology Series 4
Transduction Mechanisms of Drug Stimuli
Editors: Colpaert, Balster

perform all of the commonly studied stimulus functions, including serving as both conditioned and unconditioned stimuli in respondent conditioning paradigms and as discriminative and reinforcing stimuli in operant conditioning paradigms. The concept that drugs could serve as stimuli has had a tremendous heuristic value in behavioral pharmacology research and has helped to clarify our thinking about how learning can take place about the effects of drugs. The greatest impact of this has been on substance abuse research where drug-taking phenomena are so clearly a product of the abuser's drug history. Since most of the uses of psychoactive drugs that we as behavioral pharmacologists are most interested in involve repeated administration, learning theory is of particular importance in explaining how experience with drugs modifies their effects.

2 Drug Discrimination

2.1 Training

The stimulus function of drugs that this paper is concerned with is the discriminative stimulus function. Organisms can be readily trained to detect drug administration and use these stimulus effects to solve behavioral tasks where two or more response options are available. In its simplest form, a drug discrimination experiment is arranged in a completely analogous manner to a traditional discrimination learning experiment using more conventional exteroceptive stimuli such as lights or tones as discriminative stimuli (Schuster and Balster 1977; Stolerman 1984). Figure 1 illustrates a typical arrangement for conducting a drug discrimination experiment in laboratory animals using operant behavior. On successive days, animals are required to press levers in order to receive food pellets delivered into a food trough. On any given day, presses on only one of the levers are correct; presses on the other are incorrect and are not reinforced. Whether the left or right lever is correct is signaled only by whether or not the subject received a drug or placebo (vehicle) injection prior to the training session. In the illustration, the left lever has been designated as the drug lever. Because drug stimuli cannot be turned on and off as readily as lights and tones, it is usually necessary to alternate the drug-lever pairings on a day-to-day basis. After about 30 or so of these pairings where responding is reinforced only on the correct lever, as defined

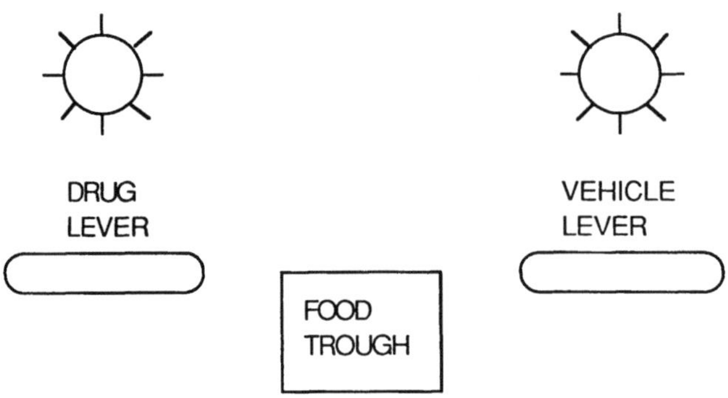

Fig. 1. Two-lever operant drug discrimination procedure

Table 1. Terminology used in discrimination learning

Discriminative stimulus	Compound stimuli
Reinforcing stimulus	Masking
Generalization	Overshadowing
Stimulus intensity	Blocking
Generalization gradient	Adaptation
Nondifferential and differential reinforcement	

by the presence or absence of drug, the drug and placebo injections come to exercise good stimulus control over lever choice. The subjects generally begin each session on the correct lever and, if tested in extinction or with responding on both levers reinforced, discriminative control of lever choice can readily be demonstrated. The only difference between this drug discrimination training procedure and discrimination procedures with lights and tones is the longer interval between changes in the stimulus conditions to allow drug effects to dissipate. To date, over 600 studies have been reported using this general approach to establish drug stimulus control [see Stolerman et al. (1982) and Stolerman and Shine (1985) for a bibliography through 1983]. Thus, when defined operationally, drugs can clearly function as discriminative stimuli.

Even a casual reading of the drug discrimination literature will reveal how the terminology from discrimination learning has been adopted by drug discrimination researchers. This ready availability of technical terminology to describe the operations in a drug discrimination experiment has greatly facilitated research in this field (Overton, this volume). Table 1 shows some of the terms that I have used already as well as some others that are often used in drug discrimination research which have analogous meaning to experimental psychologists and learning researchers.

2.2 Generalization

One of the most useful concepts derived from the stimulus control literature is that of generalization. Animals trained to discriminate a particular dose of a drug from vehicle will often respond as if they had received drug when tested with different doses of the drug or with different drugs. Tests with different doses could be viewed as testing with different stimulus intensities. Tests with different drugs could be considered a change along a qualitative dimension of the stimulus. The function relating different values of the stimulus to behavior is a *generalization gradient*. Figure 2 shows an example of a generalization gradient where drug dose is changed across a wide range. The proportion of responses on the lever which was correct on sessions following drug administration during training shows an orderly relation to drug dose or stimulus intensity. One of the interesting and hotly debated topics in both drug discrimination and stimulus control research is the true nature of such gradients. Are they quantitative or quantal functions and what is the form of the gradient when values outside the range of the training intensities are tested?

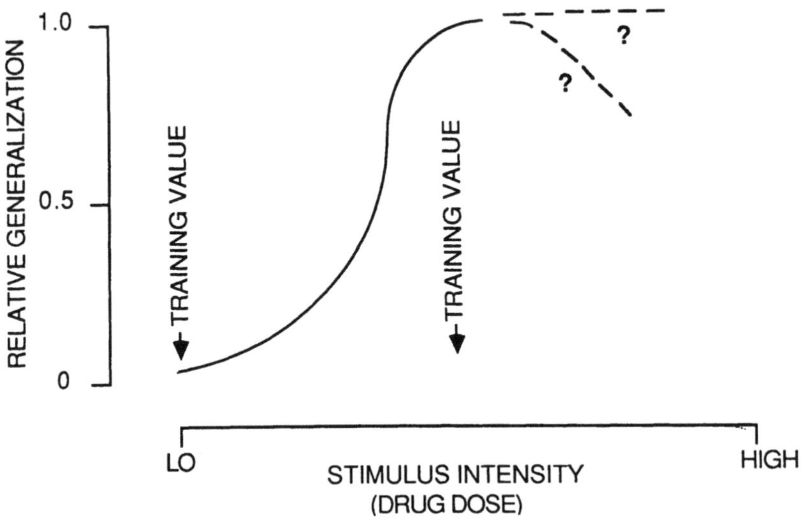

Fig. 2. Generalization gradient for drug stimuli. Shown is a model curve showing the proportion of responses on the drug lever as a function of test dose of the drug

2.3 Compound Stimuli

Another interesting and important area of research in stimulus control concerns when two or more discriminative stimuli are presented concurrently during training and/or testing, for example, a light and a tone. Such stimuli are called compound stimuli. Most drug stimuli could be thought of as compound stimuli. A pharmacologist would say that drug stimulus effects lack specificity of action, i.e., they have multiple cellular actions. Unlike with complex exteroceptive stimuli, it is more difficult to separate the component parts of compound drug stimuli. Research with compound stimuli has often centered on the phenomena of *masking*, *blocking*, and *overshadowing*. Research in some of these areas has begun using multiple drug stimuli or drug stimuli compounded with exteroceptive stimuli. The most extensive studies have been those of Järbe and his colleagues (Järbe and Johansson 1984; Järbe et al. 1983). The procedures developed by experimental psychologists for studying compound exteroceptive stimuli have been successfully applied by these investigators to the study of drug stimuli.

2.4 Adaptation

Another phenomenon from stimulus control research which may have a counterpart in drug discrimination research is that of adaptation. The corresponding phenomenon with drug stimuli may be tolerance. The analogy between adaptation and tolerance is not particularly clear since quite different experimental manipulations are typically made to study them. Adaptation refers to changes in sensitivity to the stimulus with duration of stimulation. Pharmacologists might call this acute tolerance or tachyphylaxis. Behavioral pharmacologists generally use tolerance to describe changes in sensitivity with repeated administrations, usually over a period of several days or weeks. The functional correspondence between stimulus adaptation and tolerance to the discriminative stimulus effects of drugs is unclear.

2.5 Differential Versus Nondifferential Reinforcement

One of the issues which has been of considerable interest in the study of stimulus control is that of the necessity of differential reinforcement training to establish stimulus control. In nondifferential reinforcement training, the subjects only receive training in the presence of one set of stimulus values. They do not have a history of differential reinforcement corresponding to different values of a stimulus. Nevertheless, when some aspect of the training environment is changed, responding is often affected and functions can be constructed relating the amount of behavioral disruption to the degree of stimulus change. These functions can again be referred to as generalization gradients. There is a similar phenomenon with drug stimuli where behavior acquired while subjects are treated with a drug is disrupted when the drug stimulus is removed; the reverse also can occur. This phenomenon has been referred to by Overton (1964) as *state dependent learning*. There is some debate whether or not state dependent learning is an example of drug stimulus control in the absence of discrimination training, but the experimental procedures for studying this phenomenon are completely analogous to the nondifferential reinforcement procedure used in exteroceptive stimulus control research.

2.6 Heuristic Implications

A major point to be made by the foregoing discussion is that drug effects can be studied as if they were stimuli. Conceived of in this way, drugs can enter into any of the stimulus-response operations which have been so extensively studied by behavioral scientists. This has been of significant heuristic value in behavioral pharmacology research. The research literature on stimulus control is a reservoir of ideas on how to go about studying possibly very important behavioral functions of drugs. A few more examples can be given of questions about drug discrimination which an experimental psychologist might ask.

1. Can errorless learning be established with drug stimuli and would such a discrimination display some of the same properties as errorless discrimination learning with exteroceptive stimuli?
2. What corresponds to peak shift in drug generalization testing?
3. How can thresholds be obtained for detection of drug stimuli?
4. Do the slopes of drug generalization gradients reflect a measure of strength of stimulus control as many believe them to be for exteroceptive stimuli? Since these are essentially dose-effect curves, pharmacologists might question this.
5. What corresponds to stimulus control by *concepts* such as longer, larger, and oddity when studying drug stimuli? Is multiple drug training an illustration of this?
6. What corresponds to just noticeable differences (JNDs) when working with drug stimuli?
7. Can a signal detection analysis be applied to drug stimuli?

These are just some of the possibly fruitful areas for further investigation generated by a stimulus control formulation of drug effects. How far the analogies between drug and traditional exteroceptive stimuli can be taken is not known, but it seems that we have just begun to realize the potential of this approach. Overton, in his chapter in this book, describes in considerably more detail similarities and differences in drug-produced and exteroceptive stimuli, and some of the strengths and limitations of this analogy.

3 Transduction Mechanisms for Chemical Stimuli

If drugs are stimuli, how are their stimulus effects produced? Can we carry the analogies between drug and exteroceptive stimuli at the behavioral level into research on their transduction mechanisms as well? This is the major theme of this book. Pharmacologists are familiar with asking questions about a drug's cellular mechanism of action. When the action of interest is a stimulus action, then a sensory physiologist might refer to this as a transduction mechanism. How are drug stimuli transduced?

Lest the analogy between stimulus transduction research and mechanism of action research in pharmacology seem too weak, it should be remembered that drugs are chemicals. Detection of drug stimuli could be viewed as a type of chemoreception. There are important similarities in receptor theory for drug action and current theories of olfactory and gustatory stimulus transduction. This point can be illustrated by a brief overview of the biochemistry of olfaction.

3.1 Olfactory Transduction

Olfactory stimuli act upon receptor cells, or olfactory rods, located in the olfactory mucosa. The dentritic endings of these cells form cilia which project into the mucosal layer which bathes the olfactory epithelium. Most theories of vertebrate olfactory transduction hold that chemicals activate these critical dentrites to produce graded potentials across the dentritic membrane, giving rise to action potentials in the olfactory receptor cell. How are these graded potentials produced? Figure 3 depicts one of the prominent theories for the biochemical events responsible for this odorant-cell interaction (Lancet 1986). It should look familiar to pharmacologists and neurochemists.

The odorant is believed to combine stereochemically with a receptor protein located in the lipid bilayer of the ciliary membrane. Thus, like with many drugs, the "actions" of chemical stimuli may be *receptor mediated.* In part because of the evident amplification of the sensory signal, this chemical-receptor interaction is thought to trigger a second messenger system. A good candidate for this second messenger system is the familiar cyclic nucleotide cascade (Lancet 1986). The binding of the odorant to the receptor protein has been hypothesized to activate a GTP-binding protein, or G-protein, which activates adenylate cyclase, producing the second messenger cyclic AMP. The cAMP may work directly upon ion channels or, as in synaptic transmission, may modify protein phosphorylation by activation of protein kinase C, ultimately affecting ion gating. This model of ol-

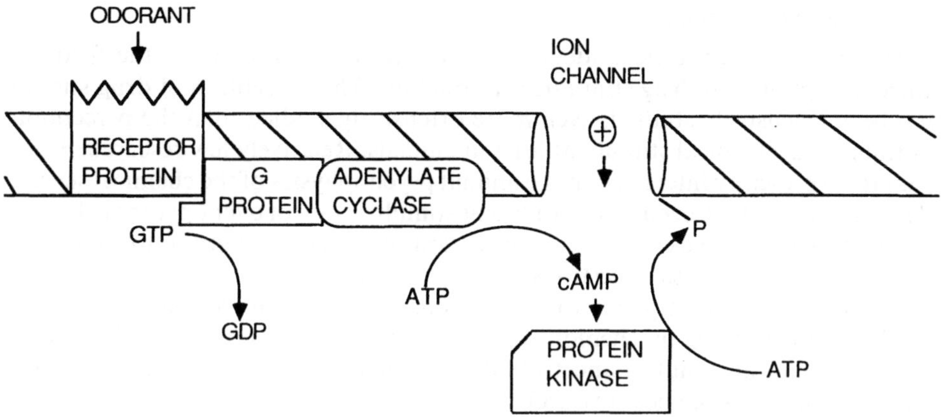

Fig. 3. Diagram of a model for olfactory stimulus transduction. (Modified from Lancet 1986)

factory stimulus transduction is essentially the same as a prevalent model of drug-receptor interaction prominent in neuropharmacology (Cooper et al. 1986). Thus, tranduction mechanisms for drug stimuli may have a great deal in common with transduction in the traditional chemical senses.

3.2 Technique for Studying Olfactory Tranduction

Some of the techniques currently being used to study olfactory transduction also have parallels in neuropharmacology. One approach to characterizing the olfactory receptor has been to use binding studies. They are commonly done using olfactory epithelium or in isolated cilia preparations. Price (1981) has described some of the special problems encountered in binding studies with odorants. Most odorants are highly lipophillic and result in large nonspecific binding. The ligands usually studied have a low affinity for the receptor, and binding to olfactory receptor proteins is often nonsaturable. Nevertheless, this approach continues to be exploited by biochemists working in this area.

Another biochemical approach has been to study the chemical composition of isolated cilia preparations (Lancet 1986). Many of the enzymes necessary for the signal transduction model discussed before such as adenylate cyclase, phosphodiesterase, and cAMP-dependent protein kinase have been found. Some of the critical receptor proteins are also beginning to be identified.

Biochemical studies of enzyme activity and of the production of putative second messengers in the presence of odorants have also been important. Historically even more important have been electrophysiological recordings of graded potentials and action potentials in olfactory receptor cells and their modulation by odorants and by biochemical manipulations of the transduction apparatus. All of these procedures have their counterparts in neurochemistry, neuropharmacology, and neurophysiology.

3.3 Olfactory Psychophysics

There has been very little evidence for cross-fertilization between the fields of chemoreception and drug stimulus transduction. The perception of drug stimuli has been extensively studied, even more extensively studied than the perception of traditional chemical stimuli, at least in animals. Relatively few researchers appear to be presently interested in the animal psychophysics of the chemical senses. Although there have been some studies of stimulus control over operant behavior by odors (e.g., Slotnick 1984), categorization of odors using generalization procedures and other basic questions in animal psychophysics with the chemical senses have not been very extensively investigated (Passe and Walker 1985). This has led those working on chemoreception in vitro to suggest the need for in vivo study of chemical stimuli (Lancet 1986), particularly in the species where the biochemical and physiological research is being done. Drug discrimination research may provide some important approaches for chemical psychophysics.

4 Subjective Effects

Finally, it may be worthwhile to speculate on what the study of transduction mechanisms for drug stimuli might tell us about normal human behavior. Receptor mechanisms for drug stimuli were not conveniently placed in our brains to mediate effects which behavioral pharmacologists would discover and exploit in the 1960s and 1970s using exogenous drugs. These transduction processes must perform a normal function in the brain and have a role in normal human behavior. What CNS functions and behaviors would these neural systems most likely be involved in? To provide some answers to these questions we should ask what drug effects in humans are most likely to correspond to discriminative stimulus effects as studied in animals.

There is good reason to believe that discriminative stimulus effects of drugs may be based entirely or in part upon their subjective effects (Schuster and Johanson, this volume). If discriminative effects are related to subjective effects, then it seems reasonable to hope that studies of the neural mechanisms for these effects may lead us toward an understanding of the neural mechanisms of some of the subjective experiences and mood states that are the basis of our perception of drug effects. Although subjective experience cannot be studied directly, psychoactive drugs give us powerful tools for directly manipulating subjective experience, mood, and feeling states and for studying their neural correlates. Useful procedures are already available for measuring and quantifying verbal reports of these uniquely personal experiences, and drug discrimination procedures for use with human subjects have been developed (Chait et al. 1984). Drug discrimination approaches in animals and in humans provide all the tools of experimental psychology and studies of transduction mechanisms bring in the rapidly expanding neurosciences. The impact that behavioral pharmacology has already had in development of drugs for the treatment of behavioral diseases and for understanding important aspects of substance-abusing behaviors may just be the beginning. Studies of the discriminative stimulus properties of drugs and their mechanisms of transduction can provide us important insights into basic brain-behavior relationships.

Acknowledgments. The preparation of this chapter was supported by research grants DA-01442 and DA-03112 from the National Institute on Drug Abuse.

References

Chait LD, Uhlenhuth EH, Johanson CE (1984) An experimental paradigm for studying the discriminative stimulus properties of drugs in humans. Psychopharmacology 82:272–274
Cooper JR, Bloom FE, Roth RH (1986) The biochemical basis of neuropharmacology, 5th edn. Oxford University Press, New York
Järbe TUC, Johansson B (1984) Interaction between drug discriminative stimuli and exteroceptive, sensory signals. Behav Neurosci 98:686–694
Järbe TUC, Svensson R, Laaksonen T (1983) Conditioning of a discriminative drug stimulus: overshadowing and blocking like procedures. Scand J Psychol 24:325–330
Lancet D (1986) Vertebrate olfactory reception. Ann Rev Neurosci 9:329–356
Overton DA (1964) State-dependent or "dissociated" learning produced with pentobarbital. J Comp Physiol Psychol 57:3–12
Passe DH, Walker JC (1985) Odor psychophysics in vertebrates. Neurosci Biobehav Rev 9:431–467
Price S (1981) Receptor proteins in vertebrate olfaction. In: Cagan RH, Kare MR (eds) Biochemistry of taste and olfaction. Academic, New York, pp 69–84
Schuster CR, Balster RL (1977) The discriminative stimulus properties of drugs. In: Thompson T, Dews PB (eds) Advances in behavioral pharmacology, vol 1. Appleton-Century-Crofts, New York, pp 85–138
Slotnick BM (1984) Olfactory stimulus control in the rat. Chemical Sens 9:157–165
Stolerman IP (1984) Internal stimulus effects of drugs. In: Sanger DJ, Blackman DE (eds) Aspects of psychopharmacology. Methuen, London, pp 57–83
Stolerman IP, Shine PJ (1985) Trends in drug discrimination research analyzed with a cross-indexed bibliography, 1982–1983. Psychopharmacology 86:1–11
Stolerman IP, Baldy RE, Shine PJ (1982) Drug discrimination procedure: a bibliography. In: Colpaert FC, Slangen JL (eds) Drug discrimination: applications in CNS pharmacology. Elsevier Biomedical, Amsterdam, pp 401–424
Thompson T, Pickens R (eds) (1981) Stimulus properties of drugs. Appleton-Century-Crofts, New York

Molecular Mechanisms

Site-Selective Serotonin Agonists as Discriminative Stimuli

R. A. GLENNON

Contents

Abstract

Various direct- and indirect-acting serotonin (5-HT) agonists serve as training drugs in tests of stimulus control of behavior; such agents include: 5-hydroxytryptophan, 5-methoxy-*N*,*N*-dimethyltryptamine, and fenfluramine. However, with the recent discovery of multiple populations of central 5-HT binding sites, the concept of site-selective serotonergic agents needs to be addressed. Certain 4-substituted 1-(2,5-dimethoxyphenyl)-2-aminopropanes such as DOM (4-methyl), DOB (4-bromo), and DOI (4-iodo) appear to be 5-HT2-selective agonists and serve as effective training drugs in rats. Stimulus generalization occurs among these agents regardless of which is used as the training drug, although stimulus generalization does not occur with 5-HT1A-selective agonists [e.g., 8-hydroxy-2-(di-*n*-propylamino)tetralin (8-OH DPAT)] or with 5-HT1B-selective agonists [e.g., 1-(3-trifluoromethylphenyl)piperazine (TFMPP)]. 8-OH DPAT and TFMPP also serve as training drugs; the 8-OH DPAT-stimulus generalizes to other 5-HT1A agonists, but not to 5-HT1B or 5-HT2 agonists, whereas the TFMPP-stimulus generalizes to other 5-HT1B agonists, but not to 5-HT1A or 5-HT2 agonists. Classical serotonin antagonists, most of which are rather selective for 5-HT2 sites, and 5-HT2-selective antagonists are able to block the stimulus effects of DOM, DOB, and DOI, but not those of 8-OH DPAT or TFMPP. The results of such studies reveal that, in rats, site-selective 5-HT agonists produce stimulus effects that are also selective; although generalization may occur with nonselective 5-HT agonists, animals trained to discriminate site-selective 5-HT agonists apparently do not recognize other 5-HT agonists that are selective for a different site. Animals trained to discriminate such agents from saline might be useful for the identification and/or investigation of novel site-selective agonists and antagonists (for example, the 8-OH DPAT-stimulus generalizes to members of a new class of anxiolytics that display high affinity for 5-HT1A binding sites), and might also aid in the overall understanding of central serotonergic mechanisms.

Department of Medicinal Chemistry, School of Pharmacy, Medical College of Virginia, Virginia Commonwealth University, Richmond, VA 23298, USA

Psychopharmacology Series 4
Transduction Mechanisms of Drug Stimuli
Editors: Colpaert, Balster

1 Introduction

There have been no reports on the use of serotonin (5-hydroxytryptamine; 5-HT) as a training drug in tests of stimulus control of behavior. It might be anticipated, because of the poor ability of serotonin to penetrate the blood-brain barrier and because 5-HT is rapidly metabolized in vivo, that a 5-HT cue would be very difficult for the animals to learn. Indeed, there are preliminary data (Colpaert, personal communication) suggesting that acquisition of a 5-HT-stimulus was difficult and that performance was unreliable. On the other hand, various direct- and indirect-acting serotonergic agonists have been used as training drugs. In the past, evidence was accumulated supporting serotonergic involvement in the mechanism of action of these agents as discriminative stimuli. However, now, with the discovery of multiple populations of central 5-HT binding sites, studies are being focussed (a) on determining which 5-HT sites might be responsible for the stimulus effects produced by these agents and (b) on developing novel site-selective serotonergic agents that might serve as site-selective training drugs. Drug discrimination studies using such agents as training drugs can help elucidate the mechanism of action of known drugs (for which mechanisms may be unknown), can aid in the development of newer and more selective 5-HT agonists and antagonists that might be of potential therapeutic value, and can contribute to our understanding of serotonergic mechanisms in general.

The 1970s saw the advent of radioligand binding studies for the investigation of central neutransmitter binding sites, tritiated 5-HT was introduced as a radioligand in the mid-1970s. Shortly thereafter, Peroutka and Snyder (1979) provided evidence for the existence of two distinct populations of central 5-HT binding sites: 5-HT1 and 5-HT2; 5-HT1 sites were labeled with [^{3}H]5-HT whereas 5-HT2 sites were labeled (frontal cortex) with [^{3}H]spiperone (SPIP). Newer radioligands have now been developed for both sites; for example, [^{3}H]ketanserin (KET) is perhaps the most widely used radioligand for labeling 5-HT2 sites. Pedigo et al. (1981) later found that tritiated SPIP possesses a high affinity for a subpopulation of 5-HT1 sites (i.e., 5-HT1A sites); 5-HT1 sites with low affinity for this ligand were termed 5-HT1B sites. More recently, tritiated 8-hydroxy-2-(di-*n*-propylamino)tetralin (8-OH DPAT) and iodocyanopindolol have been introduced as radioligands for labeling 5-HT1A and 5-HT1B sites, respectively. (Evidence suggests the possible existence of other populations and/or subpopulations of sites; these have been less well investigated and will not be discussed herein.) Although, to date, there are no site-specific agents, Table 1 lists some site-selective 5-HT agonists. It should be noted that these designations have been made on the basis of available radioligand binding data and may eventually be in need of modification as new radioligands become available and/or as new sites are identified. It should also be noted that certain agents, e.g., lysergic acid diethylamide (LSD) and 5-methoxy-*N*,*N*-dimethyltryptamine (5-OMe DMT), bind with high affinity to more than one population of sites and, for this reason, might be considered as being nonselective. With respect to antagonists, ketanserin, pirenperone, ritanserin, and, perhaps, pizotyline (pizotifen; BC-105) are quite selective 5-HT2 antagonists, whereas most other 5-HT antagonists are considerably less selective. Currently, there are no generally accepted 5-HT1-selective antagonists [although

Table 1. Selected site-selective serotonin agonists[a]

5-HT1A	8-OH DPAT	8-Hydroxy-2-(di-*n*-propylamino)tetralin HBr
	PAPP	1-[(3-Trifluoromethyl)phenyl]-4-[2-(4-aminophenyl)ethyl]-piperazine HCl
	TVX Q 7821[b]	2-[4-(4-(2-Pyrimidinyl)-1-piperazinyl)butyl]-1,2-bezisothizol-3(2H)-one 1,1-dioxide HCl
5-HT1B	TFMPP	1-[(3-Trifluoromethyl)phenyl]piperazine HCl
	mCPP	1-(3-Chlorophenyl)piperazine hydrochloride
	RU 24969	5-Methoxy-3-(1,2,3,6-tetrahydro-4-pridinyl)-1H-indole succinate
5-HT2	DOM	1-(2,5-Dimethoxy-4-methylphenyl)-2-aminopropane hydrochloride
	DOB	1-(2,5-Dimethoxy-4-bromophenyl)-2-aminopropane hydrochloride

[a] Categorized on the basis of available binding data.
[b] May (?) be a mixed agonist/antagonist. TVX Q 7821 is also known as ipsapirone.

there has been a very recent report (Nelson and Taylor 1986) of a potential 5-HT1A-selective antagonist]. For recent accounts of central 5-HT binding sites and site-selective agonists and antagonists, see reviews by Leysen (1985) and Glennon (1986a).

2 Selectivity of Serotonergic Agents Based on Discrimination Studies

Several serotonergic agents have been used in drug discrimination studies; this section deals with these agents and with attempts to understand which 5-HT sites might be involved in their stimulus mechanisms.

2.1 5-Hydroxytryptophan

A likely candidate for the investigation of serotonergic stimuli is the 5-HT precursor L-5-hydroxytryptophan (5-HTP). 5-HTP produces considerable behavioral disruption in animals, but this is believed to be a peripheral effect that can be alleviated by the coadministration of a peripheral decarboxylase inhibitor. Table 2 lists several studies where 5-HTP has been used as a training drug. The effect of low doses of 5-HTP can be potentiated by administration of the 5-HT uptake inhibitor fluoxetine and the 5-HTP stimulus can be antagonized by the centrally acting decarboxylase inhibitor NSD-105. The 5-HTP stimulus generalizes to the 5-HT releasing agent fenfluramine (Table 3), supporting the mechanistic involvement of 5-HT. Surprisingly, however, most 5-HT antagonists are ineffective in attenuating the 5-HTP stimulus (Table 3). There are conflicting reports regarding the ability of pizotyline to block the 5-HTP stimulus. Certain antagonists such as KET and pirenperone (PIR) are quite selective for 5-HT2 versus 5-HT1 sites; thus, it is possible that the stimulus is 5-HT1 mediated. However, some of the

Table 2. Studies using L-5-hydroxytryptophan as a discriminative stimulus[a]

	Schedule	Route	PSII	Dose (mg/kg)	Author's conclusion
Barrett et al. (1982)	VI-20	s.c.	30 min	30	5-HT-mediated stimulus
Friedman et al. (1983)	VI-30	s.c.	30 min	35	Unique 5-HT receptor subtype
Haenlein et al. (1983)	VI-20	s.c.	30 min	30	–
Cunningham et al. (1985a)	FR-20	i.p.	30 min	50	Mixed 5-HT1/5-HT2-mediated stimulus
Moser and Redfern (1985)[b]	(T-maze)	i.p.	30 min	35	5-HT1-mediated stimulus

[a] All studies employed male rats. A decarboxylase inhibitor (with varying dose and presession injection interval) was used in each case.
[b] Stereochemistry of 5-HTP not specified.

Table 3. Properties of the 5-hydroxytryptophan stimulus

Generalized to	*Not generalized to*
Fenfluramine[a]	Amphetamine[a]
Quipazine[c]	Quipazine[d, e]
LSD[c]	LSD[e]
5-OMe DMT[d]	
RU 24969[d]	
Antagonized by	*Not antagonized by*
Pizotyline[b]	Pizotyline[c]
	Ketanserin[c]
	Metergoline[a]
	Methiotepin[a]
	Methysergide[a]
Partially antagonized by	Mianserin[b]
Cyproheptadine[a]	Pirenperone[d]
	Xylamidine[c]

[a] Barrett et al. (1982).
[b] Friedman et al. (1983).
[c] Cunningham et al. (1985a).
[d] Moser and Redfern (1985).
[e] Friedman et al. (1985).

other antagonists are much less selective for 5-HT2 sites and have been demonstrated to antagonize various 5-HT1-mediated events (Arvidsson et al. 1986); from this standpoint, it is difficult to explain this lack of antagonism. The 5-HTP stimulus also generalizes to the nonselective 5-HT agonist 5-OMe DMT and the 5-HT1B-selective agonist RU 24969. Whereas Moser and Redfern (1985) propose that the 5-HTP stimulus is 5-HT1 mediated, Cunningham et al. (1985a) suggest that it might involve a mixed 5-HT1/5-HT2 mechanism.

2.2 5-Methoxy-*N,N*-Dimethyltryptamine

5-Methoxy-*N,N*-dimethyltryptamine (5-OMe DMT) serves as an effective discriminative stimulus in rats at doses of 1.0, 1.5, and 3.0 mg/kg; for a review, see Glennon et al. (1982). Binding data would suggest that 5-OMe DMT is a fairly nonselective serotonergic agent (Tricklebank et al. 1985; Lyon et al. 1986b); this is borne out by the results of various stimulus generalization studies. The 5-OMe DMT stimulus generalizes to DOM (but not to DOB), LSD, 8-OH DPAT, and TVX Q 7821, and partially generalizes to TFMPP and quipazine (Glennon et al. 1982; Spencer et al. 1984; Ortmann 1985). The mechanistic aspects of the stimulus may be dose related; pizotyline and cyproheptadine are effective antagonists of the stimulus when the training dose of 5-OMe DMT was 1.5 mg/kg but are without effect at a training dose of 3.0 mg/kg (Young et al. 1983). As with the 5-HTP stimulus, metergoline and methysergide were ineffective as antagonists at both training doses (Glennon et al. 1983a). It appears that 5-OMe DMT most probably produces its stimulus effects via a mixed serotonergic mechanism and that the degree of mechanistic involvement of the various 5-HT sites may be dose related.

2.3 Fenfluramine

An indirect-acting serotonergic agent that has received some, though limited, attention is fenfluramine. Both racemic (Goudie 1977; White and Appel 1981) and (−)-fenfluramine (McElroy and Feldman 1984) have been used as training drugs. The fenfluramine stimulus generalizes to quipazine and 2-(1-piperazinyl)-6-chloropyrazine (MK-212), partially generalizes to LSD, but does not generalize with (+)-amphetamine (White and Appel 1981).

Richard Young (personal communication) has also shown that the fenfluramine stimulus generalizes to quipazine, MK-212, 5-OMe DMT, and mCPP, partially generalizes to TFMPP, but does not generalize to the 5-HT1A-selective agent 8-OH DPAT (nor to buspirone or TVX Q 7821). The fenfluramine stimulus can be antagonized by the 5-HT antagonists cyproheptadine, methysergide, cinanserin, and methiothepin, but not by the dopamine antagonist haloperidol (White and Appel 1981; McElroy and Feldmann 1984). Although too few results are currently available to be able to draw any conclusions, they do suggest that fenfluramine probably produces a nonselective stimulus.

2.4 Lysergic Acid Diethylamide

The potent hallucinogen (+)-lysergic acid diethylamide LSD has seen extensive application as a training drug in drug discrimination studies; see Glennon et al. (1983a) for a recent review. Though LSD binds nearly equally well at 5-HT1 and 5-HT2 sites, the stimulus effects produced by typical doses of LSD appear to be manifested, for the most part, by a 5-HT2 mechanism. Indeed, Glennon et al. (1984b) have proposed, based on the results of radioligand binding studies, and stimulus generalization and antagonism studies, that certain phenalkylamine, indolealkylamine (including LSD), and related hallucinogenic agents produce their stimulus (and their human hallucinogenic) effects primarily via a 5-HT2-mediated mechanism. With respect to the LSD stimulus, however, there are data to suggest that the mechanism, as appears to be the case with 5-OMe DMT, may be dose related. For example, pizotyline was able to block the stimulus effects of 0.06 mg/kg, but not those of a particularly high dose (i.e., 0.18 mg/kg), of LSD (Young et al. 1983). Also, using rats trained to discriminate 0.16 mg/kg LSD from saline, Colpaert and Janssen (1983) have suggested (on the basis of a curvilinear antagonism curve with PIR) that the stimulus effects of this dose of LSD might involve multiple 5-HT sites. White and Appel (1982) have also reported some interesting differences in the stimulus effects produced by different (i.e., 0.02, 0.08, and 0.32 mg/kg) doses of LSD. Thus, even though LSD has been the subject of numerous drug discrimination studies, additional studies seem warranted.

2.5 Quipazine

Quipazine is another example of an agent that binds to several populations of 5-HT binding sites but that apparently produces its stimulus effect predominantly via a single mechanism. Table 4 lists studies using quipazine as a training drug and Table 5 presents the results of various stimulus generalization and stimulus antagonism studies. The quipazine stimulus generalizes to LSD and 5-OMe DMT. Conversely, an LSD stimulus generalizes to quipazine; however, stimulus generalization does not occur upon administration of quipazine to rats trained to discriminate either 1.0 or 3.0 mg/kg 5-OMe DMT from saline (i.e., partial generalization was obtained in both instances). Stimulus generalization studies using some of the more selective 5-HT agonists have not yet been reported. The quipazine stimulus can be antagonized by most 5-HT antagonists, and by agents that possess a high affinity both for 5-HT and dopamine binding sites, but not by dopamine (or other neurotransmitter) antagonists. Based on the ability of the 5-HT2 antagonist PIR to block DOM-stimulus generalization to quipazine, it was suggested that the stimulus effects of quipazine might involve a 5-HT2 mechanism (Glennon et al. 1983b). Yamamoto et al. (1983) independently demonstrated that PIR could antagonize the stimulus effects of quipazine in pigeons, and Friedman et al. (1984) later showed that the ability of various 5-HT2 antagonists to block the stimulus effects of quipazine in rats was significantly correlated

Table 4. Studies using quipazine as a discriminative stimulus

Study	Species/sex	Schedule	Route	PSII	Dose (mg/kg)
1. White et al. (1977)	Rats	FR-32	i.p.	30 min	2.5
2. Appel et al. (1978)	Rats (M)	FR-32	i.p.	30 min	1.0
3. Winter (1979)	Rats	FR-10	i.p.	15 min	3.0
4. White et al. (1979)	Rats (M)	FR-32	i.p.	30 min	1.0
5. Glennon et al. (1981)	Rats (M)	VI-15	i.p.	15 min	1.0
6. Glennon et al. (1982)	Rats (M)	VI-15	i.p.	15 min	1.0
7. Yamamoto et al. (1983)	Pigeons	–	i.m.	–	1.0
8. Haenlein et al. (1983)	Rats (M)	VI-20	s.c.	30 min	3.0
9. Friedman et al. (1984)	Rats (M)	VI-20	s.c.	30 min	2.0
10. Friedman et al. (1985)	Rats (M)	VI-30	s.c.	30 min	2.5

Table 5. Properties of the quipazine stimulus[a]

Generalized to		*Not generalized to*
LSD (1, 2, 3)[a]		Apomorphine (1, 4)
Psilocybin (2)		Amphetamine (1, 4, 5, 7)
Fenfluramine (4)		Methysergide (8)
5-OMe DMT (6)		
Antagonized by		*Not antagonized by*
Cinanserin (8)	Metergoline (8)	Butaclamol (3)
Clozapine (9)	Methiothepin (1, 2, 4)	Fluphenazine (1)
Cyproheptadine (1, 2, 4, 8)	Mianserin (8)	Haloperidol (1)
	Pirenperone (7, 8)	Naltrexone (7)
Desipramine (8)	Pizotyline (8)	Phentolamine (3, 4)
Ketanserin (8)		Propranolol (4)
Loxapine (8)	Methysergide (1, 2, 4, 8)[b]	Xylamidine (2, 4)

[a] Numbers in parenthesis refer to the studies in Table 4.
[b] Partial antagonism.

with their affinities for 5-HT2 sites. Indeed, using a new radioligand that presumably labels the high-affinity state of 5-HT2 binding sites, Lyon et al. (1986b) have recently found that quipazine possesses a considerably higher affinity for 5-HT2 sites ($K_i = 17$ nM) than previously thought. Thus, here is an example of the application of drug discrimination methodology to study the mechanism of action of a serotonergic agent whereby a mechanism was proposed that was initially inconsistent with binding data, but for which new binding data now support the results of the discrimination study.

3 Discrimination Studies Using Site-Selective Serotonergic Agents

Several site-selective serotonergic agents have now been employed as training drugs in drug discrimination studies (Table 6). Most of these studies used agents that had been defined as possessing site selectivity on the basis of radioligand binding data. There is one interesting group of agents that was proposed to be site-selective on the basis of drug discrimination studies – a proposal that is now supported by binding data. These agents (i.e., 5-HT2-selective agonists) will be discussed first.

3.1 5-HT2-Selective Agonists

Hallucinogenic agents have long been thought to act via a serotonergic mechanism. Various drug discrimination studies employing hallucinogens (e.g., LSD, DOM, mescaline) as training drugs arrived at similar conclusions. Shortly after the finding that KET and PIR constituted 5-HT2-selective antagonists, Colpaert et al. (1982) reported that PIR was a potent antagonist of the stimulus effects of LSD in rats. Subsequently, it was demonstrated that KET and PIR potently antagonize the stimulus properties of the phenalkyamine hallucinogen DOM and DOM-stimulus generalization to LSD and mescaline (Glennon et al. 1983b). This provided the first evidence that hallucinogens might be acting via a 5-HT2 mech-

Table 6. Site-selective serotonergic agonists as training drugs

	Schedule	Route	PSII	Dose
8-OH DPAT (5-HT1A)				
Glennon and McKenney (1985)	VI-15	i.p.	15 min	0.2
Cunningham et al. (1985b)	–	i.p.	–	0.4
Glennon (1986c)	VI-15	i.p.	15 min	0.2
Tricklebank[a] (1986)	–	s.c.	–	0.05
TFMPP (5-HT1B)				
Glennon et al. (1984a)	VI-15	i.p.	15 min	1.0
Glennon and McKenney (1985)	VI-15	i.p.	15 min	0.5
McKenney and Glennon (1986)	VI-15	i.p.	15 min	1.0
Lyon et al. (1986a)	VI-15	i.p.	15 min	0.5
Appel et al[a] (1986)	FR-20	i.p.	15 min	0.8
Schechter et al.[a] (1986)	FR-10	i.p.	15 min	1.0
DOM (5-HT2)				
Silverman and Ho (1980)	FR/VI	i.p.	15 min	1.5
Young et al. (1981)	VI-15	i.p.	15 min	1.0
Glennon et al.[b] (1983a)	VI-15	i.p.	15 min	1.0
DOI (5-HT2)				
Glennon and McKenney (1985)	VI-15	i.p.	15 min	0.5
Glennon (1986b)	VI-15	i.p.	15 min	0.5
R(−)DOB (5-HT2)				
Glennon[c] (1986)	VI-15	i.p.	15 min	0.2

[a] Personal communication.
[b] Review article.
[c] Unpublished data.

anism, and suggested that certain phenalkylamines might constitute selective 5-HT2 agonists (examples of which had not been previously reported). Radioligand binding studies have now shown that, indeed, appropriately substituted phenalkylamines possess up to a several thousand-fold selectivity for 5-HT2 versus 5-HT1 sites (Shannon et al. 1984; Glennon et al. 1984b; Titeler et al. 1985). Ketanserin and pirenperone, though selective for 5-HT2 versus 5-HT1 sites, are not necessarily specific for serotonin sites. That is, these agents also display affinity for other neurotransmitter binding sites (e.g., dopaminergic, histaminergic, adrenergic) (Leysen 1985). Silverman and Ho (1980) showed that dopamine antagonists are ineffective in blocking the DOM stimulus. We have obtained similar results with the dopamine antagonist haloperidol, and have further demonstrated that histaminergic and adrenergic antagonists are also without effect. We have also shown that various antagonists with different binding profiles, but that share a common 5-HT2 component, can antagonize the stimulus effects of DOM (Glennon 1985; Glennon and Hauck 1985). Examples of such agents include: pizotyline, cyproheptadine, ritanserin, LY 53857, tetrahydrotrazodone, and dihydroflutroline. Tests of stimulus generalization (Table 7) showed that the DOM stimulus generalizes to quipazine and fenfluramine, but not to 8-OH DPAT, TFMPP, mCPP, or RU 24969. All of these data provided evidence that DOM might be producing its stimulus effects via 5-HT2 agonism. Finally, we have dem-

Table 7. Properties of the DOM stimulus

Generalized to	*Not generalized to*	*Antagonized by*
Quipazine	8-OH DPAT	Pizotyline
Fenfluramine	8-OMe DPAT	Ketanserin
5-OMe DMT	TFMPP	Pirenperone
LSD	mCPP	Ritanserin
PAAs[a]	RU 24696	LY 53857
IAAs[a]		Dihydroflutroline
		Tetrahydrotrazodone
		1-Naphthylpiperazine

Data are from Glennon (1985), Glennon and Hauck (1985), and Glennon et al. (1983a).
[a] The DOM stimulus generalizes to a number of phenalkylamines (PAAs) and indolealkylamines (IAAs); see Glennon et al. (1983a) for a review.

onstrated that the potencies of 22 agents (i.e., ED_{50} values derived from stimulus generalization studies conducted using DOM as the training drug) are significantly correlated ($r>0.9$) with the affinities (K_i values) of these agents at [^{3}H]KET-labeled 5-HT2 sites (Glennon et al. 1984b). The use of DOM as a discriminative stimulus, and further discussion of the above studies, have been described in detail in several reviews (see Glennon et al. 1983a, 1986a; Glennon 1985).

Using [^{3}H]KET as the radioligand, competition studies revealed that DOB and DOI, the 4-bromo and 4-iodo analogs of DOM, possess a higher affinity and selectivity than DOM for 5-HT2 binding sites. Binding is stereoselective in that both optical isomers bind at 5-HT2 sites, but, in both cases, the affinity of the R(−)-isomers are several times that of their S(+)-enantiomers. As a consequence, DOI was used as a training drug in a drug discrimination study with the anticipation that it might provide a more 5-HT2-selective stimulus (Glennon and McKenney 1985; Glennon 1986b). Although our investigations with this agent have not yet been completed, we have found that the DOI stimulus (training dose, 0.5 mg/kg; $ED/_{50}$, 0.16 mg/kg) generalizes to DOM, DOB, R(−)-DOI, LSD, and to the less potent S(+)-DOI, but not to 8-OH DPAT or TFMPP. In addition, the DOI stimulus was potently antagonized by KET. In contrast to the results obtained using the somewhat less selective DOM as the training drug, the DOI stimulus did not generalize to fenfluramine. In the DOI-trained animals, fenfluramine produced saline-appropriate responding at doses of 0.3 and 0.6 mg/kg, 48% and 44% drug-appropriate responding at 0.7 ($n=6$) and 0.75 ($n=4$) mg/kg respectively, and disruption of behavior at doses of 0.8 mg/kg and 1.0 mg/kg. Variation of the presession injection interval from 5 to as much as 30 min for a dose of 0.7 mg/kg fenfluramine did not result in greater than 48% DOI-appropriate responding. A potential disadvantage of DOI as a training drug is the possibility that it might undergo a rapid deiodination in vivo to afford 1-(2,5-dimethoxyphenyl)-2-aminopropane (i.e., 2,5-DMA); 2,5-DMA, though active, is less "potent" and selective than DOI. The relative potencies obtained in the abovementioned stimulus generalization studies argue against this possibility. Furthermore, instead of being a short-acting agent (which could reflect a rapid rate of metabolism to 2,5-DMA), DOI appears to produce a stimulus that is of a fairly long duration (Fig. 1).

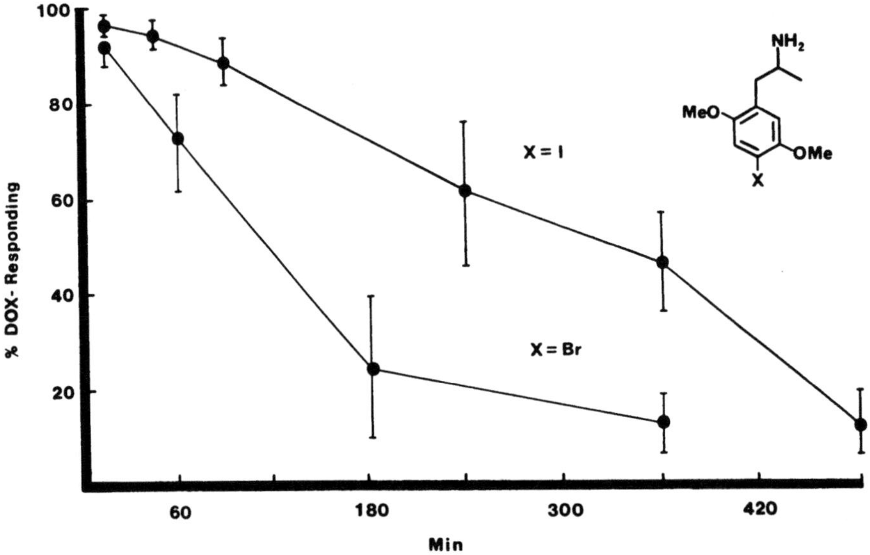

Fig. 1. Time course of action for the training dose of racemic DOI (0.5 mg/kg) in DOI-trained animals (X=I) and for R(−)-DOB (0.2 mg/kg) in R(−)-DOB-trained animals (X=Br) (n=4–6 rats at each dose). *% DOX-responding*, number of responses made on the drug-appropriate lever as a percentage of the total responses; *Min* is the presession injection interval (*psii*) in minutes. (Animals were trained using a 15-min *psii*)

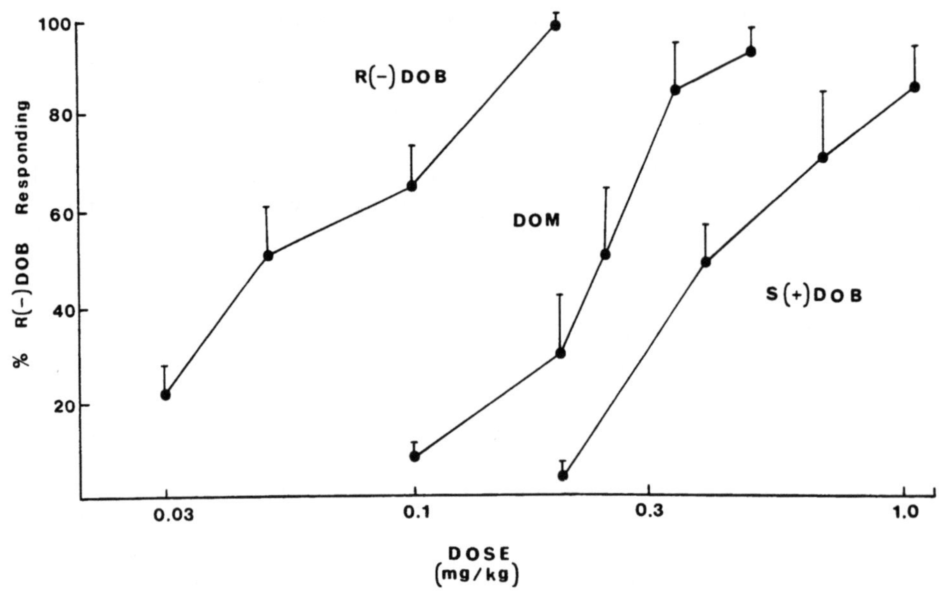

Fig. 2. Tests of stimulus generalization using animals trained to discriminate R(−)-DOB (0.2 mg/kg) from saline (n=four to five rats at each dose; *psii*=15 min)

While these studies were in progress, we were investigating the structure-activity relationships of DOB. A series of DOB analogs, systematically modified to provide information on the contribution of each of the pendant substituent groups, was synthesized and then examined in the radioligand binding assay and in the drug discrimination paradigm using DOM as the training drug (Glennon et al. 1986b). The results of this investigation revealed that R(−)-DOB was the most potent/selective of these agents; molecular modification served only to reduce potency and/or selectivity. Subsequently, [^{3}H]DOB has been prepared as a new radioligand for labeling 5-HT2 sites, and R(−)-DOB has been used as a training drug in drug discrimination studies. Here too, the discrimination studies are still in progress. However, preliminary data show that the R(−)-DOB stimulus (training dose 0.2 mg/kg; ED_{50}, 0.05 mg/kg) generalizes to DOM (ED_{50}, 0.24 mg/kg), to LSD (ED_{50}, 0.04 mg/kg), and to the less potent S(+)-DOB (ED_{50}, 0.56 mg/kg) (Fig. 2), but not to 8-OH DPAT or TFMPP. The duration of action of R(−)-DOB is considerably shorter than that of DOI (Fig. 1) and is, in fact, slightly shorter than that of DOM (Young et al. 1981). The DOB stimulus can be completely antagonized by the 5-HT2 antagonist pirenperone (ED_{50}, 0.03 mg/kg).

3.2 5-HT1A-Selective Agonists

Arvidsson et al. (1981) reported the synthesis of the novel serotonin agonist 8-OH DPAT, and shortly thereafter Middlemiss and Fozard (1983) found that this agent binds selectively at 5-HT1A sites. Gozlan et al (1983) introduced [^{3}H]8-OH DPAT and this tritiated ligand is now commercially available for labeling 5-HT1A sites. 8-OH DPAT serves as an effective training drug in rats at doses of 0.2 (Glennon and McKenney 1985) and 0.4 (Cunningham et al. 1985b) mg/kg. More recently, Tricklebank has used a training dose of 50 μg/kg (personal communication). The 8-OH DPAT stimulus does not generalize to the 5-HT1B agonists TFMPP and RU 24969, to the 5-HT2 agonists DOM and DOI, or to other less-selective agonists such as 5-OMe DMT (Table 8). Serotonin (and dopamine)

Table 8. Properties of the 8-OH DPAT stimulus

Generalized to	*Not generalized to*	*Not antagonized by*
8-OMe DPAT[a]	TFMPP[a, b]	Spiperone[a, d]
8-OH DEAT[a]	LSD[b]	Ketanserin[a]
Buspirone[a, b, c]	5-OMe DMT[a, b]	Pirenperone[b]
TVX Q 7821[a, b]	MK-212[b]	Metergoline[b]
Gepirone[a]	Propranolol[a]	Methiothepin[b]
8-OH DMAT[a, c]	Quipazine[a, b]	Haloperidol[b]
	RU 24969[b]	
	8-OH DBAT[a]	
	DOM[a]	
	DOI[a]	

[a] Glennon and McKenney (1985), Glennon (1986c), and unpublished results.
[b] Cunningham et al. (1985b).
[c] Partial generalization.
[d] Spiperone-produced disruption of behavior at low doses precluding complete investigation.

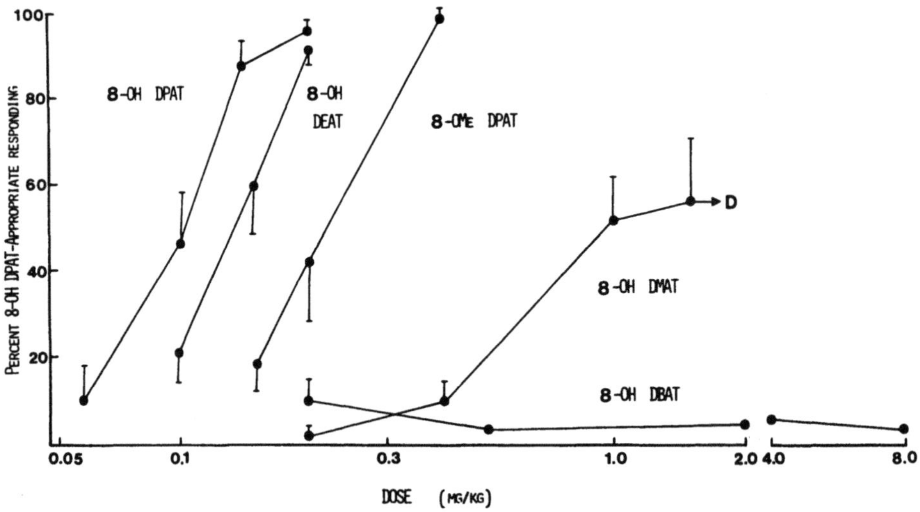

Fig. 3. Tests of stimulus generalization in rats trained to discriminate 8-OH DPAT (0.2 mg/kg) from saline (n = four to seven rats at each dose, *psii* = 15 min). [*Note:* 8-OH DMAT produced 56% 8-OH DPAT-appropriate responding at 1.5 mg/kg (n = 5), and disruption of behavior at 1.7 (n = 4) and 2.0 (n = 7) mg/kg]

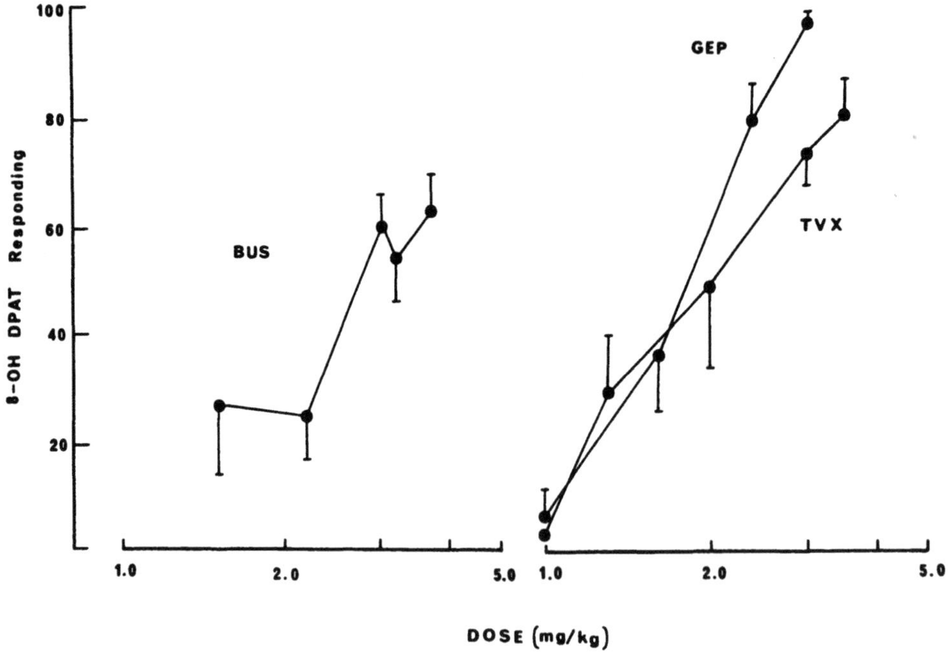

Fig. 4. Tests of stimulus generalization in rats trained to discriminate 8-OH DPAT from saline (training dose = 0.2 mg/kg, 15 min *psii*, n = 4–7 at each dose). *BUS*, buspirone; *GEP*, gepirone; *TVX*, TVX Q 7821. [*Note:* BUS produced 63% 8-OH DPAT-appropriate responding at 3.7 mg/kg (n = 5), and disruption of behavior at doses of 4.0 mg/kg (n = 5) or greater]

antagonists were ineffective in blocking the 8-OH DPAT stimulus (Table 8). Indeed, at this time, there is little evidence that the 8-OH DPAT stimulus is 5-HT mediated. The only support for a 5-HT-related mechanism is derived from studies on several 8-OH DPAT analogs. In a biochemical measure of serotonin agonism, 8-OH DPAT is similar in potency to its diethyl homolog 8-OH DEAT and significantly more potent than its dimethyl homolog 8-OH DMAT; the di-*n*-butyl analog of 8-OH DPAT (i.e., 8-OH DBAT) is essentially inactive (Arvidsson et al. 1984). Figure 3 shows that in tests of stimulus generalization, the 8-OH DPAT stimulus generalizes with 8-OH DEAT; 8-OH DMAT results in partial generalization, and 8-OH DBAT produces saline-appropriate responding at 100 times the ED_{50} dose of 8-OH DPAT (ED_{50}, 0.08 mg/kg).

Buspirone, TVX Q 7821, and gepirone are examples of a novel class of "second-generation" anxiolytic agents. Such agents possess a low affinity for benzodiazepine binding sites, and, unlike the benzodiazepines, display a high affinity for 5-HT1A sites (Gozlan et al. 1983; Traber et al. 1984; Peroutka 1985). Cunningham et al. (1985b) have reported that the 8-OH DPAT stimulus generalizes to buspirone, whereas we have found only partial generalization (Fig. 4). Nevertheless, the 8-OH DPAT stimulus does generalize to TVX Q 7821 and gepirone (Table 8). Thus, animals trained to discriminate 8-OH DPAT from saline might be useful for identifying and studying novel anxiolytic agents.

3.3 5-HT1B-Selective Agonists

Only one 5-HT1B-selective agonist has been used as a training drug in drug discrimination studies: TFMPP (Table 9). TFMPP was initially developed as a serotonin agonist by Fuller and Mason (1979); subsequent studies identified TFMPP as being selective for 5-HT1B sites. As shown in Table 9, we have found that the TFMPP stimulus (training dose, 1.0 mg/kg; ED_{50}, 0.23 mg/kg) generalizes to other 5-HT1B agonists such as mCPP (ED_{50}, 0.47 mg/kg) and RU 24969 (ED_{50},

Table 9. Results of stimulus generalization studies with TFMPP

Agonist	Generalization	No generalization
5-HT1A		8-OH DPAT[a, b]
5-HT1B	mCPP[a, b]	
	RU 24969[a, b]	
5-HT2		DOM[a]
		DOB[a]
		DOI[a]
Misc	Fenfluramine[a, b]	5-OMe DMT[a]
	2-MPP[a]	Quipazine[b, c]
	1-Naphthylpiperazine[a]	LSD[c, d]
		Apomorphine[b]
		2,5-DMA[a]

[a] Glennon et al. (includes some previous unpublished data).
[b] Appel et al.: personal communication.
[c] Glennon et al.: partial generaLIZATION.
[D] Sppel et al.: partial generalization.

0.17 mg/kg), but not to the 5-HT1A agonist 8-OH DPAT or to the 5-HT2 agonists DOM or DOI. The TFMPP stimulus also generalizes to fenfluramine (1.2 mg/kg) but not to quipazine or to the nonselective agonist 5-OMe DMT. Appel has recently obtained similar results with many of the same agents (Table 9) (personal communication). To date, all attempts to antagonize the TFMPP stimulus with 5-HT and dopamine antagonists have been unsuccessful.

The TFMPP stimulus does not generalize to either (+)-LSD or quipazine, although both result in partial generalization (i.e., 50%–60% TFMPP-appropriate responding, with disruption of behavior at slightly higher doses). Early in our studies, prior to acceptance of the concept of 5-HT1 subpopulations of binding sites, but after we had already demonstrated DOM stimulus generalization to both LSD and quipazine, we surmised that if LSD and quipazine were nonselective serotonergic agents (as was suggested by the binding data that were available at that time), administration of these agents in combination with a 5-HT2 antagonist might result in TFMPP stimulus generalization. To this extent, studies were conducted whereby doses of either LSD or quipazine (partial results shown in Fig. 5 for LSD) were administered to TFMPP-trained rats that had been pretreated with various doses of ketanserin; as shown, stimulus generalization was not evident (McKenney and Glennon, unpublished data). It may now be concluded that generalization did not occur under these conditions either because

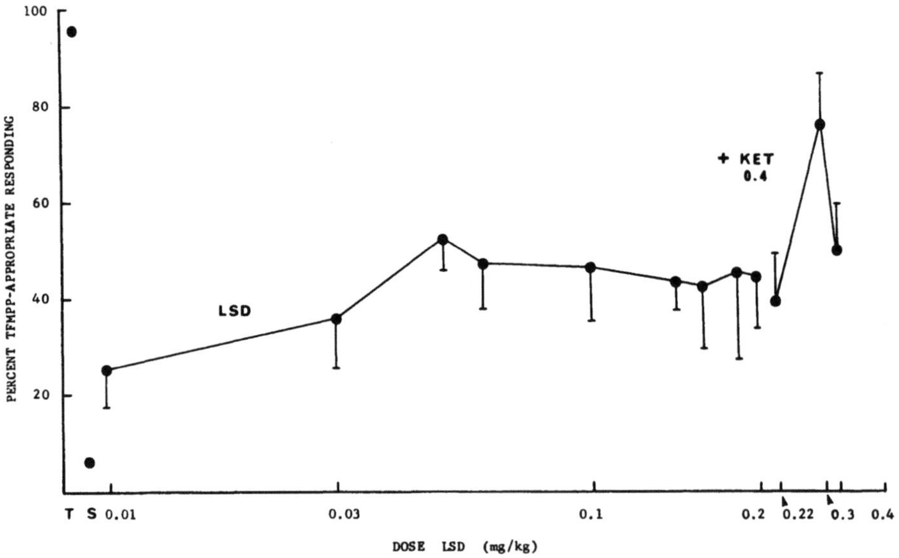

Fig. 5. Tests of stimulus generalization with (+)-LSD in rats trained to discriminate TFMPP from saline (training dose=0.5 mg/kg, *psii*=15 min, *n*=5–7 rats at each dose). *T*=effect of training dose of TFMPP, *S*=effect produced by 1.0 mg/kg saline. Doses of greater than 0.2 mg/kg LSD resulted in disruption of behavior; however, 0.22 mg/kg LSD in combination with 0.4 mg/kg ketanserin (45-min *psii*) produced 38% TFMPP-appropriate responding. Doses of greater than 0.32 mg/kg LSD in combination with ketanserin resulted in disruption of behavior. [Higher and lower doses of ketanserin were used (data not shown), but the combination never produced greater than 77% TFMPP-appropriate responding]

LSD and quipazine act at 5-HT1 sites other than (or in addition to) 5-HT1B sites, or that the stimulus effects of these agents are primarily of a 5-HT2 nature.

4 Conclusion

Certain serotonergic agents (e.g., 5-OMe DMT) lack selectivity with respect to binding at 5-HT binding sites and, as training drugs, produce a stimulus that also lacks selectivity. However, certain other agents are selective both with respect to binding and stimulus properties. 5-HT1A- (8-OH DPAT), 5-HT1B- (TFMPP), and 5-HT2-selective (DOM, DOB, DOI) serotonin agonists have now been used as training drugs and appear to produce a stimulus effect that is site selective. That is, the 8-OH DPAT stimulus generalizes with other 5-HT1A agonists, but not to 5-HT1B or 5-HT2 agonists, the TFMPP stimulus generalizes to other 5-HT1B agonists, but not to 5-HT1A or 5-HT2 agonists, and the DOM- (DOI- and DOB-) stimulus generalizes to other 5-HT2 agonists, but not to 5-HT1A or to 5-HT1B agonists. However, animals trained to discriminate a selective agonist may recognize nonselective agents; for example, the DOM stimulus generalizes to the 5-HT releasing agent fenfluramine and to the 5-HT agonist 5-OMe DMT. Thus, a certain amount of caution must be used in interpreting the results of stimulus generalization studies. Nevertheless, the present results demonstrate that animals can be trained to recognize these site-selective agents, and suggest that such animals might prove useful in future mechanistic, structure-activity, and drug development studies.

Acknowledgments. Of the investigations described above, those conducted in our laboratory were supported, in part, by funds from PHS grant DA-01642, the World Health Organization, the VCU Grant-in-Aid Program, and the A. D. Williams Fund.

References

Appel JB, White FJ, Kuhn DM (1978) The use of drugs as discriminative stimuli in behavioral pharmacodynamics. In: Colpaert FC, Rosecrans JA (eds) Stimulus properties of drugs: ten years of progress. Elsevier, Amsterdam, pp 7–27

Arvidsson LE, Hacksell U, Nilsson JLG, Hjorth S, Carlsson A, Lindberg P, Sanchez D, Wikstrom H (1981) 8-Hydroxy-2-(di-*n*-propylamino)tetralin, a new centrally acting 5-hydroxytryptamine receptor agonist. J Med Chem 24:921–923

Arvidsson LE, Hacksell U, Johansson AM, Nilsson JLG, Lindberg P, Sanchez D, Wikstrom H, Svensson K, Hjorth S, Carlsson A (1984) 8-Hydroxy-2-(alkylamino)tetralins and related compounds as central 5-hydroxytryptamine agonists. J Med Chem 27:45–51

Arvidsson LE, Hacksell U, Glennon RA (1986) Recent advances in central 5-hydroxytryptamine agonists and antagonists. In: Jucker E (ed) Progress in drug research (to be published)

Barrett RJ, Blackshear MA, Sanders-Bush E (1982) discriminative stimulus properties of L-5-hydroxytryptophan: behavioral evidence for multiple serotonin receptors. Psychopharmacology 76:29–35

Colpaert FC, Janssen PAJ (1983) A characterization of LSD-antagonist effects of pirenperone in the rat. Neuropharmacology 22:1001–1005

Colpaert FC, Niemegeers CJE, Janssen PAJ (1982) A drug discrimination analysis of lysergic acid diethylamide (LSD): in vivo agonist and antagonist effects of purported 5-hydroxytryptamine antagonists and pirenperone, a LSD antagonist. J Pharmacol Exp Ther 221:206–215

Cunningham KA, Callahan PM, Appel JB (1985a) Differentiation between the stimulus effects of L-5-hydroxytryptophan and LSD. Eur J Pharmacol 108:179–186
Cunningham KA, Callahan PM, Appel JB (1985b) Similarities in the stimulus effects of 8-hydroxy-2-(di-*n*-propylamino)tetralin (8-OH DPAT), buspirone, and TVX Q 7821: implications for understanding the actions of novel anxiolytics. Soc Neurosci Abstr 11:1106
Friedman R, Barrett RJ, Sanders-Bush E (1983) Additional evidence that L-5-hydroxytryptophan discrimination models a unique serotonin receptor. Psychopharmacology 80:209–213
Friedman RA, Barrett RJ, Sanders-Bush E (1984) Discriminative stimulus properties of quipazine: mediation by serotonin-2 binding sites. J Pharmacol Exp Ther 228:628–635
Friedman RL, Sanders-Bush E, Barrett RL (1985) Clozapine blocks disruptive and discriminative stimulus effects of quipazine. Eur J Pharmacol 106:191–193
Fuller RW, Mason NR (1979) Structure-activity relationships in the actions of 1-phenylpiperazines on brain serotonin receptors. In: Haber B, Gabay S, Issidorides MR, Alivisatos SGA (eds) Serotonin: current aspects of neurochemistry and function. Plenum, New York, pp 359–368
Glennon RA (1985) Interaction of hallucinogenic agents with serotonin receptors. In: Dahlbom R, Nilsson JLG (eds) Medicinal chemistry. Swedish Pharmaceutical, Stockholm, pp 116–129
Glennon RA (1986a) Central serotonin binding sites. In: Williams M, Glennon RA, Timmermans P (eds) Receptor pharmacology and function. Dekker, New York (to be published)
Glennon RA (1986b) Discriminative stimulus properties of the serotonergic agent 1-(2,5-dimethoxy-4-iodophenyl)-2-aminopropane (DOI). Life Sci 39:825–831
Glennon RA (1986c) Discriminative stimulus properties of the 5-HT1A agonist 8-hydroxy-2-(di-*n*-propylamino)tetralin (8-OH DPAT). Pharmacol Biochem Behav 25:135–139
Glennon RA, Hauck AE (1985) Mechanistic studies on DOM as a discriminative stimulus. Pharmacol Biochem Behav 23:937–941
Glennon RA, McKenney JD (1985) Site-selective 5-HT agonists as discriminative stimuli. Pharmacologist 27:194
Glennon RA, Rosecrans JA (1981) Speculations on the mechanism of action of hallucinogenic indolealkylamines. Neurosci Biobehav Rev 5:197–207
Glennon RA, Rosecrans JA, Young R (1982) The use of the drug discrimination paradigm for studying hallucinogenic agents. A review. In: Colpaert FC, Rosecrans JA (ed) Drug discrimination: applications in CNS pharmacology. Elsevier, Amsterdam, pp 69–96
Glennon RA, Rosecrans JA, Young R (1983a) Drug-induced discrimination: a description of the paradigm and a review of its specific application to the study of hallucinogenic agents. Med Res Rev 3:289–340
Glennon RA, Young R, Rosecrans JA (1983b) Antagonism of the effects of the hallucinogen DOM and the purported 5-HT agonist quipazine by 5-HT2 antagonists. Eur J Pharmacol 91:189–196
Glennon RA, McKenney JD, Young R (1984a) Discriminative stimulus properties of the serotonin agonist 1-(trifluoromethylphenyl)piperazine (TFMPP). Life Sci 35:1475–1480
Glennon RA, Titeler M, McKenney JD (1984b) Evidence for 5-HT2 involvement in the mechanism of action of hallucinogenic agents. Life Sci 35:2502–2511
Glennon RA, Titeler M, Young R (1986a) Structure-activity relationships and mechanism of action of hallucinogenic agents based on drug discrimination and radioligand binding studies. Psychopharmacol Bull 935–959
Glennon RA, McKenney JD, Lyon RA, Titeler M (1986b) 5-HT1 and 5-HT2 binding characteristics of 1-(2,5-dimethoxy-4-bromophenyl)-2-aminopropane analogues. J Med Chem 29:194–199
Goudie AJ (1977) Discriminative stimulus properties of fenfluramine in an operant task. An analysis of its cue function. Psychopharmacology 53:97–102
Gozlan H, El Mestikawy S, Pichat L, Glowinski J, Hamon M (1983) Identification of presynaptic serotonin autoreceptors using a new ligand: [^{3}H]PAT. Nature 305:140–142
Haenlein M, Caul WF, Barrett RJ, Michaelis RC (1983) Discrimination of serotonergic drugs is unaltered in rats prenatally exposed to ethanol. Neurobehav Toxicol Teratol 5:475–478
Leysen JE (1985) characterization of serotonin receptor binding sites. In: Green AR (ed) Neuropharmacology of serotonin. Oxford University Press, Oxford, pp 79–116

Lyon RA, Titeler M, McKenney JD, Magee PS, Glennon RA (1986 a) Synthesis and evaluation of phenyl- and benzoyl piperazines as potential serotonergic agents. J Med Chem 29:630–634
Lyon RA, Titeler M, Glennon RA (1986 b) Serotonin receptor selectivity of serotonergic drugs. Soc Neurosci Abstr (to be published)
McElroy JF, Feldman RS (1984) Discriminative stimulus properties of fenfluramine: evidence for serotonergic involvement. Psychopharmacology 83:172–178
McKenney JD, Glennon RA (1986) TFMPP may produce its stimulus effects via a 5-HT1B mechanism. Pharmacol Biochem Behav 24:43–47
Middlemiss DN, Fozard JR (1983) 8-Hydroxy-2-(di-*n*-propylamino)tetralin discriminates between subtypes of the 5-HT1 recognition site. Eur J Pharmacol 90:151–153
Moser P, Redfern PH (1985) The discriminative stimulus properties of 5-HTP are mediated by the 5-HT1 receptor. Br J Pharmacol 84:70P
Nelson DL, Taylor EW (1986) Spiroxatrine: a selective serotonin 1A receptor antagonist. Eur J Pharmacol 124:207–208
Ortmann R (1985) The 5-HT syndrome and the drug discrimination paradigm in rats: applications in behavioral studies on the central 5-HT system. Pharmacopsychiatry 18:198–201
Pedigo NW, Yamamura HI, Nelson DL (1981) Discrimination of multiple [^{3}H]5-hydroxytryptamine binding sites by the neuroleptic spiperone in rat brain. J Neurochem 36:220–226
Peroutka SJ (1985) Selective interaction of novel anxiolytics with 5-hydroxytryptamine-1A receptors. Biol Psychiatry 20:971–979
Peroutka SJ, Snyder SH (1979) Multiple serotonin receptors: differential binding of [^{3}H]5-hydroxytryptamine, [^{3}H]lysergic acid diethylamide and [^{3}H]spiroperidol. Mol Pharmacol 16:687–699
Shannon M, Battaglia G, Glennon RA, Titeler M (1984) 5-HT1 and 5-HT2 binding properties of derivatives of the hallucinogen 1-(2,5-dimethoxyphenyl)-2-aminopropane (2,5-DMA). Eur J Pharmacol 102:23–29
Silverman PB, Ho BT (1980) The discriminative stimulus properties of 2,5-dimethoxy-4-methylamphetamine (DOM): differentiation from amphetamine. Psychopharmacology 68:209–215
Spencer DG, Glaser T, Schuurman T, Traber J (1984) Behavioral and neurochemical correlates of pharmacology involving the 5-HT1 receptor. Soc Neurosci Abstr 10:1072
Titeler M, Herrick K, Lyon RA, McKenney JD, Glennon RA (1985) [^{3}H]DOB: a specific agonist radioligand for 5-HT2 serotonin receptors. Eur J Pharmacol 117:145–146
Traber J, Davies MA, Dompert WU, Glaser T, Schuurman T, Seidel PR (1984) Brain serotonin receptors as a target for the putative anxiolytic TVX Q 7821. Brain Res 12:741–744
Tricklebank MD, Forler C, Middlemiss DN, Fozard JR (1985) Subtypes of the 5-HT receptor mediating the behavioral responses to 5-methoxy-*N,N*-dimethyltryptamine in the rat. Eur J Pharmacol 117:15–24
White FJ, Appel JB (1981) A neuropharmacological analysis of the discriminative stimulus properties of fenfluramine. Psychopharmacology 73:110–115
White FJ, Appel JB (1982) Lysergic acid diethylamide and lisuride: differentiation of their neuropharmacological actions. Science 216:535–537
White FJ, Kuhn DM, Appel JB (1977) Discriminative stimulus properties of quipazine. Neuropharmacology 16:827–832
White FJ, Appel JB, Kuhn DM (1979) Discriminative stimulus properties of quipazine: direct serotonergic mediation. Neuropharmacology 18:143–151
Winter JC (1979) Quipazine-induced stimulus control in the rat. Psychopharmacology 60:265–269
Yamamoto T, Solomon RE, Woods JH (1983) Amphetamine, quipazine, and cyclazocine discrimination. Pharmacologist 25:199
Young R, Glennon RA, Rosencrans JA (1981) Discriminative stimulus properties of the hallucinogenic agent DOM. Commun Psychopharm 4:501–504
Young R, Rosecrans JA, Glennon RA (1983) Behavioral effects of 5-methoxy-*N,N*-dimethyltryptamine and dose-dependent antagonism by BC-105. Psychopharmacology 80:156–160

Discriminative Stimulus Effects of Cholinergic Agonists and the Actions of Their Antagonists

I. P. Stolerman, R. Kumar, and C. Reavill

Contents

Abstract

Both muscarinic- and nicotinic-cholinergic agonists have been used for discrimination training, but only nicotine has been studied extensively. The limited information available suggests that the discriminative stimulus effects of drugs classified as muscarinic-cholinergic agonists are blocked competitively by atropine but not by ganglion-blockers. The discriminative effects of nicotine are blocked non-competitively by ganglion-blocking drugs that penetrate into the CNS (e.g. mecamylamine), but they are not blocked by atropine. The specificity of the block is shown by the failure of mecamylamine to block several non-nicotinic drugs. The ganglion-blocking drug chlorisondamine penetrates poorly into the CNS when injected systemically; when injected intraventricularly, it is a potent and specific nicotine antagonist with a 4-week duration of effect. Haloperidol attenuates discriminative effects of nicotine but this is not a specific block; there are marked reductions in response rate, the morphine stimulus is also attenuated, and other neuroleptics have much weaker effects. The results support the view that the discriminative effect of nicotine involves predominantly cholinoceptive sites, and they suggest that it is not mediated primarily by the dopamine system. The transduction mechanisms for the nicotine stimulus may include the receptor sites that mediate many of its other CNS effects, but more information is needed about possible subtypes of nicotinic receptors before definitive conclusions are possible.

1 Introduction

Drug discrimination research has made important contributions to knowledge of the effects and mode of action of drugs acting on cholinergic systems. The discriminative stimulus effects of nicotinic-cholinergic agonists have been tested ex-

Departments of Pharmacology and Psychiatry, Institute of Psychiatry, De Crespigny Park, London SE5 8AF, UK

Psychopharmacology Series 4
Transduction Mechanisms of Drug Stimuli
Editors: Colpaert, Balster

tensively from 1969 onwards, whereas the discriminative effects of muscarinic-cholinergic agonists have suffered relative neglect. This is surprising because the overall interest of pharmacologists in brain nicotinic mechanisms has been at a very low level until recently. The recent growth in knowledge of CNS nicotinic receptors and their possible involvement in Alzheimer's disease as well as in tobacco addiction is now changing this situation. Until recently, the prevailing view has been that most important brain cholinoceptors are muscarinic, and most attention has been focussed on this subclass of receptor and its possible subtypes. Others have questioned the applicability to the brain of the muscarinic-nicotinic classification of cholinoceptors, although early drug discrimination studies by Rosecrans and his associates provided clear behavioural evidence for its validity (Schechter and Rosecrans 1971, 1972a).

This paper reviews existing studies of the discriminative stimulus effects of drugs acting on cholinoceptive mechanisms, and emphasises the actions of putative antagonists. Understanding effects of antagonists provides a key approach to elucidating the process whereby drug actions at the receptor level are transduced into behaviourally effective stimulus effects. The effects of antagonists on the pharmacologically diverse types of drug used to establish discriminations will be considered in turn, including anticholinesterases and muscarinic- and nicotinic-cholinoceptor agonists.

2 Discrimination of Anticholinesterases

Johanson and Jarbe (1976) trained rats to discriminate the effects of physostigmine (0.5 mg/kg i.p.) from saline in a T-maze shock-escape procedure. This discrimination was acquired rapidly and scores of 80% correct responses were obtained after as few as 8–12 sessions of 5 trials each. There was no generalisation to the quaternary anticholinesterase neostigmine (0.5 mg/kg), suggesting that the physostigmine discrimination may have been based on central actions of the drug. Other doses of neostigmine seem not to have been tested. Treatment with the antimuscarinic drug Ditran reduced the number of drug-appropriate responses after physostigmine to a level not appreciably different from that after saline. The effect of Ditran seemed to be dose related but it was not shown if it could be reversed by increasing the dose of physostigmine, or whether it was a specific block. Other studies using T-maze shock-escape procedures have also shown that physostigmine can be quite well discriminated from the non-drug condition, although the pharmacological nature of the stimulus was not studied (Overton 1982).

These findings raise the possibility that the physostigmine (0.5 mg/kg) cue may be mediated centrally by muscarinic receptors. However, neither nicotinic antagonists nor a standard antimuscarinic agent such as scopolamine were tested: Ditran is a mixture of isomers with complex actions. It is unlikely that physostigmine or similar anticholinesterases will be very useful as discriminative stimuli since their effects are most likely to reflect cholinergic activity at both muscarinic and nicotinic sites. Anticholinesterases other than physostigmine seem not to have been used as training stimuli.

3 Discrimination of Muscarinic Agonists

3.1 Tests with Agonists

In an early experiment, Schechter and Rosecrans (1972 b) trained rats to discriminate arecoline (0.5 mg/kg s.c.) from saline in a two-bar operant conditioning procedure with simultaneous positive reinforcement and punishment contingencies. Responding under these conditions reached at least 80% correct. Subsequently, rats were trained to discriminate arecoline (1.74 mg/kg s.c.) in a two-bar procedure with milk reinforcement only (Meltzer and Rosecrans 1981). Arecoline acquired clear stimulus control over behaviour within 40 training sessions, by which time at least 95% of responses were made on the correct bar. However, the total numbers of responses in drug training sessions were only about 25% of those in non-drug sessions, indicating that the discriminative effect of arecoline at the dose used was weak in relation to its unconditioned effects. Overton (1982) has shown that arecoline and pilocarpine can also be discriminated from the nondrug condition with T-maze shock-escape procedures, but large doses seem to have been necessary (arecoline, 10–20 mg/kg i.p.; pilocarpine, 25–75 mg/kg i.p.).

The potent muscarinic agonist oxotremorine was generalised with arecoline in a dose-related manner, with complete generalisation occurring at a dose of 0.1 mg/kg (Meltzer and Rosecrans 1981). However, the muscarinic agonist pilocarpine was not generalised convincingly, yielding a maximum of 55% drug-appropriate responses as compared with 92% after the training dose of arecoline. The effects of oxotremorine and pilocarpine were evident only at doses which greatly reduced overall response rates. There was no generalisation to nicotine, showing that the arecoline stimulus had at least some specificity. In a recent study, neither oxotremorine nor pilocarpine was generalised with arecoline (Jung et al. 1986). This study also showed that pilocarpine and oxotremorine could be discriminated from the non-drug condition, but cross-tests of generalisation yielded a complex pattern of only partly positive results.

3.2 Tests with Antagonists

The antimuscarinic drug atropine (4 mg/kg) greatly attenuated the discriminative effect of a 1.74-mg/kg training dose of arecoline (Meltzer and Rosecrans 1981). The block could be overcome by increasing the dose of arecoline, and the atropine seemed to have approximately doubled the ED_{50} for arecoline by shifting the arecoline dose-response curve to the right; these results would be expected if atropine acted as a competitive antagonist of arecoline, and they support the view that the arecoline discriminative stimulus may provide a valid index of muscarinic activity. Atropine methylnitrate, which penetrates poorly into the brain, failed to attenuate the discriminative effect of arecoline. Interestingly, both atropine and atropine methylnitrate blocked signs of peripheral, autonomic stimulation, but neither compound attenuated the response-rate decreasing effect of arecoline. Scopolamine, but not methylscopolamine, potently blocked the discriminative effects of arecoline, oxotremorine and pilocarpine (Jung et al. 1986). The effects of atropine and scopolamine may have been specific since they have failed to attenuate the discriminative stimulus effects of several compounds from other pharmacological classes (e.g. Hirschhorn and Rosecrans 1974; Bennett and Lal 1982; Browne 1982; Overton 1983).

3.3 Conclusions

The results of generalisation tests to date do not establish that an unequivocal muscarinic-agonist stimulus has been identified, but the results of experiments with antagonists are encouraging. There appears to be much scope for further work with drugs acting on muscarinic receptors. Training dose is often an important variable in determining the nature and specificity of drug-produced stimuli, but there does not seem to have been any major study of the role of training dose in discrimination of muscarinic agonists. The complex pattern of results from generalisation tests with different agonists suggests the possibility that these compounds may have differing efficacies at putative subpopulations of muscarinic receptors. Proposed selective antagonists for these subpopulations seem not to have been tested in drug discrimination experiments, but such methods may have an important role to play in the furthering knowledge of muscarinic receptor subtypes. Until more information becomes available, it will be difficult to determine whether the receptor sites mediating the discriminative effects of muscarinic agonists are similar to or different from those mediating their other behavioural effects.

4 Discrimination of Nicotinic Agonists

4.1 Tests with Agonists

Virtually all the work on discrimination of nicotinic-cholinoceptor agonists has been done with (−)-nicotine as the training stimulus. Early studies established that nicotine could be well discriminated in both two-bar operant and T-maze procedures (Morrison and Stephenson 1969; Overton 1969), and subsequently Rosecrans and his coworkers studied many aspects of the nicotine discriminative stimulus in considerable detail. More recently, work has concentrated on attempts to correlate the characteristics of the nicotine stimulus with those of high-affinity binding sites for tritiated nicotine. Several general reviews of these studies have appeared (Rosecrans and Chance 1977; Rosecrans et al. 1978; Stolerman 1987). The present article will focus on attempts to attenuate or block the nicotine discriminative stimulus with antagonists from different pharmacological classes.

4.2 Ganglion-Blockers as Antagonists

It has been firmly established that the nicotine discriminative stimulus can be potently and reliably blocked by ganglion-blocking drugs such as mecamylamine and pempidine (Morrison and Stephenson 1969; Schechter and Rosecrans 1971, 1972a; Chance et al. 1978; Romano et al. 1981). Table 1 illustrates this effect. In the peripheral nervous system, nicotine acts at cholinoceptive sites that fall into two main groups, those like the acetylcholine receptors at the neuromuscular

Table 1. Discrimination of nicotine or quipazine after administration of mecamylamine or metergoline in rats ($n=7$–8). (Abridged from Stolerman et al. 1983)

Pretreatment[a] (s.c.)	Test[b] (s.c.)	Discriminative response[c]
Saline	Saline	4.2 ± 2.2
Saline	Nicotine, 0.4 mg/kg	93.2 ± 2.4[d]*
Mecamylamine, 0.75 mg/kg	Saline	1.8 ± 0.8
Mecamylamine, 0.75 mg/kg	Nicotine, 0.4 mg/kg	4.6 ± 2.3[e]*
Saline	Saline	7.1 ± 3.7
Saline	Nicotine, 0.4 mg/kg	93.1 ± 2.5[d]*
Metergoline, 0.25 mg/kg	Saline	7.3 ± 2.4
Metergoline, 4.00 mg/kg	Saline	14.0 ± 3.2
Metergoline, 0.25 mg/kg	Nicotine, 0.4 mg/kg	91.0 ± 2.4[d]*
Metergoline, 4.00 mg/kg	Nicotine, 0.4 mg/kg	90.5 ± 1.9[d]*
Saline	Saline	7.4 ± 3.1
Saline	Quipazine, 0.5 mg/kg	97.0 ± 0.9[d]*
Mecamylamine, 0.75 mg/kg	Saline	17.0 ± 13.6
Mecamylamine, 2.25 mg/kg	Saline	22.5 ± 6.0
Mecamylamine, 0.75 mg/kg	Quipazine, 0.5 mg/kg	93.1 ± 3.9[d]*
Mecamylamine, 2.25 mg/kg	Quipazine, 0.5 mg/kg	83.9 ± 8.3[d]*
Saline	Saline	13.3 ± 4.7
Saline	Quipazine, 0.35 mg/kg	84.5 ± 6.9[d]*
Metergoline, 0,25 mg/kg	Saline	10.4 ± 1.9
Metergoline, 0.25 mg/kg	Quipazine, 0.35 mg/kg	11.0 ± 3.6[e]*

[a] Drug and dose administered (s.c.) 30 min before tests (mecamylamine) or 3 h before tests (metergoline).
[b] Drug and dose administered (s.c.) 15 min before test session.
[c] Drug bar responses as percentage of total responses ± SEM.
[d] Differs from saline pretreatment and saline test.
[e] Differs from saline pretreatment and drug test.
* $p<0.01$.

junction and those like the acetylcholine receptors in autonomic ganglia. It is logical that ganglion-blocking drugs can also block the discriminative effect of nicotine since it is more likely to be mediated through receptors of the neuronal type than through receptors like those on skeletal muscle. However, more detailed studies of the effects of the ganglion-blockers have shed further light on the nature of central nicotinic receptors.

One matter concerning the actions of the ganglion-blockers as antagonists of nicotine often causes misunderstanding. Although mecamylamine, pempidine and chlorisondamine (see below) can be very effective antagonists, there is little evidence that they act competitively, by preventing nicotine from gaining access to its binding sites. In the peripheral nervous system, many such agents act by blocking ion channels that are closely linked to receptors for acetylcholine (Ascher et al. 1979; Lingle 1983). None of these drugs inhibit high-affinity nicotine binding to rat brain membranes (Romano and Goldstein 1980; Marks and Collins 1982). Furthermore, the block of the nicotine discriminative stimulus, and of some other behavioural effects of nicotine, cannot be reversed by increasing the dose of nicotine. All this evidence suggests that the ganglion-blockers attenuate the discriminative effects of nicotine by a non-competitive mechanism, perhaps by blocking a second site in close proximity to but distinct from the recognition site for nicotine itself.

Comparisons of those ganglion-blocking drugs that penetrate well into the CNS with those that penetrate poorly have shown that only the former block the nicotine stimulus. Quaternary compounds such as hexamethonium or chlorisondamine do not block the nicotine stimulus when small doses are injected systemically; very large systemic doses have occasionally been reported to block, probably due to penetration of a small proportion of the administered dose into the CNS (Morrison and Stephenson 1969; Schechter and Rosecrans 1972a; Romano et al. 1981; Stolerman et al. 1983). This work strongly suggests that the nicotine stimulus is of exclusively central origin. Unlike the amphetamine stimulus that is mediated peripherally when small drug doses are used for training, even the smallest training doses of nicotine used to date have produced stimuli by acting on the CNS (Colpaert et al. 1976; Stolerman and D'Mello 1981; Stolerman et al. 1984).

It was observed that hexamethonium (10 µg) injected intraventricularly (i.v.t.) failed to block the nicotine stimulus (Hazell et al. 1978). This finding raised questions as to whether poor penetration into the brain was an adequate explanation for the low blocking potency of hexamethonium administered systemically. Stolerman et al. (1983) extended these negative results with hexamethonium to doses as large as 0.4 mg (i.v.t.), administered 20 min before testing responses to nicotine. It was then noted that the quaternary ganglion-blocking drug chlorisondamine, administered intraventricularly, was a potent antagonist of effects of nicotine on locomotor activity (Clarke and Kumar 1983). In these and subsequent experiments, intraventricular injections were administered with the rats placed in a stereotaxic apparatus under halothane anaesthesia; the injection volume was 1 µl.

In rats trained to discriminate nicotine (0.1 mg/kg) from saline, a single injection of chlorisondamine (5 µg i.v.t.) completely blocked the discriminative effects of nicotine administered 7 days later (Table 2). The block persisted, albeit at reduced magnitude, for about 4 weeks and the findings were replicated in a second group of rats (Garcha et al. 1985; Kumar et al. 1987). In order to obtain these data, rats were tested repeatedly in extinction over the 4-week period, without the training sessions that are usually carried out on days between extinction tests. Test doses of nicotine were twice as large as training doses to maximise baseline discrimination under these unfavourable conditions. Despite the adverse conditions, discrimination between nicotine (0.2 mg/kg) and saline persisted in control subjects that had received saline (i.v.t.), although overall response rates declined sharply as a function of the number of extinction tests.

Table 2 shows that the effect of chlorisondamine was not a general property of ganglion-blocking drugs: hexamethonium had no blocking effect under comparable conditions even when administered in doses as large as 50–200 µg (i.v.t.). Pentolinium (25–100 µg i.v.t.) slightly attenuated the discriminative effect of nicotine (0.2 mg/kg s.c.) in tests 1 week after injection, but had no effect in tests after 2–4 weeks. Substantially larger doses of hexamethonium or pentolinium could not be tested because they produced convulsive phenomena after injection (i.v.t.).

Table 2. Discrimination of nicotine, morphine or midazolam after intraventricular administration of ganglion-blocking drugs in rats (n=7–10). (Garcha et al. 1985; Kumar et al. 1987)

Pretreatment[a] (i.v.t.)	Test[b] (s.c.)	Discriminative response (%)
Saline	Saline	11.7 ± 4.2
Saline	Nicotine, 0.2 mg/kg	90.8 ± 2.6[c]**
Chlorisondamine, 5 μg	Saline	21.8 ± 7.6
Chlorisondamine, 5 μg	Nicotine, 0.2 mg/kg	26.1 ± 8.2[d]**
Saline	Saline	7.8 ± 2.8
Saline	Nicotine, 0.2 mg/kg	90.6 ± 4.2[c]**
Pentolinium, 100 μg	Saline	9.4 ± 3.5
Pentolinium, 100 μg	Nicotine, 0.2 mg/kg	56.7 ± 11.8[c, d]*
Saline	Saline	17.2 ± 9.0
Saline	Nicotine, 0.2 mg/kg	93.4 ± 2.8[c]**
Hexamethonium, 200 μg	Saline	13.3 ± 5.1
Hexamethonium, 200 μg	Nicotine, 0.2 mg/kg	93.9 ± 1.6[c]**
Saline	Saline	8.7 ± 4.0
Saline	Midazolam, 0.2 mg/kg	87.2 ± 4.0[c]**
Chlorisondamine, 5 μg	Saline	10.3 ± 2.5
Chlorisondamine, 5 μg	Midazolam, 0.2 mg/kg	88.7 ± 2.9[c]**
Saline	Saline	5.7 ± 1.9
Saline	Morphine, 6 mg/kg	77.5 ± 7.5[c]**
Chlorisondamine, 5 μg	Saline	3.6 ± 1.1
Chlorisondamine, 5 μg	Morphine, 6 mg/kg	77.1 ± 5.7[c]**

[a] Drug and dose administered (i.v.t.) 7–8 days before tests.
[b] Drug and dose administered (s.c.) before test session. All test doses were double the dose used for training.
[c] Differs from saline pretreatment and saline test.
[d] Differs from saline pretreatment and drug test.
* $p<0.05$.
** $p<0.001$.

In contrast, chlorisondamine did not produce convulsive effects either at 5 μg or at 100 μg (i.v.t.).

The potent blocking effect of chlorisondamine (i.v.t.) strongly supports earlier data interpreted as evidence that the nicotine stimulus is mediated by central sites. The failure of the other ganglion blockers to have a similar action is unexplained and suggests either that the central receptor differs to some extent from those in ganglia or that hexamethonium and pentolinium did not gain adequate access to these central sites even when injected i.v.t. Alternatively, there may be a previously unobserved difference in the receptor selectivity of these agents. Studies with microinjections of antagonists into brain regions may eventually enable a distinction to be made between the pharmacodynamic and pharmacokinetic explanations.

Finally, the value of the ganglion-blockers for probing into central nicotinic mechanisms depends largely on the specificity of their actions. Several studies have examined the effects of mecamylamine on the discrimination of non-nicotinic agents, and in no case has there been any blocking action found. The drugs used to establish these discriminations have varied widely, and have included arecoline, methysergide, morphine, pentobarbitone, phencyclidine, quipazine and scopolamine (Poling et al. 1979; Meltzer and Rosecrans 1981; Overton 1983; Stolerman et al. 1983). In the studies on the quipazine stimulus, even doses of mecamylamine several times larger than doses which completely blocked responses to nicotine had no effect on responses to quipazine (Table 1). Doses of chlorisondamine (5 µg i.v.t.) that blocked responses to nicotine had no detectable effect on responses to either midazolam (0.2 mg/kg s.c.) or morphine (6 mg/kg s.c.) in rats trained to discriminate these drugs from saline (Table 2). These results also showed that chlorisondamine (i.v.t.) did not produce a general disruption of discriminative responding. The findings to date suggest strongly that at least some ganglion-blocking drugs have sufficient specificity of action to support their continued usefulness in investigations of CNS nicotinic mechanisms.

4.3 Neuroleptics as Putative Antagonists

Evidence from a number of sources indicates that there are nicotinic receptors located presynaptically on dopamine nerve terminals in the striatum (Giorguieff-Chesselet et al. 1980; Lichtensteiger et al. 1982; Clarke and Pert 1985). The action of nicotine at these sites seems to result in an enhanced release of dopamine and, thus, dopamine could mediate some actions of nicotine. Rosecrans et al. (1978) have reported that haloperidol does not block the nicotine discriminative stimulus, but this matter has been reinvestigated with haloperidol and other drugs acting as dopamine antagonists.

In one such experiment, haloperidol appeared to block the nicotine discriminative stimulus (Table 3). However, the 0.05 to 0.1 mg/kg (s.c.) doses of haloperidol that had this effect markedly reduced overall rates of responding. In other experiments, the effect of haloperidol on nicotine discrimination was weaker (Reavill and Stolerman, unpublished results). These results contrast markedly with the findings with ganglion-blockers, which reliably abolish the nicotine stimulus at doses well below those which have any effect on total numbers of responses.

Neither pimozide (0.8 mg/kg s.c.) nor droperidol (0.025 mg/kg s.c.) significantly attenuated the discriminative effect of nicotine (Table 3). It was notable that all the neuroleptics had some tendency to attenuate the effect of nicotine to about 60% drug-appropriate responding. These results question the validity of interpreting the clearer findings with haloperidol in terms of dopamine antagonism. In addition, there was some indication that the blocking action of haloperidol lacked specificity, since this neuroleptic also attenuated the discrimination of morphine (3.2 mg/kg s.c.). This attenuation was only partial, but it may be related to switching of responding from one bar to another when neuroleptics are used in certain drug discrimination protocols (Colpaert et al. 1977; Goudie et al. 1986). All neuroleptics tested reduced total numbers of responses (Table 3), and for this reason it was not possible to test larger doses. In some cases, rats made insufficient numbers of responses for discriminative effects to be assessed, thus reducing the number of observations used to calculate means.

Table 3. Discrimination of nicotine (0.1 mg/kg) or morphine (3.2 mg/kg) after administration of different neuroleptic drugs in rats ($n=8$, except in experiment with pimozide, where $n=6$). (Reavill and Stolerman, unpublished results)

Pretreatment[a] (s.c.)	Test[b] (s.c.)	Discriminative response	Total number of responses
Vehicle	Saline	11.4 ± 3.5	449 ± 45
Vehicle	Nicotine	89.6 ± 2.5[c]**	340 ± 27[c]**
Haloperidol, 0.1 mg/kg	Saline	34.3 ± 12.4	28 ± 8[c]**
Haloperidol, 0.1 mg/kg	Nicotine	18.7 ± 9.3[d]**	20 ± 7[c]**
Vehicle	Saline	18.7 ± 6.6	390 ± 38
Vehicle	Nicotine	83.6 ± 5.4[c]**	327 ± 53
Pimozide, 0.8 mg/kg	Saline	16.7 ± 5.9	80 ± 32[c]*
Pimozide, 0.8 mg/kg	Nicotine	62.1 ± 11.7[c]*	120 ± 53[c]*
Vehicle	Saline	6.9 ± 3.5	444 ± 74
Vehicle	Nicotine	87.9 ± 4.1[c]**	372 ± 51
Droperidol, 0.025 mg/kg	Saline	6.4 ± 3.5	214 ± 68[c]*
Droperidol, 0.025 mg/kg	Nicotine	63.2 ± 7.5[c]*	326 ± 64[c]*
Vehicle	Saline	4.4 ± 1.2	347 ± 42
Vehicle	Morphine	84.1 ± 3.6[c]**	269 ± 26
Haloperidol, 0.1 mg/kg	Saline	31.4 ± 16.7	29 ± 9[c]**
Haloperidol, 0.1 mg/kg	Morphine	64.3 ± 6.3[c,d]*	53 ± 11[c,d]*

[a] Drug and dose administered subcutaneously 60 min (haloperidol and droperidol) or 3 h (pimozide) before tests.
[b] Drug and dose administered (s.c.) before test session. Nicotine dose was 0.1 mg/kg and morphine dose was 3.2 mg/kg.
[c] Differs from vehicle pretreatment and saline test.
[d] Differs from vehicle pretreatment and drug test.
* $p<0.05$.
** $p<0.01$.

4.4 Conclusions

Dopaminergic mechanisms are probably not of major importance in the CNS mediation of the nicotine stimulus. This does not preclude the inverse possibility that nicotinic mechanisms may significantly modulate responses that primarily involve the dopamine system. Drugs thought to inhibit activity at various other neurotransmitter receptors have also been found not to block the discriminative effects of nicotine. These drugs include α-methyl-*p*-tyrosine, dibenamine, metergoline, naloxone, naltrexone, *p*-chloro-phenylalanine, propanolol and reserpine (Hirschhorn and Rosecrans 1974; Rosecrans and Chance 1977; Romano et al. 1981; Overton 1983; Stolerman et al. 1983). In most cases, the inhibitory drugs have been shown to block responses to agonists at the appropriate type of receptor, as illustrated in Table 1; these results show that doses of metergoline 30 times larger than those which blocked the discriminative effect of the 5-hydroxytryptamine agonist quipazine failed to block the discriminative effect of nicotine.

The primary mechanism involved in the response to nicotine seems to be a cholinoceptive site. The prolonged block of the nicotine discriminative stimulus by chlorisondamine is an unusual effect and the reason for the 4-week duration of action is not known. It may be that chlorisondamine is trapped in the CNS after intraventricular injection, but this would imply that other quaternary antagonists might also have long durations of effect, which seems not to be the case. For example, Locke and Holtzman (1985) have shown that intraventricular injections of quaternary narcotic antagonists block the discriminative effects of morphine for only 1.5–3.5 h. The mechanism whereby ganglion-blockers such as mecamylamine and chlorisondamine inhibit responses to nicotine in the CNS is unknown, although it is thought that such compounds may block ion-channels associated with peripheral nicotinic receptors (e.g. Lingle 1983).

The probable mediation of the nicotine discriminative stimulus by cholinoceptive sites suggests that the receptors involved in its discriminative effect resemble those involved in its other behavioural effects. Further work is needed to determine whether the sites for different effects have similar neuroanatomical distributions. Nearly all of the behavioural effects of nicotine, including changes in locomotor activity, and its ability to support self-administration and conditioned taste aversions, are blocked by mecamylamine. The long-term blocking effect of chlorisondamine also does not distinguish between discriminative effects and those on locomotor activity, rates of operant responding and conditioned taste aversions (Clarke and Kumar 1983; Reavill et al. 1986). However, there have been suggestions that there are multiple subtypes of CNS nicotinic receptors (e.g. Sloan et al. 1984) and there is some recent evidence that ganglion-blockers can distinguish between them (Collins et al. 1986). The fact that the nicotine cue has different characteristics at different training doses is also compatible with the presence of receptor subtypes (Stolerman et al. 1984), although such an interpretation of the data is very speculative. The present information does not enable conclusions to be reached about whether similar subtypes of receptor may mediate the discriminative and other behavioural effects of nicotine. The results discussed raise the possibility that the CNS nicotinic receptor may take the form of a supramolecular receptor complex with two or more recognition sites for different agents.

5 Comment

A fundamental question is whether stimulus effects of drugs commonly involve receptors mediating their "traditional" pharmacological effects. This issue can probably be answered affirmatively for the majority of commonly used drug stimuli, which have been selected for study for precisely this reason. Whether these drug receptors serve as the sole sensors for behaviour controlled by drug stimuli is less obvious. If drug receptors constitute an entirely adequate and sufficient form of sensory organ for the control of discriminative responding, two questions emerge. How are these receptors linked to CNS mechanisms for the control of discriminative responding? Is it likely that such a link would have evolved if there was no chemically based discriminative process that was advantageous for the organism outside the conditions of the drug discrimination laboratory? Insofar as drug discrimination behaviour shares many characteristics with behaviour controlled by exteroceptive stimuli, it might be expected that common neural pathways are involved. However, as discussed by Overton (1983), a neurochemical process or neuronal system involved in mediating the discriminative effects of drugs (as a class of stimuli) remains to be identified.

References

Ascher P, Large WA, Rang HP (1979) Studies on the mechanism of action of acetylcholine antagonists on rat parasympathetic ganglion cells. J Physiol 295:139–170

Bennett DA, Lal H (1982) Discriminative stimuli produced by clonidine: an investigation of the possible relationship to adrenoceptor stimulation and hypotension. J Pharmacol Exp Ther 223:642–648

Browne RG (1982) Discriminative stimulus properties of phencyclidine. In: Colpaert FC, Slangen JL (eds) Drug discrimination: applications in CNS pharmacology. Elsevier, Amsterdam, pp 109–122

Chance WT, Kallman MD, Rosecrans JA, Spencer RM (1978) A comparison of nicotine and structurally related compounds as discriminative stimuli. Br J Pharmacol 63:609–616

Clarke PBS, Kumar R (1983) Characterization of the locomotor stimulant action of nicotine in tolerant rats. Br J Pharmacol 80:587–594

Clarke PBS, Pert A (1985) Autoradiographic evidence for nicotine receptors on nigrostriatal and mesolimbic dopaminergic neurons. Brain Res 348:355–358

Collins AC, Evans CB, Miner LL, Marks MJ (1986) Mecamylamine blockade of nicotine responses: evidence for two brain nicotinic receptors. Pharmacol Biochem Behav 24:1767–1773

Colpaert FC, Kuyps JJMD, Niemegeers CJE, Janssen PAJ (1976) Discriminative stimulus propertiers of a low *dl*-amphetamine dose. Arch Int Pharmacodyn 223:34–42

Colpaert FC, Niemegeers CJE, Janssen PAJ (1977) Differential haloperidol effect on two indices of fentanyl-saline discrimination. Psychopharmacology 53:169–173

Garcha HS, Kumar R, Norris EA, Reavill C, Stolerman IP (1985) Long-term blockade of nicotine cue by chlorisondamine in rats. Br J Pharmacol 85:245P

Giorguieff-Chesselet MF, Chéramy A, Glowinski J (1980) In vivo and in vitro studies on the presynaptic control of dopamine release from nerve terminals of the nigrostriatal dopaminergic neuron. In: Littauer UZ, Dudai Y, Silman I, Teichberg VI, Vogel Z (eds) Neurotransmitters and their receptors. Wiley, New York, pp 33–47

Goudie AJ, Atkinson J, West CR (1986) Discriminative properties of the psychostimulant dl-cathinone in a two-lever operant task. Lack of evidence for dopaminergic mediation. Neuropharmacology 25:85–94

Hazell P, Peterson DW, Laverty R (1978) Inability of hexamethonium to block the discriminative stimulus (S^D) property of nicotine. Pharmacol Biochem Behav 9:137–140

Hirschhorn ID, Rosecrans JA (1974) Studies on the time course and the effect of cholinergic and adrenergic receptor blockers on the stimulus effect of nicotine. Psychopharmacologia 40:109–120

Johansson JO, Jarbe TUC (1976) Physostigmine as a discriminative cue in rats. Arch Int Pharmacodyn 219:97–102

Jung M, Costa L, Shearman GT, Kelly PH (1986) Discriminative properties of muscarinic agonists. Psychopharmacology 89:S46

Kumar R, Reavill C, Stolerman IP (1987) Nicotine cue in rats: effects of central administration of ganglion-blocking drugs. Br J Pharmacol 90:239–246

Lichtensteiger W, Hefti F, Felix D, Huwyler T, Melamed E, Schlumpf M (1982) Stimulation of nigrostriatal dopamine neurones by nicotine. Neuropharmacology 21:963–968

Lingle C (1983) Blockade of cholinergic channels by chlorisondamine on a crustacean muscle. J Physiol 339:395–417

Locke KW, Holtzman SG (1985) Characterization of the discriminative stimulus effects of centrally administered morphine in the rat. Psychopharmacology 87:1–6

Marks MJ, Collins AC (1982) Characterization of nicotine binding in mouse brain and comparison with the binding of α-bungarotoxin and quinuclidinyl benzilate. Mol Pharmacol 22:554–564

Meltzer LT, Rosecrans JA (1981) Discriminative stimulus properties of arecoline: a new approach for studying central muscarinic receptors. Psychopharmacology 75:383–387
Morrison CF, Stephenson JA (1969) Nicotine injections as the conditioned stimulus in discrimination learning. Psychopharmacologia 15:351–360
Overton DA (1969) Control of T-maze choice by nicotinic, antinicotinic, and antimuscarinic drugs. Proceeding of the 77th annual convention of American Psychological Association, p 869
Overton DA (1982) Comparison of the degree of discriminability of various drugs using the T-maze drug discrimination paradigm. Psychopharmacology 76:385–395
Overton DA (1983) Test for a neurochemically specific mechanism mediating drug discriminations and for stimulus masking. Psychopharmacology 81:340–344
Poling AD, White FJ, Appel JB (1979) Discriminative stimulus properties of phencyclidine. Neuropharmacology 18:459–463
Reavill C, Stolerman IP, Kumar R, Garcha HS (1986) Chlorisondamine blocks acquisition of the conditioned taste aversion produced by (−)-nicotine. Neuropharmacology 25:1067–1069
Romano C, Goldstein A (1980) Stereospecific nicotine receptors on rat brain membranes. Science 210:647–650
Romano C, Goldstein A, Jewell NP (1981) Characterization of the receptor mediating the nicotine discriminative stimulus. Psychopharmacology 74:310–315
Rosecrans JA, Chance WT (1977) Cholinergic and non-cholinergic aspects of the discriminative stimulus properties of nicotine. In: Lal H (ed) Discriminative stimulus properties of drugs. Plenum, New York, pp 155–185
Rosecrans JA, Kallman MJ, Glennon R (1978) The nicotine cue: an overview: In: Colpaert FC, Rosecrans JA (eds) Stimulus properties of drugs: ten years of progress. Elsevier, Amsterdam, pp 69–81
Schechter MD, Rosecrans JA (1971) Behavioral evidence for two types of cholinergic receptors in the C.N.S. Eur J Pharmacol 15:375–378
Schechter MD, Rosecrans JA (1972a) Effect of mecamylamine on discrimination between nicotine- and arecoline-produced cues. Eur J Pharmacol 17:179–182
Schechter MD, Rosecrans JA (1972b) Atropine antagonism of arecoline-cued behavior in the rat. Life Sci 11:517–523
Sloan JW, Todd GD, Martin WR (1984) Nature of nicotine binding to rat brain P_2 fraction. Pharmacol Biochem Behav 20:899–909
Stolerman IP (1987) Psychopharmacology of nicotine: stimulus effects and receptor mechanisms. In: Iversen LL, Iversen SD, Snyder SH (eds) Handbook of psychopharmacology, vol 19. Plenum, New York, pp 421–465
Stolerman IP, D'Mello GD (1981) Role of training conditions in discrimination of central nervous system stimulants by rats. Psychopharmacology 73:295–303
Stolerman IP, Pratt JA, Garcha HS, Giardini V, Kumar R (1983) Nicotine cue in rats analysed with drugs acting on cholinergic and 5-hydroxytryptamine mechanisms. Neuropharmacology 22:1029–1037
Stolerman IP, Garcha HS, Pratt JA, Kumar R (1984) Role of training dose in discrimination of nicotine and related compounds by rats. Psychopharmacology 84:413–419

Stimulus Properties of Dopaminergic Drugs: Comparisons Involving Selective Agonists and Antagonists

J. B. APPEL [1], R. T. WEATHERSBY [1], K. A. CUNNINGHAM [2], P. M. CALLAHAN [1], and R. L. BARRETT [1]

Contents

Abstract

Rats were trained to discriminate the putatively selective dopamine (DA) receptor agonists SKF 38393 (10 mg/kg) or Ly 171555 (0.025 mg/kg) from saline in a two-lever situation involving fixed-ratio (FR 20), extinction schedules of water reinforcement. During substitution tests, no dose of any compound [apomorphine, Ly 171555, lisuride, LSD, amphetamine, cocaine, (−) 3-PPP, or SKF 82526] mimicked SKF 38393, the effects of which were blocked by the D_1 antagonist Sch 23390 but not by haloperidol. Postsynaptic and DA "autoreceptor" agonists [apomorphine, (−) 3-PPP], as well as dopaminergic ergot derivatives (bromocriptine, lergotrile, lisuride) and Sch 23390, substituted for Ly 171555, a partial ergoline which has behavioral effects that are blocked by haloperidol and molindone, but not by either Sch 23390 or serotonin (5-HT) antagonists (ketanserin, pizotifen). Amphetamine and cocaine did not substitute for either SKF 38393 or Ly 171555. These results suggest that the stimulus properties of a variety of neuropharmacologically important and clinically useful compounds are transduced at (pre- or postsynaptic) D_2 receptors. However, this conclusion must be tempered by evidence that actions at D_1 and D_2 receptors may not be entirely independent. The behavioral effects of abused psychomotor stimulants probably involve mechanisms other than "direct" agonist activity at either D_1 or D_2 receptors.

1 Introduction

Dopaminergic neuronal systems are biologically as well as clinically important and are almost certainly involved in the actions of many psychoactive drugs. These include: (1) substances that are used in the *treatment* of psychiatric (e.g.,

[1] Behavioral Pharmacology Laboratory, Department of Psychology, University of South Carolina, Columbia, SC 29208, USA.
[2] *Present address:* Department of Pharmacology and Toxicology, University of Texas Medical Branch, Galveston, TX 77550, USA.

Psychopharmacology Series 4
Transduction Mechanisms of Drug Stimuli
Editors: Colpaert, Balster

chlorpromazine, haloperidol) and neuromuscular [l-dihydroxyphenylalanine (*l*-DOPA), bromocriptine] disorders and (2) drugs that have severe abuse potential and can *cause* psychotic-like symptoms, particularly when administered (or self-administered) chronically; that is, CNS stimulants (e.g., amphetamine and cocaine) and compounds that have both stimulant and hallucinogenic (or psychotomimetic) properties, e.g., MDA [1-(3,4-methylenedioxyphenylisopropylamine)] and MDMA [*N*-methyl-1-(3,4,-methylenedioxyphenylisopropylamine)].

Unfortunately, elucidating the mechanisms by means of which activity at dopaminergic neurons is transduced into alterations in behavior is complicated by several anatomical, biochemical, and physiological considerations. For example, radioligand binding and enzymatic experiments have characterized as many as four dopamine (DA) receptor subtypes (Seeman 1980), although the existence of only two such subtypes has been firmly established: D_1 receptors activate adenylate cyclase while D_2 receptors either act independently of, or inhibit, the activity of this enzyme (Stoof and Kebabian 1984). DA receptors can be differentiated further by their anatomical localization; that is, receptors in the mesolimbocortical (A_{10}) DA system may be involved in the therapeutic effects of neuroleptic drugs (Chiodo and Bunney 1983; White and Wang 1983) while nigrostriatal DA receptors may mediate locomotor and other "side" effects of many of these agents (Bartholini 1976; Costall et al. 1977). In addition, different types of (presynaptic) "autoreceptors" appear to regulate up to as many as three distinct functional events in DA-containing neurons: generation of the nerve impulse, release of transmitter at the terminal, and activity of the rate-limiting step of catecholamine biosynthesis (Galloway et al. 1986). Unhappily, the relationship of these (anatomically defined) autoreceptors to (biochemically and enzymatically defined) D_1 and D_2 DA receptor subtypes is not yet clear, although it is being investigated intensely (White and Wang 1986).

In this paper, drug discrimination (DD), a sensitive and specific behavioral assay which has proven to be useful in delineating the mechanisms subserving many centrally active drugs (Colpaert and Slangen 1982), was used, first to describe the stimulus properties of putatively selective D_1 and D_2 agonists, and second, to compare these properties to those of several clinically or neuropharmacologically important agents as well as two psychomotor stimulants. Our hope was that such studies would shed light on the relevance of activity at different DA receptor subtypes to in vivo mechanisms underlying both therapeutic efficacy and abuse potential. Thus, animals were trained to discriminate D_1 or D_2 agonists and were given substitution (generalization) and combination (antagonism) tests with dopaminergic, serotonergic, and related substances. More specifically, one group of rats was trained to discriminate 10 mg/kg of the D_1 agonist SKF 38393 (2,3,4,5-tetrahydro-7,8-dihydroxy-1-phenyl-1H-3-benzazepine) from saline. This substance stimulates dopamine-activated adenylate cyclase (Setler et al. 1978) and potently displaces the binding of specific ligands to D_1 but not to D_2 receptors (O'Boyle and Waddington 1984).

A second group was trained in a similar manner to detect 0.025 mg/kg Ly 171555 [*trans*-4,4a,5,6,7,8,8a-9-octahydro-5-propyl-1H-pyrazolol (3,4-g)quinoline; quinpirole]. This "partial ergoline," like apomorphine and lisuride, is said to act by stimulating (D_2) receptors (Appel et al. 1984; Kamien and Woolverton

1985; Seeman 1980; Stoof and Kebabian 1984; Titus et al. 1981; Woolverton et al. 1985) because it has behavioral effects that are blocked by haloperidol as well as more selective D_2 antagonists such as (−) sulpiride, molindone, and metoclopramide but, unlike SKF 38393 (Molloy and Waddington 1984), are not affected by the D_1 antagonist Sch 23390 [R-(+)-8-chloro-2,3,4,5-tetrahydro-3-methyl-phenyl-1H-3-benzazepine-7-ol) (Iorio et al. 1983).

2 Materials and Methods

2.1 Training Procedures

The methods have been described in detail elsewhere (Cunningham et al. 1985; Weathersby and Appel 1986). Briefly, male Sprague-Dawley rats were trained to discriminate 10 mg/kg SKF 38393 ($N=12$) or 0.025 mg/kg Ly 171555 ($N=24$) from equivalent volumes of saline (0.9% NaCl), administered intraperitoneally (i.p.), 15 min prior to each experimental session. For half of the animals, responses on the left lever were reinforced (with water) following drug while responses on the right lever were reinforced following saline; conditions were reversed for the remaining animals. Drug or saline injections were given in random order with the restriction that neither condition prevail for more than three consecutive sessions. Training began under a fixed-ratio (FR 1) schedule of reinforcement; that is, each lever press resulted in the presentation of the water-filled dipper (0.1 ml/reinforcer). As response rates stabilized, the FR requirement was increased until all rats were responding reliably under an FR 20 schedule following either drug or saline.

Both levers were then presented simultaneously. Rats were required to respond on the drug- or saline-appropriate lever to receive reinforcement under FR 20; there were no programmed consequences for incorrect responses. (Thus, a concurrent FR 20, extinction schedule was in effect.) To control for the development of position cues based upon olfactory stimuli, a random relationship was maintained between the lever programmed to deliver reinforcement for each consecutive subject run in the same experimental chamber. Training sessions were conducted at the same time each day (5–6 days/week) and were 30 min long.

Once animals demonstrated mean individual accuracies (as defined below) of at least 90% for ten consecutive sessions, testing was begun. These sessions were conducted one to two times/week; training continued during intervening days to maintain discriminative stimulus control. During test sessions, the rat was removed from the chamber without reinforcement as soon as a total of 20 responses occurred on either lever.

2.2 Testing Procedures

Three kinds of tests were conducted: In *dose-response* determinations, rats were tested with a wide range of doses of SKF 38393 or Ly 171555. In *substitution* (generalization) tests, novel drugs were given 15–30 min prior to testing instead of the training drug and, in *combination* (antagonism) tests, rats were given various doses of putatively selective neurotransmitter antagonists 60 min prior to testing as well as the training drug (SKF 38393 or Ly 171555), 15–30 min prior to testing.

2.3 Data Analysis

Accuracy (percentage correct) during training sessions was defined as the number of correct responses divided by the total number of responses (× 100) prior to the delivery of the first reinforcer and, during test sessions, as the number of drug-appropriate responses divided by the total number of responses. Rate (response per minute) was defined as the total number of responses emitted before either: (1) 20 responses were completed on the correct lever (training sessions) or (2) 20 responses occurred on either lever (test sessions), divided by the number of minutes taken to complete the first 20 responses.

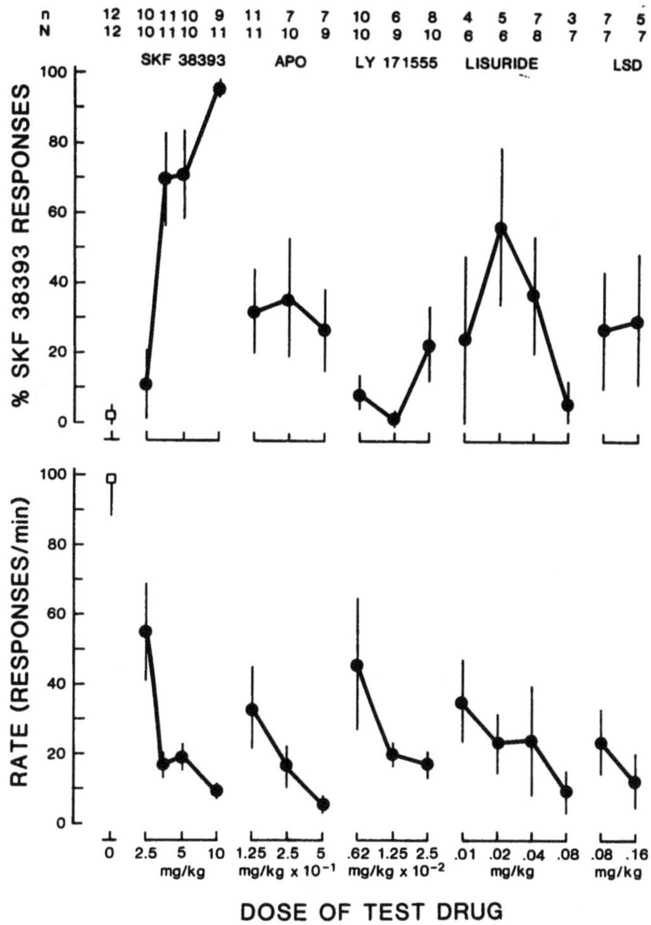

Fig. 1. Results of dose-response and substitution tests with the direct DA agonists apomorphine and Ly 171555, and lisuride and the serotonin agonist LSD in animals (N=12) trained to discriminate SKF 38393 (10 mg/kg) from saline. Mean percentages of training drug-appropriate responding (± SEM) and rates (± SEM) are plotted as functions of test drug dosages in the *upper* and *lower panels*, respectively; a single determination was made at each dose. The number of subjects responding at least 20 times (n) and the total number of animals tested at each dose (N) are shown at the *top of the figure*

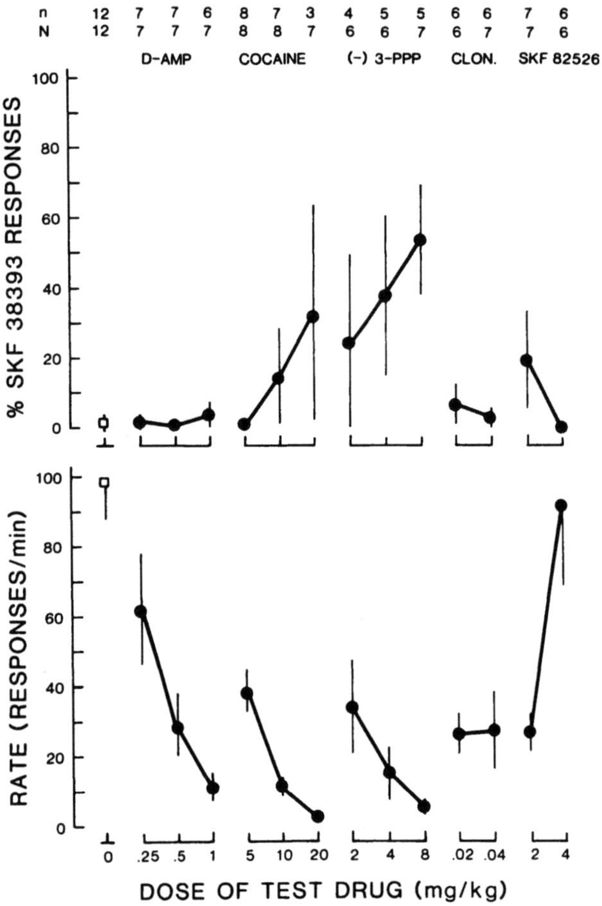

Fig. 2. Results of substitution tests with *d*-amphetamine *(D-AMP)*, cocaine, *(−) 3-PPP*, clonidine (*CLON*), and SKF 82526 in animals trained to discriminate SKF 38393 (10 mg/kg) from saline

Table 1. Effects of putatively selective dopamine antagonists (given 60 min prior to testing) in rats trained to discriminate SKF 38393 (10 mg/kg) from saline

Compound	Dose (mg/kg)	% Drug-lever responses (±SEM)	Responses/min (±SEM)	*n/N*
Saline		2± 2	99±11	12/12
Haloperidol (alone)	0.25	0	70±17	10/10
Haloperidol	0.125	98± 1	9± 3	8/12
(+SKF 38393)	0.25	89±10	14± 8	7/12
SCH 23390 (alone)	0.25	11± 5	35± 7	10/11
Sch 23390	0.031	97± 1	9± 2	10/12
(+SKF 38393)	0.063	68±14	11± 4	10/12
	0.125	29±12	33± 7	11/11

n/N, number of subjects completing the first FR 20 (*n*) out of the total number of subjects tested (*N*).

3 Results

3.1 SKF 38393-Saline Discrimination

As reported elsewhere (Cunningham et al. 1985), animals were able to discriminate SKF 38393 (10 mg/kg) from saline, even though this compound, unlike Ly 171555, suppressed responding severely (Fig. 1). In dose-response tests the percentage of drug-lever responding following different doses of SKF 38393 increased in a dose-dependent manner to a level greater than 90% (Fig. 1).

In substitution tests, apomorphine, Ly 171555, lisuride, and LSD (*d*-lysergic acid diethylamide) elicited primarily saline-lever responding (Fig. 1); neither amphetamine, cocaine, clonidine (an α_2-adrenergic agonist), (−) 3-PPP [3-(3-hydroxyphenyl)-*N*-*n*-propylpiperidine], nor the peripherally active DA agonist SKF 82596 (Kamien and Woolverton 1985) induced significant amounts of responding on the drug-appropriate lever, although there is some suggestion that (−) 3-PPP may partially and dose-dependently mimic SKF 38393 (Fig. 2).

When administered in combination with the training drug Sch 23390, but not haloperidol, blocked the SKF 38393 cue (Table 1); neither haloperidol nor Sch 23390 mimicked SKF 38393 when given alone (Table 1).

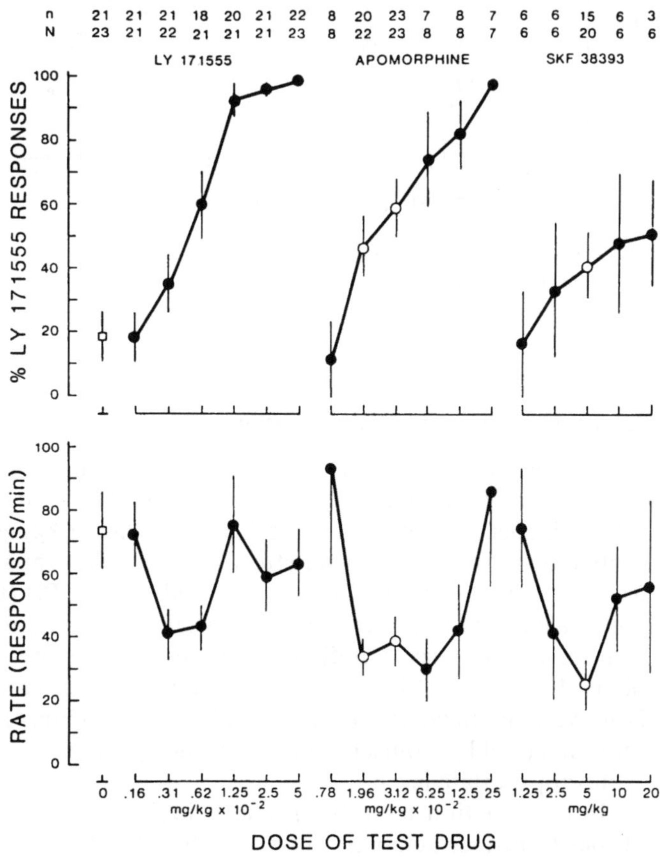

Fig. 3. Results of dose-response and substitution tests with the direct DA agonists apomorphine and SKF 38393 in animals (N=23) trained to discriminate Ly 171555 (0.025 mg/kg) from saline. *Open circles* denote the mean of three determinations (± SEM); as in all the figures, *closed circles* represent the average performance (± SEM) during a single determination at each of the doses indicated on the abscissae

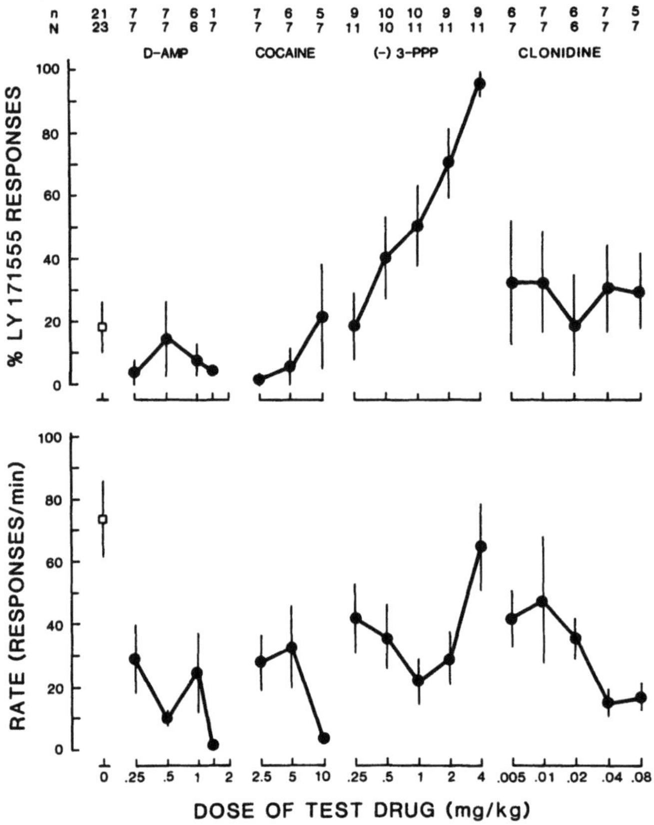

Fig. 4. Results of substitution tests with *d*-amphetamine (*D-AMP*), cocaine, (−) 3-PPP, and clonidine (*CLON*) in animals trained to discriminate Ly 171555 (0.025 mg/kg) from saline

3.2 Ly 171555-Saline Discrimination

As was the case with SKF 38393, animals rapidly learned to discriminate Ly 171555 from saline at a dose (0.025 mg/kg) that did not disrupt behavior (Fig. 3); indeed, higher doses of Ly 171555 (0.0125–0.05 mg/kg) were, if anything, *less* disruptive than lower doses – an effect that was also apparent during substitution tests with apomorphine, SKF 38393, and (−) 3-PPP (Figs. 3, 4). More importantly, a large variety of DA agonists mimicked the training drug in a dose-dependent manner at doses that did not suppress responding. These included apomorphine (Fig. 3), (−) 3-PPP (Fig. 4), bromocriptine, lergotrile, and lisuride (Fig. 5). However, amphetamine, cocaine, and clonidine (Fig. 4) as well as LSD and the serotonin (5-HT) agonist quipazine (Table 3) elicited responding primarily on the saline-appropriate lever.

In combination tests, both haloperidol and molindone blocked Ly 171555 (Table 2); neither ketanserin nor pizotifen (BC 105) had similar antagonistic ef-

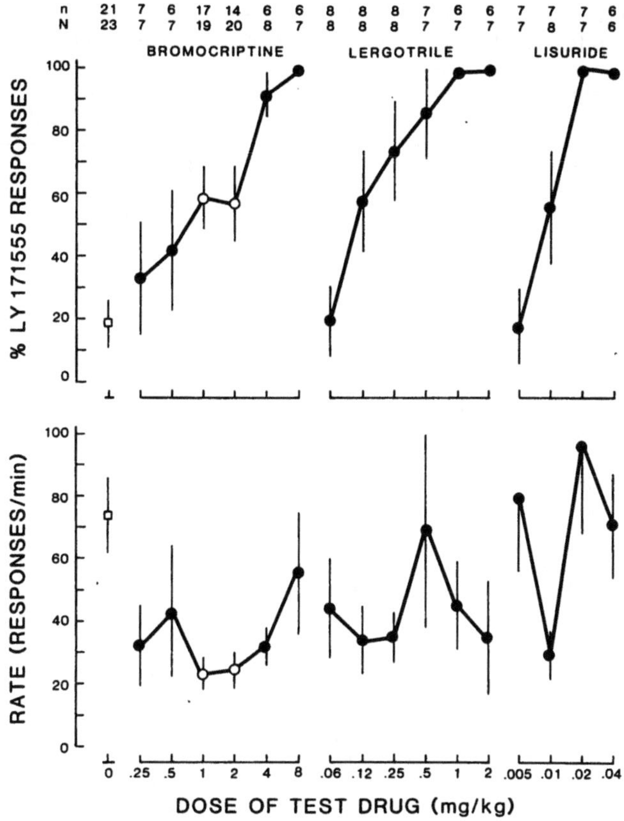

Fig. 5. Results of substitution tests with the dopaminergic ergot derivatives bromocriptine, lergotrile, and lisuride in animals trained to discriminate Ly 171555 (0.025 mg/kg) from saline

fects and both of these agents induced some measure of drug-appropriate responding when given alone (Table 3). Sch 23390 did not block the stimulus effects of Ly 171555 at doses that did not also disrupt lever-pressing behavior; indeed, when given alone, doses up to and including 0.25 mg/kg of this substance appeared to substitute for the training drug (Table 2).

Table 2. Effects of putatively selective dopamine antagonists (given 60 min prior to testing) in rats trained to discriminate Ly 171555 (0.025 mg/kg) from saline

Compound	Dose (mg/kg)	% Drug-lever responses (±SEM)	Responses/min (±SEM)	*n*/*N*
Saline		88± 7	76± 13	21/22
Domperidone (+Ly 171555)	2.0	82±16	59± 26	6/7
Haloperidol (alone)	0.0625	46±22	37± 20	5/6
	0.125	12± 8	32± 22	3/6
	0.25			0/6
Haloperidol (+Ly 171555)	0.0625	72±17	49± 17	5/6
	0.125	67±18	51± 13	7/8
	0.25	1± 1	66± 18	7/7
	0.5			0/6
Molindone (alone)	0.5	13±11	32± 16	4/7
	1.0	35±26	3	3/7
	2.0	29±29	3	2/7
Molindone (+Ly 171555)	0.5	93± 7	50± 14	7/7
	1.0	42±17	39± 10	8/8
	2.0	35±17	59± 18	7/8
	4.0	16±14	62± 20	3/7
	8.0	7± 5	46± 25	3/7
Sch 23390 (alone)	0.03125	33±19	43± 23	6/6
	0.0625	54±19	50± 13	6/6
	0.125	55±24	18± 7	3/7
	0.25	93± 7	13± 5	5/8
	0.5	23	4	1/6
Sch 23390 (+Ly 171555)	0.25	100	145±112	4/7
	0.5	66±19	42± 20	4/8

n/*N*, number of subjects completing the first FR 20 (*n*) out of the total number of subjects tested (*N*).

Table 3. Effects of serotonergic agents in rats trained to discriminate Ly 171555 (0.025 mg/kg) from saline

Compound	Dose (mg/kg)	% Drug-lever responses (±SEM)	Responses/min (±SEM)	*n*/*N*
Saline		88± 7	76±13	21/22
LSD	0.01	20±16	41±18	6/7
	0.02	14± 9	25±13	5/8
	0.04	19±11	52±20	8/8
	0.08	58±23	47±38	4/8
Quipazine	0.5	7± 4	60±17	7/7
	1.0	21±12	72±29	8/8
	1.5	100	21	1/6
	2.0	0	20	1/8
	4.0	100	2	1/8

Table 3 continued

Compound	Dose (mg/kg)	% Drug-lever responses (±SEM)	Responses/min (±SEM)	*n*/*N*
Ketanserin	4.0	45±28	10± 6	3/5
(alone)	8.0	41±20	17±10	4/6
Ketanserin	2.0	100	16± 8	5/7
(+Ly 171555)	4.0	99± 1	60±28	6/6
	8.0	96± 3	10± 3	7/7
Pizotifen (BC 105)	4.0	6± 6	21± 3	3/5
(alone)	8.0	78±20	24±14	3/5
Pizotifen (BC 105) (+Ly 171555)	2.0	83±17	59±25	6/6
(1st test)	4.0	44±17	71±18	7/7
(2nd test)	4.0	95± 5	44±16	7/7
(3rd test)	4.0	100	60±25	3/5
Mean	4.0	74±10	58±10	17/19
	8.0	87±11	47±16	7/7

Note: LSD and quipazine were given 15 min prior to testing; ketanserin and pizotifen (BC 105) were given 60 min prior to testing.
n/*N*, number of subjects completing the first FR 20 (*n*) out of the total number of subjects tested (*N*).

4 Discussion

As reported elsewhere, the results indicate that the D_1 agonist SKF 38393 and the D_2 agonist Ly 171555 have discriminable effects that differ both from saline and from each other (Cunningham et al. 1985; Kamien and Woolverton 1985; Weathersby and Appel 1986). While the dose of SKF 38393 required to maintain the discrimination (10 mg/kg), unlike that of Ly 171555 (0.025 mg/kg), disrupted behavior, the inability of SKF 82526 (which crosses the blood-brain barrier only with considerable difficulty) to substitute for SKF 38393 suggests that the SKF 38393 cue is mediated centrally (Kamien and Woolverton 1985).

Among all the drugs tested, neither directly acting DA (D_2) agonists (apomorphine, Ly 171555, lisuride), autoreceptor agonists such as (−) 3-PPP (Hjorth et al. 1981), centrally acting catecholaminergic releasing and uptake-inhibiting agents (amphetamine and cocaine), nor primarily serotonergic substances (LSD) substituted for SKF 38393 at doses that did not completely eliminate behavior in most animals; moreover, D_2 antagonists such as haloperidol did not antagonize the SKF 38393 cue while Sch 23390 did. It is therefore likely that the stimulus properties of SKF 38393 are transduced by D_1 receptors. While these receptors may have little pharmacological interest in that they do not appear to be involved in the behavioral effects of most dopaminergic substances, they alter the activity of D_2-mediated events (below) and, in addition, appear to be functionally significant in other assays. For example, they alter motor activity by increasing the spontaneous activity of cells in the *zona reticulata* (but not in the *zona compacta*) of the substantia nigra (Matthews and German 1986).

In animals trained to discriminate Ly 171555 from saline, D_2 agonists such as apomorphine and (−) 3-PPP, but not the D_1 agonist SKF 38393, mimicked the training drug. Similarly, D_2 antagonists such as haloperidol and molindone but not the D_1 antagonist Sch 23390 blocked the Ly 171555 cue. These results support those reported previously and imply that Ly 171555 is a relatively selective D_2 agonist (e.g., Titus et al. 1981). However, this conclusion must be tempered by findings that both 5-HT and D_1 antagonists *mimic* the Ly 171555 cue (Tables 2 and 3). It is interesting that similar results also occur when animals are trained to discriminate another D_2 agonist, lisuride, from saline (Appel et al. 1984). These effects could mean that D_1 and D_2 receptors do not function independently (Creese et al. 1985; Mailman 1985; Waddington et al. 1985); for example, D_1 blockade may influence behaviors initiated by D_2 stimulation (Pugh et al. 1985). Alternatively, they might indicate that drugs such as SKF 38393 and Ly 171555 act upon other neuronal systems; for example, the generalization of the Ly 171555 cue to both ketanserin and pizotifen suggests that 5-HT systems play a role in the actions of at least one purportedly selective DA agonist.

Dopaminergic (bromocriptine, lergotrile, lisuride) but not serotonergic (LSD) ergot derivatives (Appel et al. 1985) appear to have in vivo effects which strongly resemble those of Ly 171555. These data confirm previous reports that D_2 receptor subtypes mediate the effects of these (as well as other) clinically useful compounds. For this reason, Ly 171555 could be a valuable preclinical screening device as well as an important research "tool," a conclusion which is reinforced by the fact that neuroleptics such as haloperidol, molindone, and probably additional D_2 antagonists potently antagonize the discriminable effects of Ly 171555.

Amphetamine and cocaine, which are thought to act primarily by inhibiting the reuptake or causing the release of catecholamines (Feldman and Quenzer 1985) do not mimic either SKF 38393 or Ly 171555, at least when these psychomotor stimulants are administered acutely (i.p.). This is encouraging since it suggests that the mechanisms underlying the reinforcing states induced by abused substances are unlike those that occur following treatment with compounds that alleviate the symptoms of neuropsychiatric disorders or, stated otherwise, that clinically useful agents (e.g., DA agonists and antagonists) are not likely to be abused *because* they do not induce amphetamine- or cocaine-like states. It is less than encouraging in that the present analysis of the stimulus properties of DA agonists has shed little, if any, light on the nature of these mechanisms. Perhaps the ability to activate DA (as well as other catecholamine-containing) systems indirectly does not translate into ability to mimic cues that are the result of selective D_1 or D_2 receptor stimulation. Interestingly, however, in animals trained to discriminate amphetamine from saline, D_1 and D_2 agonists mimic partially, and D_1 as well as D_2 antagonists effectively block, the amphetamine cue (Nielsen and Jepsen 1985).

5 Conclusions

In spite of the complexities that have arisen in attempts to characterize the nature of dopaminergic neuronal systems and how these systems transduce excitatory or inhibitory stimulation into alterations in behavior, it is possible to draw at least six conclusions from the present results:

1. Rats can discriminate SKF 38393 (10 mg/kg) and Ly 171555 (0.025 mg/kg) both from saline and from each other.
2. The SKF 38393 cue is mediated (centrally) by DA receptors of the D_1 subtype while the Ly 171555 cue is mediated by D_2 receptors.
3. A variety of dopaminergic agents, some of which are used clinically, antagonize the Ly 171555 cue; thus, their actions also are likely to involve D_2 receptors.
4. Both the second and third conclusions must be tempered by the possibility that D_1 and D_2 receptors may not act independently; that is, D_1 antagonists such as Sch 23390 appear to have stimulus properties that mimic those of D_2 agonists. If this is correct, it may be necessary to reformulate current ideas about how different biochemically and enzymatically defined DA "receptor" subtypes function in vivo.
5. While the mechanisms subserving the behavioral and subjective effects of amphetamine and cocaine remain poorly understood, it can be concluded that these psychomotor stimulants are discriminably different from compounds that act as agonists or antagonists at either D_1 or D_2 sites; hence, it is likely that these abused substances have different, presumably more complex, actions than those of more "direct" DA agonists.
6. Finally, the present results, however they may be interpreted, demonstrate again that drug discrimination can be used to accurately describe the different behavioral (subjective) consequences of different DA receptor events.

Acknowledgments. This research was supported by USPHS Research Grant 2 RO1 DA 02543 from the National Institute on Drug Abuse. We wish to thank the following companies for their generous gifts of drugs: Astra Lakemedel AG, Sodertalje, SWEDEN (3-PPP); Boehringer Ingelheim ltd., Ridgefield, CT (clonidine); Janssen Pharmaceutica, New Brunswick, NJ (ketanserin); Eli Lilly & Co, Indianapolis, IN (lergotrile, Ly 171555); Sandoz Pharmaceuticals, East Hanover, NJ (pizotifen, bromocriptine); Schering Corp, Bloomfield, NJ (lisuride, Sch 23390); Smith Kline and French, West Point, PA (SKF 38393, SKF 82596). Cocaine and LSD were obtained from the National Institute on Drug Abuse. We also thank Sue Hilfer and the staff of the Department of Medical Illustration (USC School of Medicine) for preparing the figures.

References

Appel JB, Cunningham KA, Callahan PM (1984) Stimulus properties of lisuride revisited: new evidence for the involvement of D_2 receptors. Paper presented at the meeting of the Society for Neuroscience (Satellite session of the Society for the Stimulus Properties of Drugs), Anaheim

Appel JB, Cunningham KA, West KB (1985) Monoaminergic involvement in the behavioral effects of ergot derivatives. In: Lal H, LaBella F, Lane J (eds) Endocoids. Liss, New York, pp 51–62

Bartholini G (1976) Differential effects of neuroleptic drugs on dopamine turnover in the extrapyramidal and limbic systems. J Pharmacol 28:429–432

Chiodo LA, Bunney BS (1983) Typical and atypical neuroleptics: differential effects of chronic administration on the activity of A9 and A10 midbrain dopaminergic neurons. J Neurosci 8:1607–1619

Colpaert FC, Slangen JL (eds) (1982) Drug discrimination: applications in CNS pharmacology. Elsevier Biomedical, Amsterdam

Costall B, Marsden CD, Naylor RJ, Pycock CJ (1977) Stereotyped behavior patterns and hyperactivity induced by amphetamine and apomorphine after discrete 6-hydroxydopamine lesions of extrapyramidal and mesolimbic nuclei. Brain Res 123:89–97

Creese I, Hess E, Le H (1985) The effect of chronic Sch 23390 treatment on the biochemical and behavioral properties of D-1 and D-2 dopamine receptors. Paper presented at the meetings of the American College of Neuropsychopharmacology, Maui

Cunningham KA, Callahan PM, Appel JB (1985) Dopamine D_1 receptor mediation of the discriminative stimulus properties of SKF 38393. Eur J Pharmacol 119:121–125

Feldman RS, Quenzer LF (1984) Fundamentals of neuropsychopharmacology. Sinauer and Assoc, Sunderland

Galloway MP, Wolf NF, Roth RH (1986) Regulation of dopamine synthesis in the medial prefrontal cortex is mediated by release modulating autoreceptors: studies in vivo. J Pharmacol Exp Ther 236:689–698

Hjorth S, Carlsson A, Wikstrom H, Lindberg P, Sanchez D, Hacksell U, Arvidsson L-E, Svensson U, Nilsson JLG (1981) In: Gessa GL, Corsini GU (eds) Apomorphine and other dopaminomimetics, vol 1: basic pharmacology. Raven, New York, pp 235–260

Iorio LC, Barnett A, Leitz FH, Houser VP, Korduba CA (1983) Sch 23390, a potential benzazepine antipsychotic with unique interactions on dopaminergic systems. J Pharmacol Exp Ther 226:462–468

Kamien JB, Woolverton WL (1985) The D_1 dopamine agonist SKF 38393 functions as a discriminative stimulus in rats. Psychopharmacology 87:368–370

Mailman R (1985) Defining the D_1 dopamine receptor: its linkage to adenylate cyclase and psychopharmacological effects. Paper presented at the meetings of the American College of Neuropsychopharmacology, Maui

Matthews RT, German DC (1986) Evidence for a functional role of dopamine type-1 (D-1) receptors in the substantia nigra of rats. Eur J Pharmacol 120:87–93

Molloy AG, Waddington JL (1984) Dopaminergic behavior stereospecifically promoted by the D_1 agonist R-SK & F 38393 and selectively blocked by the D_1 antagonist Sch 23390. Psychopharmacology 82:409–410

Nielsen EB, Jepsen SA (1985) Antagonism of the amphetamine cue by both classical and atypical antipsychotic drugs. Eur J Pharmacol 111:167–176

O'Boyle KM, Waddington JL (1984) Selective and stereospecific interactions of R-SK & F 38393 with ^{3}H-piflutixol but not ^{3}H-spiperone binding to striatal D-1 and D-2 dopamine receptors: comparisons with SCH 23390. Eur J Pharmacol 98:433–436

Pugh MT, O'Boyle KM, Molloy AG, Waddington JL (1985) Effects of the putative D-1 antagonist Sch 23390 on stereotyped behavior induced by the D-2 agonist Ru 24213. Psychopharmacology 87:308–312

Seeman P (1980) Brain dopamine receptors. Pharmacol Rev 32:229–313

Setler PE, Sarau HM, Zirkle CL, Saunders HL (1978) The central effects of a novel dopamine agonist. Eur J Pharmacol 50:419–430

Stoof JC, Kebabian JW (1984) Two dopamine receptors: biochemistry, physiology and pharmacology. Life Sci 35:2281–2296

Titus RD, Kornfeld EC, Jones ND, Clemens JA, Smalstig EB, Fuller RW, Hahn RA, Hynes MD, Mason NR, Wong DT, Foreman MM (1981) Resolution and absolute configuration of an ergoline-related dopamine agonist, *trans*-4,4a,5,6,7,8,8a,9-octahydro-5-propyl-1H(or 2H)-pyrazolo(3,4-g)quinoline. J Med Chem 26:1112–1116

Waddington JL, O'Boyle K, Molloy A (1985) Structural analogues of Sch 23390 as new probes for behavioral interactions between D-1 and D-2 dopaminergic function. Paper presented at the meetings of the American College of Neuropsychopharmacology, Maui

Weathersby RT, Appel JB (1986) Dopamine D_2 receptor mediation of the discriminative stimulus properties of Ly 171555. Eur J Pharmacol 132:87–91

White FJ, Wang RY (1983) Differential effects of classical and atypical antipsychotic drugs on A9 and A10 dopamine neurons. Science 221:1054–1057

White FJ, Wang RY (1986) Electrophysiological evidence for the existence of both D-1 and D-2 dopamine receptors in the rat nucleus accumbens. J Neurosci 6:274–280

Woolverton WL, Kamien JB, Goldberg LI (1985) Effects of selective dopamine receptor agonists in rats trained to discriminate apomorphine from saline. Pharmacol Biochem Behav 22:577–581

Central Nervous System Stimulants: Neuropharmacological Mechanisms

E. B. Nielsen [1] and J. Scheel-Krüger [2]

Contents

Abstract

The mechanisms underlying CNS-stimulant drug discrimination are discussed. Although different doses of CNS stimulants may produce qualitatively different cues, it appears that a relatively low dose of *d*-amphetamine (e.g., 1 mg/kg) elicits a "general" CNS-stimulant cue. Presynaptically, this cue may primarily depend on release of endogeneous dopamine whereas inhibition of dopamine reuptake, per se, is insufficient to elicit the cue. Postsynaptically, the involvement of both dopamine D-1 and D-2 receptors is implicated. Furthermore, in the drug discrimination situation, D-1/D-2 receptors may be coupled differently than in dopamine-dependent locomotor activation. Anatomically, CNS-stimulant drug discrimination may depend primarily on mesolimbic dopamine systems.

1 Introduction

The relative inaccessibility of "internal events" for experimental analysis has led to an emphasis on the study of (unconditioned) motor behavior as a way to analyze the effects of drugs. The history of research on CNS-stimulant drug effects is illustrative in this respect. Thus, the effects of amphetamines were initially analyzed by means of their ability to cause stereotyped motor behavior (e.g., Lyon and Robbins 1975). However, the discriminative and reinforcing properties of CNS stimulants appear to be of at least equal importance for understanding, for

[1] NOVO Industri A/S, Pharmaceuticals R & D, Department of Pharmacology, DK-2880 Bagsvaerd, Denmark.
[2] Sct. Hans Mental Hospital, Psychopharmacological Research Laboratory, DK-4000 Roskilde, Denmark.

Psychopharmacology Series 4
Transduction Mechanisms of Drug Stimuli
Editors: Colpaert, Balster

example, the addictive and psychotomimetic properties of the drugs. Furthermore, as will be later discussed, there may even be different neuropharmacological mechanisms involved in the motor-activating and the discriminative/reinforcing stimulus properties of CNS stimulants.

In the present review, we will attempt to overview the neuropharmacological basis for the discriminative stimulus effects of CNS stimulants and, whenever relevant, to point out how such effects may differ in their mechanism from other behavioral effects (e.g., changes in motor behavior) of the drugs. The review will focus mostly on amphetamine and cocaine as prototypical CNS stimulants. The pharmacology of the cueing effects of direct DA agonists (when used as training drugs) will only be mentioned briefly as these drugs are discussed elsewhere in this book.

2 The Generality of CNS-Stimulant Cues

The cues produced by "classical" CNS stimulants (e.g., amphetamine, cocaine, methylphenidate) are apparently interchangeable, indicating the existence of a general CNS-stimulant cue (see review by Silverman and Ho 1977, and references therein; Aceto et al. 1984; Schechter 1977). However, subtle qualitative differences may exist between different CNS-stimulant drugs. For example, D'Mello and Stolerman (1977) found increased ED_{50} values for cocaine and amphetamine when animals trained to discriminate these drugs were subjected to a "cross-over" procedure involving retraining on the alternative drug. Further it has been found that the extent to which amphetamine and cocaine can be blocked by haloperidol sometimes differs (see, for example, Colpaert et al. 1978; Cunningham and Appel 1982). However, these issues are at present relatively unexplored.

Cathionine, the major psychoactive component of Khat leaves, has a discriminable similarity to amphetamine (Glennon et al. 1984a, b). Other drugs which produce a CNS stimulant cue include certain sympathomimetics (e.g., ephedrine). Their effect is potentiated by caffeine (Holloway et al. 1985), which alone, however, does not substitute for amphetamine (Holloway et al. 1985). Phenylethylamine (PEA), claimed as an endogenous amphetamine, substitutes partially for amphetamine (Goudie 1983). When given in conjunction with an MAO inhibitor, PEA substitutes completely for amphetamine (Huan and Ho 1974). It has been speculated the endogenous PEA may be involved in the cueing effect of both cocaine and amphetamine (Huan and Ho 1974; Colpaert et al. 1980). Finally, Δ^9-tetrahydrocannabinol was found to substitute for amphetamine (Bueno et al. 1976). However, the mechanism of action of cannabinoids remains obscure (Browne and Weismann 1981) although an involvement of dopamine (DA) has been suggested (Sakurai et al. 1985).

There is evidence to suggest that the training dose of a drug determines the pharmacological specificity of the discrimination, an issue that was first studied using opiate discrimination (see overview by Colpaert 1982). That this may apply also to CNS-stimulant discrimination is suggested by two observations: (1) the ability of apomorphine, a mixed D-1/D-2 receptor agonist, to substitute for *d*-amphetamine relates inversely to the training dose of amphetamine (e.g., apomorphine substitutes for a "high-dose" (1.6 mg/kg) amphetamine cue, but not for a "low-dose" (0.4 mg/kg) amphetamine cue (Stolerman and D'Mello 1981); (2) it has been suggested that a "high-dose" (30 mg/kg) cocaine cue involves an "anxiogenic" pharmacology, while a "low-dose" (1.25 mg/kg) cocaine cue depends on DA systems (Shearman and Lal 1981; Emmett-Oglesby et al. 1983). The results with amphetamine may parallel the observation that the drug preferentially produces locomotor activation at low doses while stereotyped behavior predominates at higher doses (Lyon and Robbins 1975). Similarly, different DA systems have been implicated in locomotor and stereotypic behavior, respectively (Arnt 1983, 1985; Costall et al. 1977; Robbins and Everett 1982).

Thus, although a general CNS-stimulant cue may exist, the dose of the CNS stimulant may be important since qualitatively different cues may perhaps exist at different doses. Furthermore, they may even involve different pharmacological mechanisms.

3 Presynaptic Mechanisms Involved in CNS-Stimulant Cues

3.1 DA/NE Uptake Inhibition

Until recently, it appeared that by inhibiting the synaptic uptake of DA with, for example, bupropion or nomifensine, an amphetamine-like stimulus could be elicited (Porsolt et al. 1982, Table 1). However, the role of a possible concurrent release of DA by at least nomifensine cannot be excluded (Braestrup and Scheel-Krüger 1976); apparently, bupropion is unable to release DA as determined from in vitro studies (Ferris et al. 1981). Neither nomifensine nor bupropion is a selective DA-uptake inhibitor (Hyttel and Larsen 1985). As a matter of fact, it is more

Table 1. Uptake inhibitors: behavioral correlates. Ability of various uptake inhibitors (1) to produce stereotyped behavior rated on a scale from 0–6 according to Nielsen (1981) and (2) to elicit the stimulus properties of amphetamine in animals trained to discriminate this drug (1 mg/kg, i.p., *t*, 15 min) from saline (Nielsen, unpublished).

Compound	ED_{50} (mg/kg)		
	A	B	A/B
	Stereotyped behavior	Amph-cue	
d-Amphethamine	4.5	1	4.5
Bupropion	40	22	1.8
Diclofensine	> 60	3	>20
GBR 12909	140	–	–
Imipramine	> 64	>20	–
Methylphenidate	10	5	2
Nisoxetine	> 30	>20	–
Nomifensine	11	4	2.8

Drugs were injected s.c., *t*, 120 min, except bupropion and nisoxetine (i.p., *t*, 30 min).

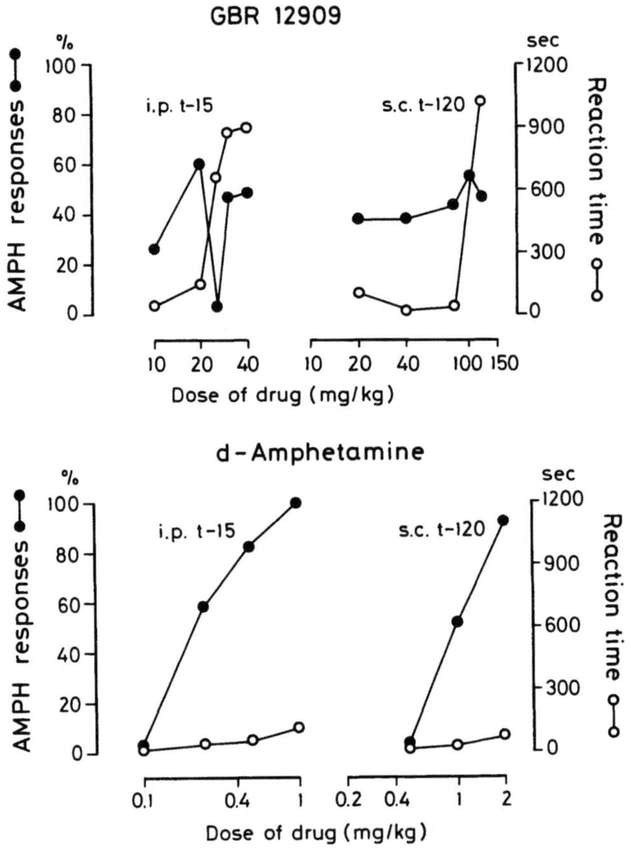

Fig. 1. Ability of i.p. (*t*, 15 min) or s.c. (*t*, 120 min) administered GBR 12909 and amphetamine to elicit the cueing effect of amphetamine in animals trained to discriminate 1 mg/kg of this drug (i.p., *t*, 15 min) from saline. Reaction time is the time to complete the first 32 responses. AMPH responses is the percentage of responses on the amphetamine lever in test sessions which were terminated after the completion of 32 responses on either lever Nielsen, unpublished).

appropriate to consider at least nomifensine a selective norepinephrine (NE) uptake inhibitor (Hyttel and Larsen 1985; Heikkila and Manzino 1984). However, inhibition of NE uptake cannot per se explain that nomifensine substitutes for amphetamine since nisoxetine, a potent and selective NE-uptake inhibitor (Hyttel and Larsen 1985), fails to substitute in the present experiments [substitution for amphetamine by high doses of nisoxetine in mice has, however, been reported by Snoddy and Tessel (1983) (Table 1)]. Thus, the basis for the ability of the drugs to substitute for amphetamine is unclear.

GBR 12909, a phenyl-substituted piperazine derivative, however, exhibits an extraordinary selectivity for, and potency at DA-uptake inhibition (Heikkila and Manzino 1984). GBR 12909 failed to substitute for amphetamine or did so only partially (Fig. 1) although GBR 12909 does induce stereotyped behavior (see Table 1 for data for GBR 12909 and representative uptake inhibitors). It is noteworthy that there are widely different ratios between amphetamine cue-induction and stereotypye induction, suggesting different mechanisms of action of the compounds. Indeed, GBR 12909 causes stereotyped behavior without eliciting the amphetamine cue. In contrast, diclofensine, an isoquinoline derivative developed as an antidepressant (Keller et al. 1982), fails to produce stereotyped behavior although it readily substituted for the amphetamine cue. Similar to bupropion, diclofensine inhibits the uptake of both DA and NE. However, concurrent inhibition of the uptake of DA and NE is unable to account for the amphetamine-like cueing effect of the drugs since the combination of an NA-uptake inhibitor (talsopram) and a DA-uptake inhibitor (GBR 13069) fails to elicit the amphetamine cue (J. Arnt 1986, personal communication). Thus, inhibition by a drug of the uptake of DA, NE, or 5-HT fails to predict per se the resulting behavioral effect (e.g., amphetamine cue substitution or locomotor activation). Rather, it can be speculated that release of DA, at least, plays a critical role for CNS-stimulant cue-induction.

What may be the basis for the behavioral profile of GBR 12909? It can be speculated that drugs which act via the DA-uptake mechanism may exert their effect according to the ongoing physiological turnover/uptake rate of endogenous DA. This rate may differ in different parts of the CNS. As a matter of fact, it has generally been observed that the levels of DA metabolites are higher in striatal areas than in mesolimbic areas (e.g., Wiesel and Sedvall 1975). A further interesting point is the anatomical selectivity of the DAergic neurotoxin, MPTP. This compound is metabolized in glial cells into MPP^+, which is subsequently taken up via the DA-uptake mechanism (Langston 1985). MPP^+ then eventually reaches the DA cell body, which is killed by interference of MPP^+ with the cell's energy metabolism (Langston 1985). A remarkable feature of MPTP's neurotoxicity is that it primarily involves the nigrostriatal DA-pathway (Langston 1985), consistent with the hypothesis of regional selectivity of drugs acting via the DA-uptake mechanism. Thus, GBR 12909 may exert a regional selectivity, resulting in a preferential activation of striatal DA systems. Finally, this possible striatal selectivity of GBR 12909 parallels the contention that CNS-stimulant cues primarily involve nonstriatal (e.g., mesolimbic) DA systems.

3.2 Other Presynaptic Mechanisms

As previously discussed, inhibition of DA uptake per se apparently may not increase DA levels sufficiently to produce a CNS-stimulant cue. However, relatively limited information exists regarding presynaptic and other mechanisms involved in CNS-stimulant cues. Several investigators have found that pretreatment of amphetamine-trained animals with α-methyl-*p*-tyrosine (α-MPT) antagonized this drug cue, suggesting involvement of "newly" synthesized DA (Ho and Huang 1975; Schechter and Cook 1975; Kuhn et al. 1974). McKenna and Ho (1979) found that α-MPT had no effect on the cocaine cue, although it was blocked by reserpine, suggesting involvement of the "storage pool" of DA. Thus, similar neuronal mechanisms of action of cocaine and amphetamine, which have been determined for the motor actions of the compounds (Scheel-Krüger 1971, 1977), appear to be involved in the cueing effects of the drugs.

Stereoselectivity for elicitation of amphetamine's cueing effect has been demonstrated (Jones et al. 1974; Schechter 1978). Thus, *d*-amphetamine is approximately five times more potent than *l*-amphetamine in eliciting a *d*-amphetamine cue. This suggests that at least the cueing effect of *d*-amphetamine may depend on DA since the same potency ratio has been obtained in studies of DA release/uptake inhibition by the two isomers (Louilot et al. 1985).

Structure-activity studies have revealed that the α-methyl group of amphetamine and of cathionine plays an important role for the drugs' cueing effects (Glennon et al. 1984a, b).

It has been reported that inhibition of monoamine oxidase (MAO) type B could promote a cue similar to that produced by 5 mg/kg cocaine (Colpaert et al. 1980). These authors also speculated that endogenous PEA could be involved in the cueing effect of cocaine. MAO inhibition does not, however, seem to play a role in amphetamine discrimination (Porsolt et al. 1984).

An interesting finding was reported by Ho and McKenna (1978): propranolol (a β-adrenergic antagonist) was able to substitute for the cocaine cue. Colpaert et al. (1979), however, found the cocaine/propranolol generalization to be partial and to go both ways. This is further of interest since it has been shown that both compounds have similar actions as local anesthetics on calmodulin (Volpi et al. 1981; Memo et al. 1981), suggesting that nerve-membrane processes are implicated in the cocaine cue. Propranolol does not, however, substitute for amphetamine (unpublished results). Thus amphetamine and cocaine may differ with respect to the neuronal mechanisms involved in their respective cues. Further work is needed to clarify whether the effect of propranol (and cocaine) is mediated by DAergic or other mechanisms (e.g., calmodulin).

4 Postsynaptic Mechanisms Involved in CNS-Stimulant Cues

4.1 Antagonism Studies

It was reported early that the prototypical DA antagonist haloperidol blocks the cueing effects of amphetamine (e.g., Schechter and Cook 1975). Thorough characterization of the cue-antagonistic effect of DA antagonists was later reported by Schechter (1979) and by Nielsen and Jepsen (1985). Of particular interest was the findings by Nielsen and Jepsen (1985) that so-called "atypical" antipsychotic drugs, in addition to classical DA antagonists, antagonize the amphetamine cue. Atypical antipsychotic compounds are relatively unable to inhibit DA-dependent stereotyped behavior; yet the drugs are clinically efficacious (Gerlach et al. 1984). It has been speculated that the atypical nature of clozapine and thioridazine may involve their strong anticholinergic properties, which is thought to outbalance striatal but not mesolimbic DA-antagonism (e.g., Arnt 1983, 1985; Creese 1983). That atypical antipsychotic drugs may have regional selectivity has further been established by electrophysiological studies of firing rates of DA neurons (White and Wang 1983; Chiodo and Bunney 1983), by in vivo receptor binding techniques (e.g., Köhler et al. 1981), and by DA-turnover studies (e.g., Bartholini 1976). These observations provide support for an involvement of mesolimbic DA systems in the ability of DA-antagonists to block the cueing effect of CNS stimulants.

The results from the antagonism studies suggest further that both D-1 and D-2 receptors are involved in the amphetamine cue since it is antagonized by both a D-1-selective antagonist (SCH 23390) and D-2-selective antagonists (e.g., spiroperidol and sulpiride) as well as by nonselective drugs (e.g., *cis*-flupenthixol)

(Nielsen and Jepsen 1985). The cueing effects of CNS stimulants are remarkably pharmacologically specific since they, apparently, are only counteracted by DA antagonists. Drugs which have no antagonistic effect on a CNS-stimulant cue include α- and β-adrenergic agonists/antagonists and 5-HT-, opioid-, and cholinergic antagonists (see reviews by Silverman and Ho 1977; Ho and McKenna 1978; see also Ho and Huang 1975; Järbe 1978; McKenna and Ho 1979; Hernandez et al. 1978; Nielsen and Jepsen 1985).

Finally, it is of interest to note that cocaine and amphetamine may differ with respect to their involvement of DA receptors. Thus, the D-2 antagonist haloperidol apparently is less able to block completely the cueing effect of cocaine (Colpaert et al. 1978). However, relatively little work has been done with different types of DA antagonists (e.g., D-1/D-2 selective) against the cocaine cue.

4.2 Substitution Studies

Direct DA agonists have been tested for their ability to substitute for CNS stimulants. Following the scheme of Stoof and Kebabian (1981) of only two types of DA receptors, DA agonists fall into three groups: (1) D-1 agonists (e.g., fenoldopam, SKF 38393, and SKF 75670), (2) mixed D-1/D-2 agonists (e.g., apomorphine), and (3) D-2 agonists (e.g., pergolide, LY 171555, and CH 29717). The benzazepines SKF 38393 and SKF 75670 do not, even in high doses, substitute for amphetamine (Fig. 2). However, none of these compounds are full D-1 ago-

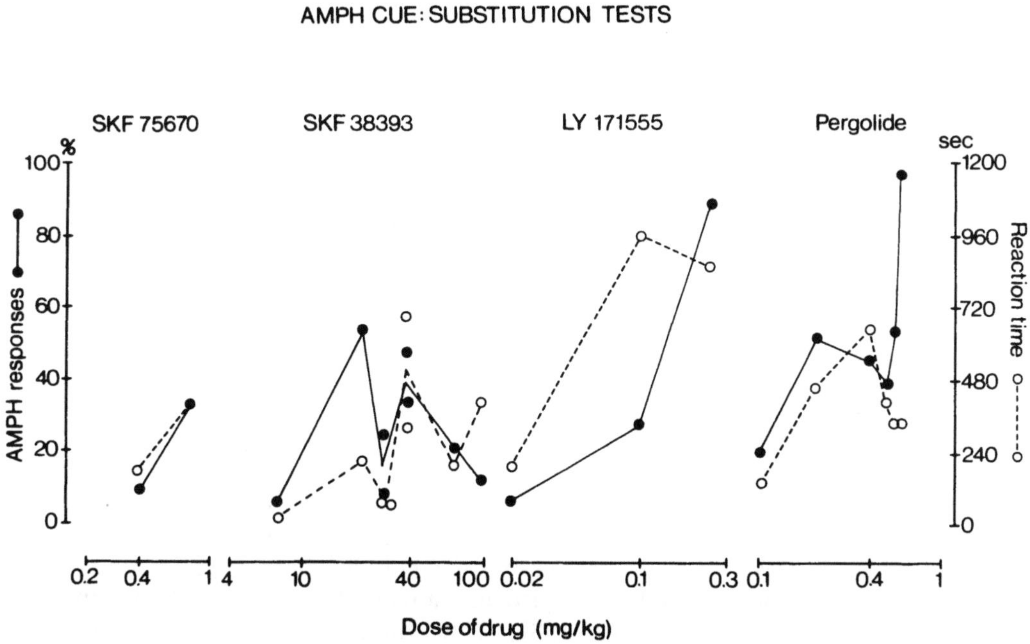

Fig. 2. Ability of D-1 and D-2 agonists to substitute for *d*-amphetamine sulfate (1 mg/kg, i.p., *t*, 15 min). See legend to Fig. 1 for other details

Table 2. D-1 receptor ligands: affinity, selectivity, and efficiency

Compound	[^{3}H]SCH 23390 binding K_i (nM)	[^{3}H]Spiperone binding K_i (nM)	D-2/D-1 ratio	Adenylate cyclase K_i/K_m (nM)	Relative efficacy
Dopamine	380	660	1.7	43000	1
Fenoldopam	3.1	170	55	800[a]	0.65
SKF 38393	18	9300	517	190[a]	0.45
SKF 75670	1.9	840	442	30[a]	0.30
SCH 23390	0.14	895	6400	40	0

[a] Stimulation of adenylate cyclase activity did not conform to Michaelis-Menten kinetics. Data from Andersen et al. (1985) and unpublished values.

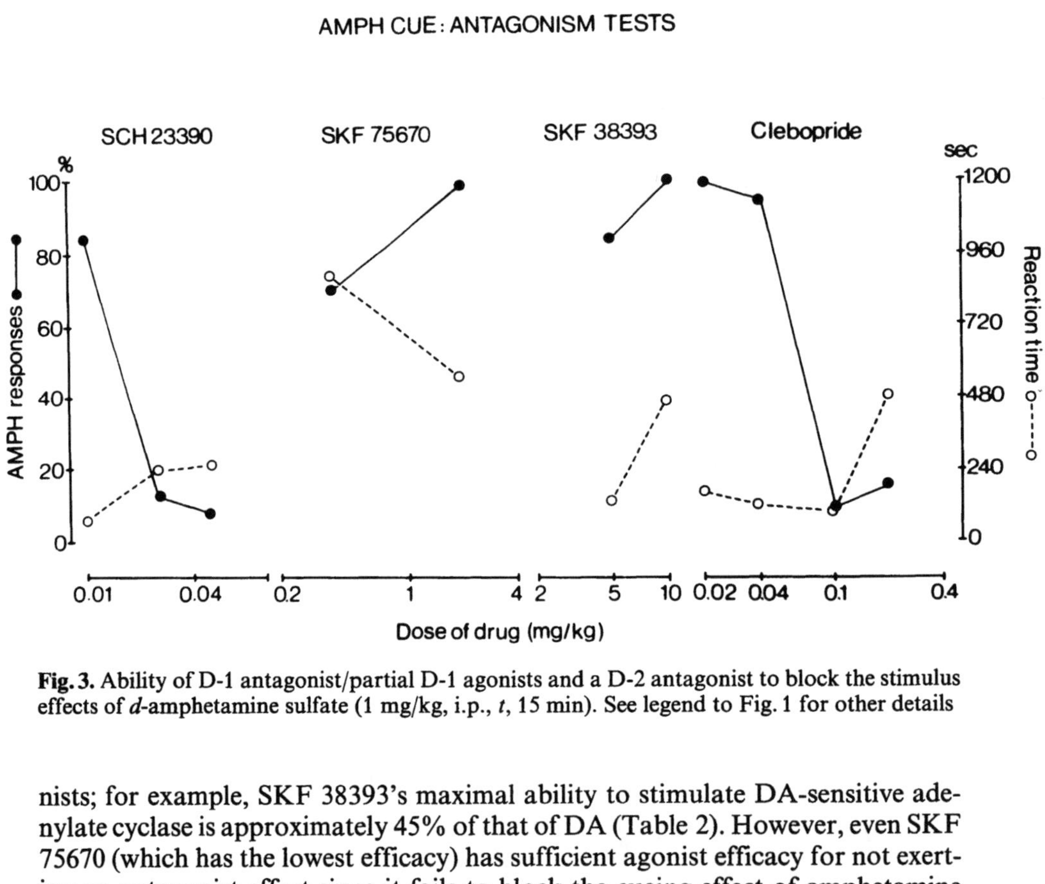

Fig. 3. Ability of D-1 antagonist/partial D-1 agonists and a D-2 antagonist to block the stimulus effects of *d*-amphetamine sulfate (1 mg/kg, i.p., *t*, 15 min). See legend to Fig. 1 for other details

nists; for example, SKF 38393's maximal ability to stimulate DA-sensitive adenylate cyclase is approximately 45% of that of DA (Table 2). However, even SKF 75670 (which has the lowest efficacy) has sufficient agonist efficacy for not exerting an antagonist effect since it fails to block the cueing effect of amphetamine (as does SCH 23390) (Fig. 3). However, it is also possible that there exist subtypes of the D-1 receptor (e.g., cyclase coupled and noncoupled receptors; Andersen et al. 1985; Andersen and Braestrup 1987). Thus, the role of D-1 receptors in the cueing effects of CNS stimulants is difficult to evaluate due to the lack of selective and potent agonists.

As noted previously, the ability of apomorphine to substitute for amphetamine, at least, depends on the training dose of amphetamine (Stolerman and D'Mello 1981). Thus apomorphine does not substitute for a low-dose amphetamine cue (produced, for example, by 1 mg/kg). This is contrasted by the fact that specific D-2 agonists (e.g., pergolide and LY 171555) substitute completely for amphetamine. At present it is unclear whether this lack of substitution by apomorphine may reflect a regional selectivity of the compound (Grome and McCulloch 1983) or whether its D-1 agonist effect is related to its lack of amphetamine-like stimulus properties. With respect to the motor effects of these compounds, it is interesting to note that apomorphine does not readily produce locomotor stimulation but has a relatively selective stereotypy-inducing effect (Randrup et al. 1980). This may parallel its ability to substitute only for higher "stereotypic" doses of amphetamine (above). Pergolide, on the other hand, induces locomotor activation to a degree comparable to that of amphetamine (unpublished results).

Another direct DA agonist, lisuride, also fails to substitute for amphetamine (Nielsen and Jepsen 1985); however, lisuride has many other discriminable effects (e.g., α_2-agonist; Colpaert and Janssen 1985), which may "overshadow" its D-2 agonist properties.

Thus, specific D-2 agonists substitute completely for amphetamine; coupled with the fact that specific D-2 antagonists block the cue (Nielsen and Jepsen 1985), a crucial role can be ascribed to D-2 receptors in the amphetamine cue. Furthermore, D-1 receptors also appear to be involved since SCH 23390 blocks completely the amphetamine cue (Nielsen and Jepsen 1985). In further experiments we have explored this possible coupling of D-1 and D-2 receptors in mediating the amphetamine cue.

4.3 D-1/D-2 Receptor Coupling

In these (unpublished) experiments we have studied pergolide's ability to substitute for amphetamine. The D-1 antagonist, SCH 23390, was unable to block this effect of pergolide although SCH 23390 completely blocked the motor-activating properties of pergolide (Figs. 4, 5). The substitution for amphetamine by pergolide was, however, completely blocked by raclopride, a specific D-2 receptor antagonist of the substituted benzamide class. The lack of effect of SCH 23390 against the pergolide substitution for amphetamine may indicate independency of D-1 and D-2 receptors in the DA systems involved in amphetamine discrimination. SKF 38393 potentiated the motor-activating properties of pergolide (Table 3). However, when tested in amphetamine-trained animals, the combination of SKF 38393 and pergolide decreased the rate of responding to the extent that only one out of eight animals responded (unpublished). This may indicate the occurrence of non-cue-specific effects (e.g., stereotypy).

Thus, it appears that (1) D-1 and D-2 receptors may be directly coupled in (some) DA systems controlling motor function and (2) D-1 and D-2 receptors may be indirectly coupled in mesolimbic DA systems possibly underlying amphetamine discrimination. It can be further speculated that at some point in the chain of synapses involved in CNS-stimulant cues, D-1 receptors may be preced-

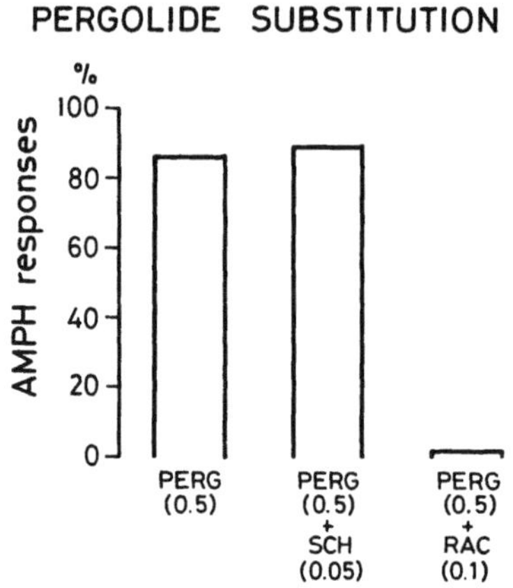

Fig. 4. Ability of pergolide (0.5 mg/kg) to substitute for amphetamine in animals trained to discriminate 1 mg/kg of this drug (i.p. *t*, 15 min). Pergolide (0.5 mg/kg) was given alone or in conjunction with the D-1 antagonist SCH 23390 or the D-2 antagonist raclopride

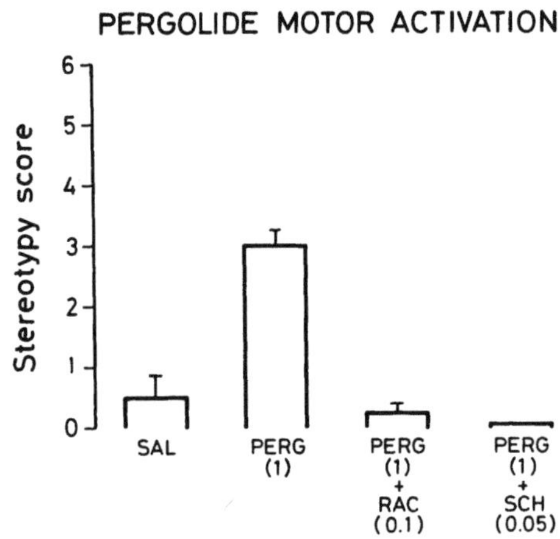

Fig. 5. Ability of the D-1 antagonist SCH 23390 and the D-2 antagonist raclopride to block the locomotor activation induced by a low dose of pergolide (1 mg/kg)

Table 3. SKF 38393-pergolide locomotor activation. Ability of SKF 38393 to potentiate the locomotor response by 1 mg/kg pergolide

Treatment	Stereotype score	*N*
Saline	1.8±0.6	8
SKF 38393 (5)	1.0±0	6
Pergolide (1)	2.8±0.3*	6
Perg+SKF	3.8±0.2*	6

* $P<0.05$.

Behavior was rated according to a conventional continuum from 0–6 (0, inactivity; 1, awake, grooming, eating, drinking; 2, locomotor activity; 3, sniffing over large area of cage; 4, sniffing in a restricted area; 5, licking, biting, forepaw treading; 6, abnormal crouched posture). Doses in milligrams per kilogram in parentheses

ing D-2 receptors. This is based on the observation that the cue can be elicited by a D-2 agonist despite blockade of D-1 receptors by SCH 23390 (which is otherwise capable of antagonizing amphetamine; above). Indeed, it has been found that low concentrations of amphetamine release sufficient DA to activate D-1-regulated cyclic AMP production (Kelly and Nahorski 1986).

D-1 and D-2 receptors are unevenly distributed in the brain (Kebabian et al. 1986; Dawson et al. 1985). This may indicate that there are differences in the extent to which selective compounds acting at one receptor may affect the converse receptor function. However, based on the present complexity of putative D-1/D-2 interactions, it appears premature to try to diagram the possible coupling of the receptors; for example, it has been shown that in the nucleus accumbens both separate D-1- and D-2- as well as combined D-1/D-2-responsive cells exist (White and Wang 1986). Finally, it should be noted that although the coupling of D-1 and D-2 receptors may be "direct" regarding motor function, it nevertheless is dependent on a functional DA-input since its removal by 6-OHDA or reserpine has been shown to "decouple" D-1/D-2 receptor function (Arnt and Hyttel 1984; Arnt 1985; Breese and Mueller 1985).

In summary, the present results show that the coupling of D-1 and D-2 receptors may depend on the particular behavioral function in question; this dependency may in turn reflect regional differences in D-1/D-2 receptor coupling.

5 Intracerebral Drug Studies

Early work (D'Mello 1981) attempted to determine whether electrical stimulation of mesolimbic DA A10 cells could promote discriminable similarity to the cueing effect of amphetamine. The findings indicated a partial similarity. A more direct approach was taken by Nielsen and Scheel-Krüger (1986), who trained animals to discriminate amphetamine. Drugs were then microinjected into discrete areas of the brain. Most importantly, it was found that low doses of amphetamine injected into the accumbens produced a discriminable similarity (ED_{50}, 0.24 μg) to i.p.-injected amphetamine (the training condition) (Fig. 6). Amphetamine injected into two striatal areas failed to produce the amphetamine cue (Nielsen and Scheel-Krüger 1986). The intraaccumbens amphetamine generalization to systemic amphetamine was blocked with low doses (50–100 ng) of (−)-sulpiride, coinjected into the accumbens (Nielsen and Scheel-Krüger 1986). These findings represent the first direct evidence of site specificity of a drug stimulus using intracerebral drug application.

In further (unpublished) experiments, we have studied the ability of specific D-1 and D-2 agonists, injected intracerebrally, to substitute for amphetamine. Briefly, the D-2 agonist CH 27717 (1 μg) substituted completely for amphetamine. Partial (max ≈ 50% amphetamine-lever responses) effect was obtained with the D-1 agonists fenoldopam, SKF 38393, and SKF 75670 (1–10 μg). However, it should be recalled that none of these D-1 agonists are full-efficacy agonists when evaluated with respect to their stimulatory effect on DA-sensitive adenylate cyclase. Future plans involve similar experiments with newly developed compounds which appear to be full D-1 agonists (Truex et al. 1985).

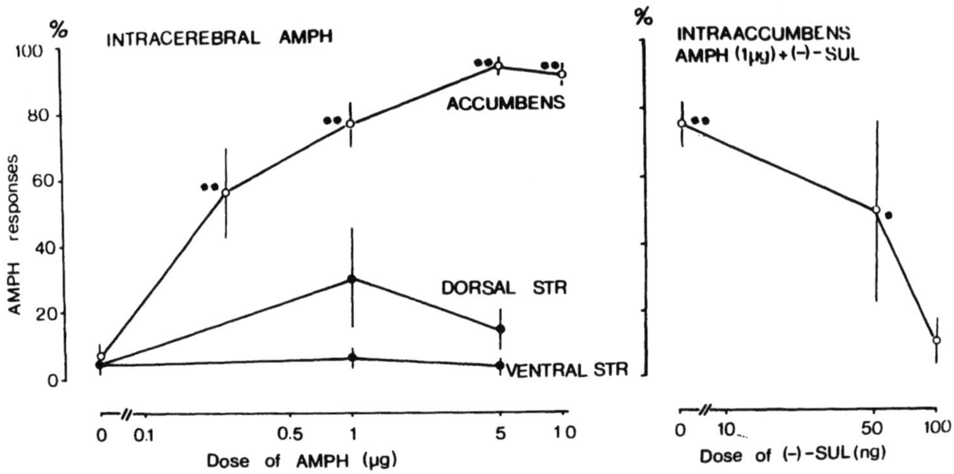

Fig. 6. *Left* Ability of *d*-amphetamine injected into the nucleus accumbens, the anterior dorsomedial striatum, or the anterior ventrolateral striatum to mimic the stimulus effect of systemically administered *d*-amphetamine sulfate (1 mg/kg i.p., *t*, 15 min; the training condition). *Right* Dose-dependent ability of (−)-sulpiride, when injected into the nucleus accumbens, to antagonize the cueing effect of *d*-amphetamine (1 µg), coinjected in the same structure. Reproduced from Nielsen and Scheel-Krüger (1986) with permission from Elsevier Science Publishers, Amsterdam, and the authors

The intracerebral data support the contention that the motor-activating properties of amphetamine and its stimulus effects are mediated by separate mechanisms. This may have relevance for understanding some of the actions of atypical antipsychotic compounds. Thus clozapine and thioridazine, molindone, and sulpiride have been claimed to exert a preferential mesolimbic action. The fact that these compounds do not readily block the stereotyped behavior induced by amphetamine, while blocking the cueing effect of the drug, parallels the preferential mesolimbic DA-involvement in this latter effect. For this reason, future work in this area may be highly important for discovering neuronal systems which may point to new strategies for drug treatment of schizophrenia and related disorders.

6 Summary

Central nervous system-stimulant drugs appear to produce similar cueing effects in animals. Different cues may, however, be produced at different doses of at least cocaine and amphetamine. At relatively low doses of CNS stimulants, this general CNS stimulant cue is mediated by endogenous DA, released from presynaptic stores. The CNS stimulant cue is primarily mediated by activation of mesolimbic DA systems. In this respect, it differs from the motor effects of the drugs which may, in addition or primarily, involve striatal DA systems. Postsynaptically, evidence for involvement of both DA D-1 and D-2 receptors exist. The arrangement of the D-1/D-2 links in the chain of synaptic connections involved in CNS-stimulant discrimination is at present complex and not well understood.

Acknowledgments. Dr. Peter H. Andersen is thanked for providing the biochemical data. Mrs Judith L. Knutsen is thanked for preparing the manuscript.

References

Aceto MD, Rosecrans JA, Young R, Glennon RA (1984) Similarity between (+)-amphetamine and amfonelic acid. Pharmacol Biochem Behav 20:635–637

Andersen P, Braestrup C (1987) Evidence for different states of the dopamine D1 receptor. J Neurochem 48:1667–1672

Andersen PH, Grønvald FC, Janssen JA (1985) A comparison between dopamine-stimulated adenylate cyclase and ^{3}H-SCH 23390 binding in rat striatum. Life Sci 37:1971–1983

Arnt J (1983) Neuroleptic inhibition of 6,7-ADTN-induced hyperactivity after injection into the nucleus accumbens. Specificity and comparison with other models. Eur J Pharmacol 90:47–55

Arnt J (1985) Antistereotypic effects of dopamine D-1 and D-2 antagonists after intrastriatal injection in rats. Pharmacological and regional specificity. Naunyn-Schmiedeberg's Arch Pharmacol 330:97–104

Arnt J, Hyttel J (1984) Differential inhibition by dopamine D-1 and D-2 antagonists of circling behaviour induced by dopamine agonists in rats with unilateral 6-hydroxydopamine lesions. Eur J Pharmacol 102:349–354

Bartholini G (1976) Differential effect of neuroleptic drugs on dopamine turnover in the extrapyramidal and limbic system. J Pharm Pharmacol 28:429–433

Braestrup C, Scheel-Krüger J (1976) Methylphenidate-like effect of the antidepressant drug nomifensine (HOE 984). Eur J Pharmacol 38:305–312

Breese GR, Mueller RA (1985) SCH 23390 antagonism of a D-2 dopamine agonist depends upon catecholaminergic neurons. Eur J Pharmacol 113:109–114

Browne RG, Weissman A (1981) Discriminative stimulus properties of tetrahydrocannabinol: mechanistic studies. J Clin Pharmacol 21:2275–2345

Bueno OFA, Carlini EA, Finkelfarb E, Suzuki JS (1976) 9-Tetrahydrocannabinol, ethanol and amphetamine as discriminative stimuli – generalization tests with other drugs. Psychopharmacologia 46:235–243

Chiodo LA, Bunney BS (1983) Typical and atypical neuroleptics: differential effects of chronic administration on the activity of A9 and A10 midbrain dopaminergic neurons. J Neurosci 8:1Å07–1Å19

Colpaert FC (1982) The pharmacological specificity of opiate-drug discrimination. In: Colpaert FC, Janssen PA (eds) Drug discrimination: applications in CNS pharmacology. Elsevier Biomedical, Amsterdam, pp 3–16

Colpaert FC, Janssen PA (1985) Discriminative stimulus properties of xylazine in rat: discriminability and effects of putative alpha-2-adrenoceptor agonists and antagonists. J Pharmacol Exp Ther 235:521–527

Colpaert FC, Slangen JF (1982) Drug discrimination: applications in CNS pharmacology. Elsevier Biomedical, Amsterdam, pp 3–16

Colpaert FC, Niemegeers CJE, Janssen PAJ (1978) Discriminative stimulus properties of cocaine and *d*-amphetamine, and antagonism by haloperidol: a comparative study. Neuropharmacology 17:937–942

Colpaert FC, Niemegeers CJE, Janssen PAJ (1979) Discriminative stimulus properties of cocaine: neuropharmacological characteristics as derived from stimulus generalization experiments. Pharmacol Biochem Behav 10:535–546

Colpaert FC, Niemegeers CJE, Janssen PAJ (1980) Evidence that a preferred substrate for type B monoamine oxidase mediates stimulus properties of MAO inhibitors: a possible role for β-phenylethylamine in the cocaine cue. Pharmacol Biochem Behav 13:513–517

Costall B, Marsden CD, Naylor RJ, Pycock CJ (1977) Stereotyped behavior patterns and hyperactivity induced by amphetamine and apomorphine after discrete 6-hydroxydopamine lesions of extrapyramidal and mesolimbic nuclei. Brain Res 123:89–111

Creese I (1983) Receptor interactions of neuroleptics. In: Coyle JT, Enna SJ (eds) Neuroleptics: neurochemical, behavioral and clinical perspectives. Raven, New York, pp 183–222

Cunningham KA, Appel JB (1982) Discriminative stimulus properties of cocaine and phencyclidine: similarities in the mechanism of action. In: Colpaert FC, Slangen JL (eds) Drug discrimination: applications in CNS pharmacology. Elsevier Biomedical, Amsterdam, pp 181–192

Dawson TM, Gehlert DR, Yamamura HI, Barnett A, Wamsley JK (1985) D-1 dopamine receptors in the rat brain: autoradiographic localization using [^{3}H]SCH 23390. Eur J Pharmacol 108:323–325

D'Mello GD (1981) A comparison of some behavioral effects of amphetamine and electrical brain stimulation of the mesolimbic dopamine system in rats. Psychopharmacology 7:184–192

D'Mello GD, Stolerman IP (1977) Comparison of the discriminative stimulus properties of cocaine and amphetamine in rats. Br J Pharmacol 61:415–422

Emmett-Oglesby MW, Wurst M, Lal H (1983) Discriminative stimulus properties of a small dose of cocaine. Neuropharmacology 22:97–101

Ferris RM, White HL, Cooper BR, Maxwell RA, Tang FLM, Beaman OJ, Russell A (1981) Some neurochemical properties of a new antidepressant, bupropion hydrochloride (Wellbutrin). Drug Dev Res 1:21–35

Gerlach J, Korsgaard S, Lund Laursen A, Noring U, Paulsen UJ (1984) New drugs and perspectives in antipsychotic treatment. Clin Neuropharmacol 7:550–551

Glennon RA, Schechter MD, Rosecrans JA (1984 a) Discriminative stimulus properties of S(−)- and R(+)-cathinone, (+)-cathine and several structural modifications. Pharmacol Biochem Behav 21:1–3

Glennon RA, Young R, Hauck AE, McKenny JD (1984 b) Structure-activity studies on amphetamine analogs using drug discrimination methodology. Pharmacol Biochem Behav 21:895–901

Goudie AJ (1983) Discriminative stimulus properties in an operant task of β-phenylethylamine. In: Colpaert FC, Slangen JL (eds) Drug discrimination: applications in CNS pharmacology. Elsevier Biomedical, Amsterdam, pp 165–180

Grome JJ, McCulloch J (1983) The effects of apomorphine upon local cerebral glucose utilization in conscious rats and in rats anesthesized with chloral hydrate. J Neurochem 40:569–576

Heikkila RE, Manzino L (1984) Behavioral properties of GBR 12909, GBR 13069 and GBR 13098: specific inhibitors of dopamine uptake. Eur J Pharmacol 103:241–248

Hernandez LL, Holohean AM, Appel JB (1978) Effects of opiates on the discriminative stimulus properties of dopamine agonists. Pharmacol Biochem Behav 9:459–461

Ho BT, Huang J-T (1975) Role of dopamine in *d*-amphetamine-induced discriminative responding. Pharmacol Biochem Behav 3:1085–1092

Ho BT, McKenna M (1978) Discriminative stimulus properties of central stimulants. In: Ho BT, Richards DW III, Chute DL (eds) Drug discrimination and state dependent learning. Academic, New York, pp 67–77

Holloway FA, Michaelis R, Huerta PL (1985) Caffeine-phenylephylamine combinations mimic the amphetamine discriminative cue. Life Sci 36:723–730

Huan J-T, Ho BT (1974) The effect of pretreatment with iproniazid on the behavioral activities of β-phenylethylamine in rats. Psychopharmacologia 35:77–81

Hyttel J, Larsen J-J (1985) Neurochemical profile of Lu 19-005, a potent inhibitor of uptake of dopamine, noradrenaline, and serotonin. J Neurochem 44:1615–1622

Järbe TUC (1978) Cocaine as a discriminative cue in rats: interactions with neuroleptics and other drugs. Psychopharmacology 59:183–187

Jones CN, Hill HF, Harris RT (1974) Discriminative response control by *d*-amphetamine and related compounds in the rat. Psychopharmacologia 36:347–356

Kebabian JW, Agui T, van Oene JC, Shigematsu K, Saavedra JM (1986) The D_1 dopamine receptor: new perspectives. Trends Pharmacol Sci 7:96–99

Keller HH, Schaffner R, Carruba MO, Burkard WP, Pieri M, Bonetti EP, Scherschlicht R, Do Prada M, Maefely WE (1982) Diclofensin (Ro 8-4650) – a potent inhibitor of monoamine uptake. In: Costa E, Racagni G (eds) Typical and atypical antidepressants: molecular mechanisms. Raven, New York, pp 249–263
Kelly E, Nahorshi SR (1986) Functional activation of DA D-1 and D-2 receptors by endogenous released dopamine in the rat striatum. Br J Pharmacol 88:305
Köhler C, Haglund L, Ögren S-O, Ängeby T (1981) Regional blockade by neuroleptic drugs of in vivo ^{3}H-spiperone binding in the rat brain. Relation to blockade of apomorphine induced hyperactivity and stereotypies. J Neural Transm 52:163–173
Kuhn DM, Appel JB, Greenberg I (1974) An analysis of some discriminative properties of *d*-amphetamine. Psychopharmacologia 39:57–66
Langston JW (1985) Mechanism of MPTP toxicity: more answers, more questions. Trends Pharmacol Sci 6:375–378
Louilot A, Gonon F, Buda M, Simon H, Le Moal M, Pujol JF (1985) Effects of *d*- and *l*-amphetamine on dopamine metabolism and ascorbic acid levels in nucleus accumbens and olfactory tubercle as studied by in vivo differential pulse voltammetry. Brain Res 336:235–263
Lyon M, Robbins TW (1975) The action of central nervous system stimulant drugs: a general theory concerning amphetamine effects. In: Essman WB, Valzelli L (eds) Current developments in psychopharmacology, vol 2. Spectrum, New York, pp 80–163
McKenna ML, Ho BT (1979) The role of dopamine in the discriminative stimulus properties of cocaine. Neuropharmacology 19:297–303
Memo M, Pradhan S, Hanbauer I (1981) Cocaine-induced supersensitivity of striatal dopamine receptors: role of endogenous calmodulin. Neuropharmacology 20:1145–1150
Nielsen EB (1981) Rapid decline of stereotyped behavior in rats during constant one-week amphetamine administration via implanted osmotic ALZET minipumps. Pharmacol Biochem Behav 15:161–165
Nielsen EB, Jepsen SA (1985) Antagonism of the amphetamine cue by both classical and atypical antipsychotic drugs. Eur J Pharmacol 111:167–176
Nielsen EB, Scheel-Krüger J (1986) Cueing effects of amphetamine and LSD: elicitation by direct microinjection of the drugs into the nucleus accumbens. Eur J Pharmacol 125:85–92
Porsolt RD, Pawalec C, Jalfre M (1982) Use of a drug discrimination procedure to detect amphetamine-like effects of antidepressants. In: Colpaert FC, Slangen JL (eds) Drug discrimination: applications in CNS pharmacology. Elsevier Biomedical, Amsterdam, pp 193–202
Porsolt RD, Pawelec C, Roux S, Jalfre M (1984) Discrimination of the amphetamine cue. Effects of A, B and mixed type inhibitors of monoamine oxidase. Neuropharmacology 23:569–573
Randrup A, Kjellberg B, Schiørring E, Scheel-Krüger J, Fog R, Munkvad I (1980) Stereotyped behavior and its relevance for testing neuroleptics. In: Hoffmeister F, Stille G (eds) Psychotropic agents, vol 1. Springer, Berlin Heidelberg New York, pp 97–110
Robbins TW, Everett BJ (1982) Functional studies of the central catecholamines. Int Rev Neurobiol 23:303–365
Sakurai Y, Hisashi O, Shimazoe T, Kataoka Y, Fujiwara M, Ueki S (1985) Δ^9-Tetrahydrocannabinol elicited ipsilateral circling behavior in rats with unilateral nigral lesion. Life Sci 37:2181–2185
Schechter MD (1977) Amphetamine discrimination as a test for anti-Parkinsonism drugs. Eur J Pharmacol 44:51–56
Schechter MD (1978) Stimulus properties of *d*-amphetamine as compared to *l*-amphetamine. Eur J Pharmacol 47:461–464
Schechter MD (1979) Effect of neuroleptics and tricyclic antidepressants upon D-amphetamine discrimination. Pharmacol Biochem Behav 12:1–5
Schechter MD, Cook PG (1975) Dopaminergic mediation of the interoceptive cue produced by *d*-amphetamine in rats. Psychopharmacologia 42:185–193
Scheel-Krüger J (1971) Comparative studies of various amphetamine analogues demonstrating different interactions with the metabolism of the catecholamines in the brain. Eur J Pharmacol 14:47–59
Scheel-Krüger J, Braestrup C, Nielsen M, Golembiowska K, Mogilnicka E (1977) Cocaine: discussion on the role of dopamine in the biochemical mechanism of action. In: Ellingwood EH, Kilbey MM (eds) Cocaine and other stimulants. Advances in behavioral biology. Plenum, New York, pp 373–407

Shearman GT, Lal H (1981) Discriminative stimulus properties of cocaine related to an anxiogenic action. Prog Neuro-psychopharm 5:57–63
Silverman PB, Ho BT (1977) Characterization of discriminative response control by psychomotor stimulants. In: Lal H (ed) Discriminative stimulus properties of drugs. Plenum, New York, pp 107–119
Snoddy AM, Tessel RE (1983) Nisoxetine and amphetamine share discriminative stimulus properties in mice. Pharmacol Biochem Behav 19:205–210
Stolerman IP, D'Mello GD (1981) Role of training conditions in discrimination of central nervous system stimulants by rats. Psychopharmacology 73:295–303
Stoof JC, Kebabian JW (1981) Opposing roles for D-1 and D-2 dopamine receptors in efflux of cyclic AMP from rat neostriatum. Nature 294:366–368
Truex LL, Foreman MM, Rigs RM, Nichols DE (1985) Effects of modifications of the 4-(3,4-dihydroxyphenyl)-1,2,3,4-tetrahydro-isoquinoline structure on dopamine sensitive rat retinal adenylate cyclase activity. Soc Neurosci 11:1094
Volpi M, Shaafi RI, Epstein PM, Andrenyak DM, Feinstein MB (1981) Local anaesthetics, mepacrine, and propranolol are antagonists of calmodulin. Proc Natl Acad Sci USA 78:795–799
White FJ, Wang RY (1983) Differential effects of classical and atypical antipsychotic drugs on A9 and A10 dopamine neurons. Science 221:1054–1057
White FJ, Wang RY (1986) Electrophysiological evidence of both D-1 and D-2 dopamine receptors in the rat nucleus accumbens. J Neurosci 6:1–7
Wiesel F-A, Sedvall G (1975) Effect of antipsychotic drugs on homovanillic acid levels in striatum and olfactory tubercle of the rat. Eur J Pharmacol 30:364–367

Discriminative Stimulus Properties of Anxiolytic and Sedative Drugs: Pharmacological Specificity

D. J. SANGER

Contents

Abstract

In the first set of experiments rats were trained to discriminate a dose of 5 mg/kg chlordiazepoxide from saline. The chlordiazepoxide cue was antagonized by flumazepil (Ro 15-1788) and by CGS 8216, and generalized to a variety of anxiolytic and sedative drugs including the benzodiazepine receptor ligands zopiclone, suriclone, CL 218,872, CGS 9896, and ZK 91296. The novel imidazopyridine hypnotic, zolpidem, which also displaces benzodiazepines from their binding sites, failed to produce high levels of responding on the chlordiazepoxide-associated lever except at a dose which greatly reduced rates of lever pressing. In further experiments rats were trained to discriminate a dose of 2 mg/kg zolpidem from saline. This dose produced reductions in response rates but an attempt to establish a lower dose of zolpidem as a discriminative stimulus was largely unsuccessful. Zolpidem-appropriate responding was produced by pentobarbital, chlordiazepoxide, triazolam, CL 218,872, clorazepate, lorazepam, quazepam, and zopiclone but only at doses which reduced response rates. The zolpidem cue was antagonized by flumazepil, CGS 9896, and ZK 91296. While the discriminative stimulus produced by chlordiazepoxide may be related to its anxiolytic action, the zolpidem stimulus is probably more closely associated with sedation. It was also tentatively concluded that the stimulus properties of chlordiazepoxide and zolpidem are produced by activity at different subtypes of benzodiazepine receptors.

1 Introduction

Some of the earlier studies of state-dependent learning and drug dissociation investigated the actions of barbiturates (Bindra et al. 1965; Overton 1966) and more recent research has confirmed that these drugs can act effectively as discriminative stimuli in experimental animals (Ator and Griffiths 1983; Herling and Shan-

Laboratoires d'Etudes et de Recherches Synthélabo (L.E.R.S.), 31 Av. P.V. Couturier, 92220 Bagneux, France.

Psychopharmacology Series 4
Transduction Mechanisms of Drug Stimuli
Editors: Colpaert, Balster

non 1982; Jarbe 1976; Johansson and Jarbe 1975a; Oka et al. 1980; Schechter 1984). Benzodiazepines, the drugs most widely used clinically as sedatives and anxiolytics, have also been studied in similar experiments. Following the study described by Colpaert and his colleagues (1976) in which oral administration of a dose of 5 mg/kg chlordiazepoxide served effectively as a discriminative stimulus in rats, many experiments have dealt with the stimulus properties of benzodiazepines. In addition to a number of other studies using chlordiazepoxide (De Vry and Slangen 1986a–c; McElroy and Feldman 1982; Sanger et al. 1985; Schechter 1984) discriminative stimulus control has been demonstrated with diazepam (Haug 1983; Haug and Gotestam 1982; Johansson and Jarbe 1975b; Shannon and Herling 1983b), lorazepam (Ator and Griffiths 1983, 1985, 1986), midazolam (Garcha et al. 1985; Spealman 1985; Stolerman et al. 1986), clorazepate (Dantzer and Perio 1982), and oxazepam (Hendry et al. 1983).

In general, the results of these studies present a coherent picture of the stimulus properties of benzodiazepines. Benzodiazepine-induced stimulus control is readily acquired and almost certainly is mediated by activity at benzodiazepine binding sites as it is antagonized by the benzodiazepine antagonists flumazepil (Ro 15-1788) and CGS 8216 (Ator and Griffiths 1983, 1985; De Vry and Slangen 1986a; Herling and Shannon 1982; Schechter 1984; Spealman 1985; Stolerman et al. 1986). At least with flumazepil, however, the extent of this antagonism depends upon the training dose of benzodiazepine (Dantzer and Perio 1982; De Vry and Slangen 1986a). There is cross generalization between different benzodiazepines and most studies suggest that differences observed between different members of this drug class are quantitative (Shannon and Herling 1983a). Also, cross generalization is usually observed between benzodiazepines and barbiturates or meprobamate, although there are some exceptions to this (Ator and Griffiths 1983) and it has been argued that there may be qualitative differences between the discriminative stimuli produced by chlordiazepoxide and pentobarbital (Barry and Krimmer 1979).

Therefore, a great deal is now known about the characteristics of the stimulus properties of benzodiazepines and other sedative and anxiolytic drugs. What has not yet been established, however, are the behavioral and biochemical mechanisms underlying these effects. Benzodiazepines have a pharmacological profile characterized by anticonvulsant, anxiolytic, muscle relaxant and sedative effects, according to the dose administered (Randall et al. 1974). However, the extent to which these actions are important for generating the stimulus properties of these drugs remains unknown. Similarly, although the involvement of benzodiazepine receptors in the stimulus properties of these drugs is well established, the similarities between benzodiazepines and barbiturates, which do not bind to these sites, indicates that other biochemical mechanisms are also important. This consideration has given rise to a number of studies of the role of GABAergic mechanisms in the stimulus properties of anxiolytic drugs but, as yet it does not seem possible to draw any very clear conclusions from this research (Nielsen et al. 1983; Sanger 1985; Stolerman et al. 1986).

One of the most interesting recent developments in the pharmacology of anxiolytics and sedatives has been the discovery of novel compounds which, while differing chemically from benzodiazepines, bind to the same receptors in the brain. A number of such compounds have been described (Williams 1984), some of which have pharmacological properties identical to those of benzodiazepines. Others, however, differ in some respects from benzodiazepines and it is research with these compounds which has recently been providing important information which may eventually result in a more thorough understanding of the behavioral pharmacology of benzodiazepine-like drugs.

The present paper deals with the discriminative stimulus properties of chlordiazepoxide and zolpidem in rats. The results obtained with chlordiazepoxide provide a generally accurate representation of what is known of the stimulus properties of benzodiazepines. Zolpidem is a non-benzodiazepine which displaces benzodiazepines from their binding sites and has a pharmacological profile which, although similar to the benzodiazepines, shows some important differences. In particular, zolpidem appears to have preferential sedative properties (Arbilla et al. 1985; Depoortere et al. 1986) and is known to be an effective hypnotic in man (Nicholson and Pascoe 1986). As will become clear, there are both similarities and differences between the stimulus properties of chlordiazepoxide and those of zolpidem.

2 Discriminative Stimulus Properties of Chlordiazepoxide

2.1 Methods

Male Wistar rats were trained to discriminate between saline and a dose of 5 mg/kg chlordiazepoxide injected intraperitoneally. The procedure involved food-rewarded lever pressing maintained by an FR 10 schedule; further methodological details are given elsewhere (Sanger et al. 1985). The discrimination was acquired rapidly, most rats reaching a preestablished criterion of correct discrimination in between 30 and 40 sessions. Once this criterion had been reached tests of generalization and antagonism were carried out.

2.2 Results of Generalization Tests

In tests of generalization a variety of compounds gave rise to responding on the chlordiazepoxide-associated lever. Figure 1 shows dose-response curves for several benzodiazepines, all of which produced dose-related responding on the chlordiazepoxide lever. Other drugs to which the chlordiazepoxide discriminative stimulus generalized included the sedatives pentobarbital and meprobamate, the antiepileptic sodium valproate, and the benzodiazepine receptor ligands zopiclone, suriclone, CL 218,872, CGS 9896, and ZK 91296. Predominantly saline lever responding was produced by a number of other drugs, some of which are listed in Table 1. It is interesting to note that these inactive compounds include agents which act as GABA agonists or antagonists of serotonin. These data, therefore, do not provide evidence for a role for either GABA or serotonin in the stimulus properties of chlordiazepoxide.

The data presented in Fig. 1 show that the relative potencies of the different benzodiazepines in producing chlordiazepoxide-appropriate responding were

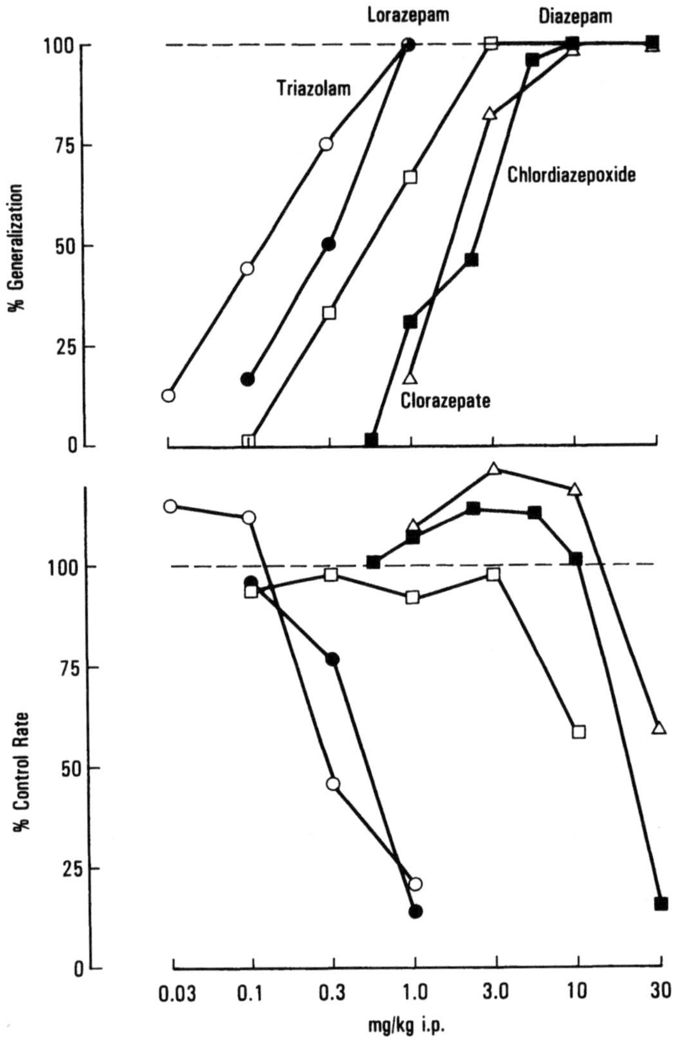

Fig. 1. Results of generalization tests with rats trained to discriminate a dose of 5 mg/kg chlordiazepoxide from saline. Rats were injected with the test drug and were reinforced for responding on the first lever on which they emitted ten responses. Data are shown as the percentage of those animals which made a choice which responded on the chlordiazepoxide-associated lever. Also shown are effects on rates of lever pressing expressed as percentages of rates after saline injection. Each drug was tested in at least six rats

similar to their relative potencies in other pharmacological tests (triazolam > lorazepam > diazepam > chlordiazepoxide = clorazepate). Also shown in Fig. 1 are the effects of the benzodiazepines on rates of lever pressing during the generalization tests. It is apparent that, with diazepam, chlordiazepoxide, and clorazepate, reductions in response rates occurred only at doses higher than those at which all animals responded on the chlordiazepoxide lever. In contrast, with

Table 1. Drugs which failed to produce chlordiazepoxide-lever responding in rats

Compound	Pharmacological action	Doses tested (mg/kg)
Haloperidol	Antipsychotic	0.1–0.3
Clozapine	Antipsychotic	0.3–3.0
Cyproheptadine	5-HT antagonist	5, 10
Pirenperone	5-HT antagonist	0.03–1.0
Etifoxine	Tranquilizer	10–30
d-Amphetamine	Stimulant	0.5, 1.0
Progabide	GABA agonist	25–100
SL 75.102	GABA agonist	50, 100

All drugs were administered by the intraperitoneal route, 30 min before sessions.

Table 2. Generalization and reduction of response rates in rats trained with chlordiazepoxide (5 mg/kg)

Drug	ED_{50}		Ratio
	Generalization	Rate reduction	
	(mg/kg)		
Chlordiazepoxide	1.84	19	10.3
Pentobarbital	9.4	14.1	1.5
Sodium valporate	123.2	235.8	1.9
Suriclone	1.1	2.58	2.4
CL 218,872	3.79	9.51	2.5
Lorazepam	0.23	0.63	2.7
Quazepam	7.23	24.36	3.4
Triazolam	0.11	0.44	4.0
Meprobamate	22.8	111.6	4.9
Midazolam	1.87	12.9	6.9
ZK 91,296	3.0	> 30.0	>10.0
CGS 9896	12.2	>200	>16.4
Diazepam	0.56	10.0	17.9
Zopiclone	1.69	31.1	18.4
Clorazepate	1.70	49.3	29.0

Generalization data were the percentage of rats which chose the drug-associated lever and response rate data were converted to percentages of control (saline) rates. ED_{50} values were calculated using probit analysis.

triazolam and lorazepam, chlordiazepoxide-lever responding was produced only with doses which markedly reduced response rates. These observations confirm the view that the chlordiazepoxide cue is not associated with response rate-reducing effects (Colpaert et al. 1976) and this point is further emphasized by the results shown in Table 2. This table presents a list of the drugs which gave rise to dose-related responding on the chlordiazepoxide lever. ED_{50} values are shown for drug-lever responding, together with ED_{50}s for reductions in response rates. When the ratios between these two ED_{50}s are considered it is clear that they can take a wide range of values.

In general the data presented in Table 2 correspond well with what is known of the other pharmacological properties of the same drugs. Pentobarbital is very sedative, for example, and probably shows little separation between anxiolytic and sedative doses. Some benzodiazepines, such as triazolam and quazepam, are better known as hypnotics than as anxiolytics while others, such as diazepam and clorazepate, are generally used as anxiolytics and seem to show a wider separation between anxiolytic and sedative doses. CGS 9896 has pharmacological properties suggesting that it may be an effective anxiolytic without producing sedation or muscle relaxation (Bennett and Petrack 1984; Sanger et al. 1985; Yokoyama et al. 1982) although clinical results confirming this possibility have yet to become available. Thus, these results show that the discriminative stimulus produced by 5 mg/kg chlordiazepoxide is probably not associated with depressant (i.e., muscle relaxant and sedative) effects and may be associated with an anxiolytic action.

2.3 Effects of Zolpidem

Although chlordiazepoxide-appropriate responding was produced by a variety of non-benzodiazepines which act at the same binding site, the results obtained with zolpidem were somewhat different. When rats trained with chlordiazepoxide were injected with zolpidem few of them responded on the chlordiazepoxide lever except with the highest dose of zolpidem which could be tested (3 mg/kg). At this dose ten rats were tested, three failed to respond at all, and of the remaining seven, five responded on the chlordiazepoxide lever. Because of the very marked depressant effects of this dose it is difficult to know whether zolpidem had, in fact, given rise to a chlordiazepoxide-like discriminative stimulus. Certainly, if it did do so, it was active only at the high dose (Sanger and Zivkovic 1986).

3 Discriminative Properties of Zolpidem

In order to investigate the discriminative stimulus properties of zolpidem in more detail a second series of experiments was carried out in which rats were trained to discriminate between zolpidem and saline. A protocol identical to that of the chlordiazepoxide study was used and, in the first experiment, an attempt was made to train rats to discriminate between saline and a dose of 1 mg/kg zolpidem. This dose is known to be pharmacologically active in rats (Depoortere et al. 1986) and in this experiment produced reductions in response rates of approximately 25%. However, after 40 sessions only two of eight rats had reached a satisfactory criterion of discrimination. The dose of zolpidem was therefore increased to 2 mg/kg for all rats and a second group of animals was also trained with the higher dose. With 2 mg/kg zolpidem most rats acquired the discrimination quite readily although this dose greatly reduced overall response rates (Sanger and Zivkovic 1986). It is interesting to note that response rates after zolpidem injection remained substantially reduced below saline values even after many months of drug administration.

When rats had acquired the zolpidem discrimination, tests of antagonism and generalization were carried out. The zolpidem cue was completely blocked by flumazepil even with doses of the antagonist as low as 1 mg/kg. The mixed agonist-antagonists CGS 9896 and ZK 91296 (Petersen et al. 1984) also produced dose-related antagonism of the zolpidem cue (Sanger and Zivkovic 1987).

Results of generalization tests with several drugs are shown in Fig. 2, which shows that dose-related responding on the zolpidem lever was produced by triazolam, zopiclone, and chlordiazepoxide but not by CGS 9896. The zolpidem cue also generalized to pentobarbital, CL 218,872, clorazepate, lorazepam, and quazepam but not to haloperidol or clonidine. What was most striking about the results of these generalization tests, however, and contrasted sharply with the re-

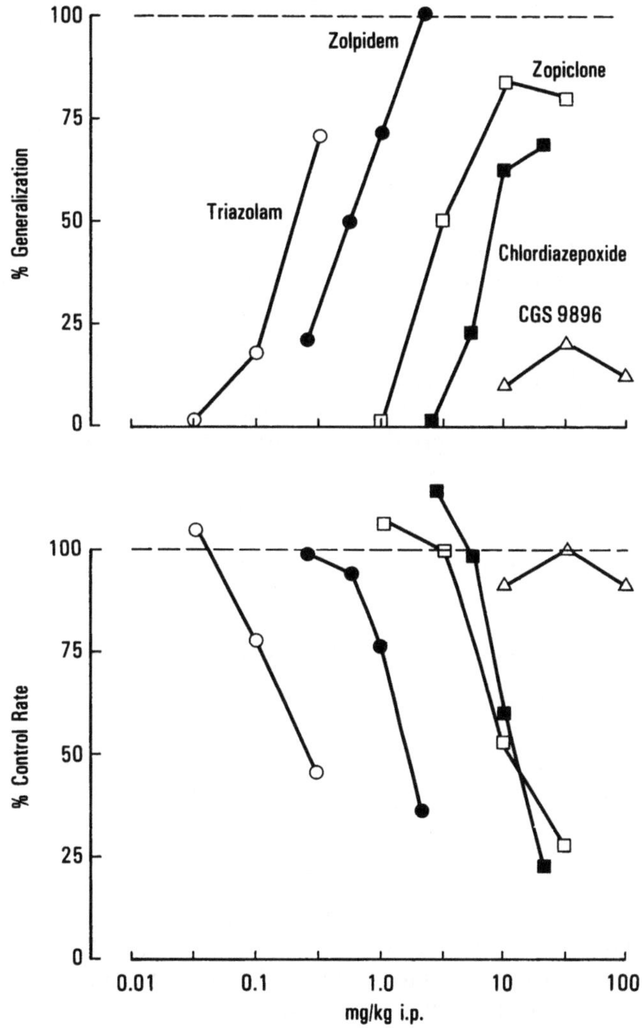

Fig. 2. Results of generalization tests in rats trained to discriminate zolpidem (2 mg/kg) from saline. See Fig. 1 for further explanation

Table 3. Generalization and reduction of response rates in rats trained with zolpidem (2 mg/kg)

Drug	ED_{50}		Ratio
	Generalization	Rate reduction	
	(mg/kg)		
Zolpidem	0.51	1.7	3.3
Pentobarbital	14.0	14.5	1.0
Chlordiazepoxide	12.1	12.9	1.1
Triazolam	0.20	0.24	1.2
CL 218,872	8.7	10.1	1.2
Clorazepate	30.0	40.0	1.3
Lorazepam	0.41	0.82	2.9
Quazepam	9.0	25.0	2.8
Zopiclone	4.6	15.1	3.3

Generalization data were the percentage of rats which chose the drug-associated lever and response rate data were converted to percentages of control (saline) rates. ED_{50} values were calculated using probit analysis.

sults described above for animals trained to discriminate chlordiazepoxide, was that zolpidem-lever responding was produced only by drug doses which reduced rates of responding. This is shown clearly in Table 3, which presents ED_{50} values for generalization and rate-reducing effects of several drugs. The ratios between these two values are generally small. If the data shown in Table 3 are compared with those of Table 1 it is clear that, although the doses of these drugs which reduced rates of responding were similar in the two studies, much higher doses were needed to produce the zolpidem cue than to give rise to the chlordiazepoxide cue. Thus, although as argued in an earlier section, the chlordiazepoxide cue is not related to depressant effects, the zolpidem cue seems to be associated with such actions which may represent the preferential sedative properties of this drug.

4 Comparison of the Discriminative Stimuli Produced by Chlordiazepoxide and Zolpidem

The foregoing descriptions of the discriminative stimuli produced by chlordiazepoxide at 5 mg/kg and zolpidem at 2 mg/kg demonstrate both similarities and differences between these two cues. Both are mediated by activity at benzodiazepine-binding sites and there is at least partial cross-generalization between the two drugs. A variety of other drugs are generalized by rats trained with either chlordiazepoxide or zolpidem although higher doses are necessary in the latter case. There are, however, at least two compounds (CGS 9896 and ZK 91296) which can produce the chlordiazepoxide cue but antagonize the cue produced by zolpidem. At the behavioral level, the zolpidem stimulus is associated with depressant, possibly sedative, effects whereas the chlordiazepoxide stimulus is not and may, therefore, be related to anxiolytic action.

The question which arises at this point is whether the chlordiazepoxide and zolpidem cues are quantitatively different, as seems to be the case for different doses of a single benzodiazepine or for different benzodiazepines (Shannon and Herling 1983 a), or whether the differences are qualitative as has been proposed for the stimulus properties of chlordiazepoxide and pentobarbital (Barry and Krimmer 1979). In more mechanistic terms, this question can be rephrased to ask whether the cues produced by chlordiazepoxide and zolpidem result from activity at a single benzodiazepine receptor or at different subtypes of receptor.

If the differences noted above are quantitative it should be possible, by more closely equating doses, for training with lower doses of zolpidem to give rise to a cue exactly equivalent to that seen with 5 mg/kg chlordiazepoxide and, conversely, for training with higher chlordiazepoxide doses to produce a cue more closely equivalent to that seen with zolpidem at 2 mg/kg. As noted above, rats did not readily acquire stimulus control with a lower dose of zolpidem and thus it is not clear that studies using different doses of this drug would be particularly fruitful. Studies involving a higher dose of chlordiazepoxide are underway (cf. De Vry and Slangen, 1986 a, b) and preliminary results suggest that the stimulus so produced is not identical to that seen with zolpidem.

The classical pharmacological technique for distinguishing between different receptor mechanisms involves a search for compounds which act as selective agonists or antagonists at different receptor subtypes. Unfortunately, as yet, compounds highly selective for different subtypes of benzodiazepine receptors, whether these receptors are defined biochemically or pharmacologically, have not been described. In the present studies, CGS 9896 and ZK 91296 were found to give rise to the chlordiazepoxide cue while antagonizing the cue produced by zolpidem. These results might indicate either that CGS 9896 and ZK 91296 are partial agonists at a single receptor type, antagonizing only effects produced by high levels of receptor occupancy (as might be the case for the zolpidem cue), or that these compounds are mixed agonist-antagonists, acting as agonists at the receptor subtype mediating the chlordiazepoxide cue and as antagonists at another receptor subtype responsible for the zolpidem cue (Bernard et al. 1985).

The data presently available do not allow a firm conclusion to be drawn as to which of these possibilities is correct. However, several findings suggest that at least two receptor subtypes need to be postulated. Thus, although CGS 9896 is generalized by chlordiazepoxide-trained rats and does not antagonize the chlordiazepoxide cue, it does antagonize the effect of identical doses of chlordiazepoxide in rats trained with zolpidem (Sanger and Zivkovic 1987). The most parsimonious explanation of this result would seem to be that the effects of chlordiazepoxide in chlordiazepoxide-trained or zolpidem-trained rats are largely mediated by activity at different receptor subtypes. Similarly, as has been argued previously (Depoortere et al. 1986), it does not seem possible to explain the differences between the pharmacological effects of zolpidem and those of benzodiazepines without proposing the existence of two receptor subtypes. However, it should be noted that if such receptor subtypes do exist they may not be identical to the different categories of binding sites which have been defined by the results of biochemical studies. Thus, although zolpidem has been found in binding studies to have preferential affinity for the BZI receptor subtype, its pharmacological effects, including stimulus properties as described above, are not identical to those of other compounds such as CL 218,872 and quazepam which have similar preferential affinities (Klepner et al. 1979; Sieghart and Schuster 1984).

5 Conclusions

Studies of the discriminative stimulus properties of chlordiazepoxide and zolpidem lead to the conclusion that these effects may be, at least partially, mediated by activity at different subtypes of benzodiazepine receptors. The receptor mediating the zolpidem cue may be associated with sedative activity while that mediating the chlordiazepoxide cue may be related to anxiolytic drug action. If this conclusion is correct it is possible that novel compounds may soon be discovered with greater selectivity as either agonists or antagonists at one or other of these receptors. Such compounds would not only be valuable experimental tools but might also represent important advances in drug therapy.

Acknowledgments. The skilled technical assistance of Mrs. C. Leonardon is gratefully acknowledged. Drugs were donated by the following organizations and individuals: quazepam, Schering Corporation, courtesy of Dr. A. Barnett; ZK 91296, Schering AG, courtesy of Dr. D. N. Stephens; pirenperone, Janssen Pharmaceutica, courtesy of Dr. F. C. Colpaert; CL 218,872, American Cyanamid Company, courtesy of Dr. J. W. Epstein; triazolam, Laboratoires Upjohn, courtesy of Dr. J.-P. Paturaud. Other drugs were obtained from commercial sources or synthesized at the Chemistry Department, LERS.

References

Arbilla S, Depoortere H, George P, Langer SZ (1985) Pharmacological profile of zolpidem at benzodiazepine receptors and electrocorticogram in rats. Naunyn Schmiedeberg's Arch Pharmacol 330:248–251

Ator NA, Griffiths RR (1983) Lorazepam and pentobarbital drug discrimination in baboons: cross-drug generalization and interaction with Ro 15-1788. J Pharmacol Exp Ther 226:776–782

Ator NA, Griffiths RR (1985) Lorazepam and pentobarbital discrimination: interactions with CGS 8216 and caffeine. Eur J Pharmacol 107:169–181

Ator NA, Griffiths RR (1986) Discriminative stimulus effects of atypical anxiolytics in baboons and rats. J Pharmacol Exp Ther 237:393–403

Barry H III, Krimmer EC (1979) Differential stimulus attributes of chlordiazepoxide and pentobarbital. Neuropharmacology 18:991–998

Bennett DA, Petrack B (1984) CGS 9896: a nonbenzodiazepine, nonsedating potential anxiolytic. Drug Dev Res 4:75–82

Bernard PS, Bennett DA, Pastor G, Yokoyama N, Liebman JM (1985) CGS 9896: agonist-antagonist benzodiazepine receptor activity revealed by anxiolytic, anticonvulsant and muscle relaxation assessment in rodents. J Pharmacol Exp Ther 235:98–105

Bindra D, Nyman K, Wise J (1965) Barbiturate-induced dissociation of acquisition and extinction: role of movement-initiating processes. J Comp Physiol Psychol 60:223–228

Colpaert FC, Desmedt LKC, Janssen PAJ (1976) Discriminative stimulus properties of benzodiazepines, barbiturates and pharmacologically related drugs: relation to some intrinsic and anticonvulsant effects. Eur J Pharmacol 37:113–123

Dantzer R, Perio A (1982) Behavioral evidence for partial antagonistic properties of Ro 15-1788, a benzodiazepine receptor antagonist. Eur J Pharmacol 81:655–658

Depoortere H, Zivkovic B, Lloyd KG, Sanger DJ, Perrault G, Langer SZ, Bartholini G (1986) Zolpidem, a novel non-benzodiazepine hypnotic. I. Neuropharmacological and behavioral effects. J Pharmacol Exp Ther 237:649–658
De Vry J, Slangen JL (1986a) Effects of chlordiazepoxide training dose on the mixed agonist-antagonist properties of benzodiazepine receptor antagonist Ro 15-1788, in a drug discrimination procedure. Psychopharmacology 88:177–183
De Vry J, Slangen JL (1986b) Effects of training dose on discrimination and cross-generalization of chlordiazepoxide, pentobarbital and ethanol in the rat. Psychopharmacology 88:341–345
De Vry J, Slangen JL (1986c) Differential interactions between chlordiazepoxide, pentobarbital and benzodiazepine antagonists Ro 15-1788 and CGS 8216 in a drug discrimination procedure. Pharmacol Biochem Behav 24:999–1005
Garcha HS, Rose IC, Stolerman IP (1985) Midazolam cue in rats: generalization tests with anxiolytic and other drugs. Psychopharmacology 87:233–237
Haug T (1983) Neuropharmacological specificity of the diazepam stimulus complex: effects of agonists and antagonists. Eur J Pharmacol 93:221–227
Haug T, Gotestam KG (1982) The diazepam stimulus complex: specificity in a rat model. Eur J Pharmacol 80:225–230
Hendry JS, Balster RL, Rosecrans JA (1983) Discriminative stimulus properties of buspirone compared to central nervous system depressants in rats. Pharmacol Biochem Behav 19:97–101
Herling S, Shannon HE (1982) Ro 15-1788 antagonizes the discriminative stimulus effects of diazepam in rats but not similar effects of pentobarbital. Life Sci 31:2105–2112
Jarbe TUC (1976) Characteristics of pentobarbital discrimination in the gerbil: transfer and antagonism. Psychopharmacology 49:33–40
Johansson JO, Jarbe JUC (1975a) Antagonism of pentobarbital induced discrimination in the gerbil. Psychopharmacologia 41:225–228
Johansson JO, Jarbe JUC (1975b) Diazepam as a discriminative cue: its antagonism by bemegride. Eur J Pharmacol 30:372–375
Klepner CA, Lippa AS, Benson DI, Sano MC, Beer B (1979) Resolution of two biochemically and pharmacologically distinct benzodiazepine receptors. Pharmacol Biochem Behav 11:457–462
McElroy JF, Feldman RS (1982) Generalization between benzodiazepine- and triazolopyridazine-elicited discriminative cues. Pharmacol Biochem Behav 17:709–713
Nicholson AN, Pascoe PA (1986) Hypnotic activity of an imidazo-pyridine (zolpidem). Br J Clin Pharmacol 21:205–211
Nielsen EB, Valentine JD, Holohean AM, Appel JB (1983) Benzodiazepine receptor mediated discriminative cues: effects of GABA-ergic drugs and inverse agonists. Life Sci 33:2213–2220
Oka M, Yamada K, Yoshida K, Shimizu M (1980) Avoidance enhancement and discriminative response control by anxiolytics with drugs acting on the GABA system. Jpn J Pharmacol 30:325–336
Overton DA (1966) State-dependent learning produced by depressant and atropine-like drugs. Psychopharmacologia 10:6–31
Petersen EN, Jensen LH, Honore T, Braestrup C, Kehr W, Stephens DN, Wachtel H, Seidelman D, Schmiechen R (1984) ZK 91296, a partial agonist at benzodiazepine receptors. Psychopharmacology 83:240–248
Randall LO, Schallek W, Sternbach LH, Ning RY (1974) Chemistry and pharmacology of the 1,4-benzodiazepines. In: Gordon M (ed) Psychopharmacological agents, vol 4, part III. Academic, New York, pp 175–281
Sanger DJ (1985) GABA and the behavioral effects of anxiolytic drugs. Life Sci 36:1503–1513
Sanger DJ, Zivkovic B (1986) The discriminative stimulus properties of zolpidem, a novel imidazopyridine hypnotic. Psychopharmacology 89:317–322
Sanger DJ, Zivkovic B (1987) Discriminative stimulus properties of chlordiazepoxide and zolpidem. Agonist and antagonist effects of CGS 9896 and ZK 91296. Neuropharmacology 26:499–505

Sanger DJ, Joly D, Zivkovic B (1985) Behavioral effects of non-benzodiazepine anxiolytic drugs: a comparison of CGS 9896 and zopiclone with chlordiazepoxide. J Pharmacol Exp Ther 232:831–837
Schechter MD (1984) Specific antagonism of the behavioral effects of chlordiazepoxide and pentobarbital in the rat. Prog Neuro-Psychopharm Biol Psychiatry 8:359–364
Shannon HE, Davis SL (1984) CGS 8216 noncompetitively antagonizes the discriminative effects of diazepam in rats. Life Sci 34:2589–2596
Shannon HE, Herling S (1983a) Discriminative stimulus effects of diazepam in rats: evidence for a maximal effect. J Pharmacol Exp Ther 227:160–166
Shannon HE, Herling S (1983b) Antagonism of the discriminative effects of diazepam by pyrazoloquinolines in rats. Eur J Pharmacol 92:155–157
Sieghart W, Schuster A (1984) Affinity of various ligands for benzodiazepine receptors in rat cerebellum and hippocampus. Biochem Pharmacol 33:4033–4038
Spealman RD (1985) Discriminative stimulus effects of midazolam in squirrel monkeys: comparison with other drugs and antagonism by Ro 15-1788. J Pharmacol Exp Ther 235:456–462
Stolerman IP, Garcha HS, Rose IC (1986) Midazolam cue in rats: effects of flumazepil (Ro 15-1788) and picrotoxin. Psychopharmacology 89:183–188
Williams M (1984) Molecular aspects of the action of benzodiazepine and non-benzodiazepine anxiolytics: a hypothetical allosteric model of the benzodiazepine receptor complex. Prog Neuro-Psychopharm Biol Psychiatry 8:209–247
Yokoyama N, Ritter B, Neubert A (1982) 2-Arylpyrazolo[4,3-c]-quinolin-3-ones: novel agonist, partial agonist and antagonist of benzodiazepines. J Med Chem 25:337–339

Comparison of Discriminative Stimuli Produced by Full and Partial Benzodiazepine Agonists: Pharmacological Specificity

D. A. Bennett

Contents

Abstract

Benzodiazepines produce discriminative stimuli that are stereospecific and antagonized by specific benzodiazepine receptor antagonists. The potency of these stimuli correlate with the ability of these compounds to bind to the benzodiazepine receptor complex. These data indicate that benzodiazepine stimuli are transduced via the benzodiazepine receptor. The underlying basis of these stimuli is unclear. Results with novel compounds that produce preclinical anxiolytic effects without the sedation and muscle relaxation of the classical benzodiazepines suggest that muscle relaxation may contribute to these stimuli. A direct comparison of discriminative stimuli established on a classical benzodiazepine agonist with the stimuli established on a partial benzodiazepine agonist supports the possibility that the classical benzodiazepine cue is mediated by a muscle relaxant effect, while the partial agonist cue is related to the anxioselectivity of the compound.

1 Introduction

Benzodiazepines are known to produce reliable discriminative stimuli. Various compounds have been used to establish drug discriminations as shown in Table 1. These experiments vary according to training drug and dose, drug pretreatment intervals, species, experimental contingencies (i.e., food reinforcement, shock avoidance, fixed ratio or variable interval reinforcement schedules, etc.), dependent variable assessed (i.e., percentage of animals selecting the drug lever or percentage of responding on the drug lever), etc. Some of these differences do not seem to affect the interpretation of results from different laboratories, while other variables might have important consequences. We have found, for example, that data analyzed by either percentage of rats selecting the drug lever or percentage of responses on the drug lever yield comparable ED_{50} values in generalization ex-

Research Department, Pharmaceuticals Division, CIBA-GEIGY Corporation Summit, NJ 07901, USA.

Psychopharmacology Series 4
Transduction Mechanisms of Drug Stimuli
Editors: Colpaert, Balster

Table 1. Discriminative stimuli established on benzodiazepine drugs

Training drug	Species	Ref.
Diazepam	Rat	Overton (1976), Haug and Gotestam (1982a, b), Haug (1983), Shannon and Herling (1983a, b), Nielsen et al. (1983), Stephens et al. (1984)
	Gerbil	Johansson and Järbe (1975)
Chlordiazepoxide	Rat	Overton (1966), Colpaert et al. (1976), Barry and Krimmer (1978a, b), McElroy and Feldman (1982), Stephens et al. (1984), Schechter (1984), Sanger et al. (1985)
Midazolam	Monkey	Spealman (1985), Garcha et al. (1985)
Lorazepam	Baboon	Ator and Griffiths (1983, 1985)
Oxazepam	Pigeon	De la Garza et al. (1983)
Clorazepate	Rat	Dantzer and Perio (1982)

periments (unpublished data). On the other hand, other investigators have shown that different training doses of the same drug can lead to different results (Overton 1966, 1971; Barry and Krimmer 1978a; Shannon and Holtzman 1979; Stolerman and D'Mello 1981; Shannon and Herling 1983a). The role of training dose in the results reviewed below is unclear but may be a factor in some of the inconsistent findings.

The purpose of this review is to consider what is known about the transduction mechanisms by which benzodiazepine stimuli are mediated. Furthermore, the transduction mechanisms of the classical benzodiazepine compounds will be compared with the discriminative stimulus effects of some of the newer benzodiazepine partial agonist compounds.

2 Classical Benzodiazepine Discriminative Stimuli

Whatever the differences between benzodiazepine discriminative stimuli, there are also robust similarities. It is known that specific benzodiazepine receptors mediate the therapeutic (and also side effect) actions of these drugs (Braestrup and Nielsen 1983) and that radiolabeled benzodiazepine ligands bind to these receptors (Braestrup and Squires 1977; Mohler and Okada 1977a). There is also ample evidence to indicate that benzodiazepine discriminative stimuli are mediated by an interaction of these drugs with the benzodiazepine macromolecular receptor complex.

Shannon and Herling (1983a) compared the potency with which various benzodiazepines generalized to diazepam discriminative stimuli with the potency of these drugs to displace [^{3}H]diazepam binding, as reported by Mohler and Okada (1977b, 1978). There appears to be a good correlation ($r = 0.78$) between these two measures, suggesting a relationship between benzodiazepine receptor binding and benzodiazepine discriminative effects.

The discriminative stimuli produced by diazepam have also been shown to be stereospecific (Shannon and Herling 1983a). The dextro isomer (Ro 11-6896) produces dose-dependent generalization to the diazepam interoceptive cue, with near complete generalization reported at the 0.03-mg/kg dose. The levo isomer (Ro 11-6893), however, produces no degree of generalization even at the 3.0-mg/kg dose. Higher doses could not be evaluated because of general operant response suppression.

The third piece of evidence which indicates that benzodiazepine discriminative stimuli are being mediated by the benzodiazepine macromolecular receptor complex comes from antagonism studies. The selective benzodiazepine receptor antagonists CGS 8216 (Boast et al. 1983) and Ro 15-1788 (Bonetti et al. 1982) reliably antagonize discriminative stimuli produced by a variety of benzodiazepines, including diazepam (Herling and Shannon 1982; Nielsen et al. 1983; Shannon and Herling 1983a, b; Stephens et al. 1984; Bennett et al. 1985), chlordiazepoxide (Schechter 1984), lorazepam (Ator and Griffiths 1983), oxazepam (De la Garza et al. 1983), and midazolam (Spealman 1985). Drugs from a variety of other classes either do not antagonize these stimuli or show inconsistent results depending upon the experimental situation in which they are tested (Table 2).

The antagonism of benzodiazepine stimuli by the receptor antagonists is specific to the benzodiazepines. Although barbiturate drugs generalize to discriminative cues established on the benzodiazepines, the benzodiazepine antagonists do

Table 2. Drugs that do not, or inconsistently, antagonize benzodiazepine discriminative stimuli

Drug	Antagonism	Ref.
Pentylenetetrazol	Yes	Haug (1983)
	No	Colpaert (1977)
Bicuculline	Yes	Nielsen et al. (1983)
	No	Haug (1983)
Bemegrid	Yes	Johansson and Järbe (1975)
	Yes	Colpaert (1977)
	No	Schechter (1984)
Isoniazid	Yes	Haug and Gotestam (1982a)
	No	Colpaert (1977)
Nicotine	Yes	Colpaert (1977)
	No	Nielsen et al. (1983)
3-Mercaptopropionic acid	No	Haug 1983)
Strychnine	No	Haug (1983)
	No	Colpaert (1977)
FG 7142	No	Nielsen et al. (1983)
B-CCM	No	Nielsen et al. (1983)
Harmaline	No	Nielsen et al. (1983)
Naltrexone	No	Nielsen et al. (1983)
Caffeine	No	Colpaert (1977)
DL-Amphetamine	No	Colpaert (1977)
Apomorphine	No	Colpaert (1977)
Cocaine	No	Colpaert (1977)
Mescaline	No	Colpaert (1977)
Picrotoxin	No	Colpaert (1977)
Tryptamine	No	Colpaert (1977)
Carbachol	No	Colpaert (1977)
Neostigmine	No	Colpaert (1977)
Physostigmine	No	Colpaert (1977)

not antagonize this generalization (Shannon and Herling 1983a; Herling and Shannon 1982; Ator and Griffiths 1983). Therefore, even though other drugs may mimic the interoceptive effects produced by the benzodiazepines, the mechanism by which these effects are produced differs from benzodiazepine actions which are selectively mediated directly through benzodiazepine receptors.

These three pieces of evidence – the correlation of discriminative stimuli with benzodiazepine receptor binding, the stereospecificity of the diazepam discrimination and the selective antagonism of benzodiazepine stimuli by specific antagonists of the benzodiazepine receptor – show that the discriminative stimuli established using the classical benzodiazepine drugs are transduced by a direct interaction with the benzodiazepine receptor.

The underlying basis of the classical benzodiazepine-mediated interoceptive cues is more difficult to interpret. As these compounds are known to produce sedation, muscle relaxation, and abuse potential, in addition to the antianxiety and anticonvulsive actions, it remains possible that the underlying interoceptive effects may be influenced by any of these actions.

In Table 3 results from various benzodiazepine discrimination studies, in which drugs from different classes were evaluated for generalization, are presented. The benzodiazepines and barbiturates reliably generalize to benzodiazepine stimuli. There is some controversy in the literature as to whether the discriminative stimuli produced by the benzodiazepines (and by the barbiturates) are mediated by anxiolytic actions of the drugs or are related to possible side effects (i.e., sedation/muscle relaxation) also shared by these drugs.

Colpaert et al. (1976) reported on the intercorrelations between the discriminative stimulus properties of benzodiazepines (the ability of various benzodiazepines to generalize to chlordiazepoxide discriminative stimuli) and the intrinsic behavioral effects of ataxia and loss of righting reflex. While there is no significant correlation between potencies for loss of the righting reflex and benzodiazepine discriminative stimuli ($r = -0.02$), there is a high correlation between ataxia and the stimuli ($r = 0.92$). However, Colpaert does not attribute the underlying basis of the benzodiazepine cue to ataxia, as two drugs (lorazepam and nitrazepam) produce generalization to the chlordiazepoxide stimulus at doses much lower than those that induce ataxia. In addition, the muscle relaxant, baclofen, does not produce drug lever selection in these animals, even though it induces ataxia. Sedation is also not the basis for these stimuli, as in many instances Colpaert reported that the doses of benzodiazepines that generalize to the chlordiazepoxide stimuli are actually increasing rather than decreasing the rates of responding.

Alternatively, in another drug discrimination in which rats were trained to discriminate diazepam, the muscle relaxants baclofen and methocarbamol both produce dose-dependent generalization to the diazepam stimuli (Bennett and Amrick 1986), suggesting that muscle relaxant activity is an important factor underlying these discriminative stimuli. The discrepancy between the Colpaert et al. (1976) study and these results may be based upon either the training drug (chlordiazepoxide versus diazepam) or the training doses employed. In support of Colpaert's findings, sedation does not appear to be an important factor in the Bennett and Amrick investigation, as diazepam itself produces increases in responding at doses that also produce the discriminative stimulus (unpublished data).

There is also a question as to whether GABA has a role in defining the benzodiazepine discriminative stimulus. A study by Haug (1983) indicates that GABA agonists do not generalize to benzodiazepine stimuli. Nielsen et al. (1983), however, showed that the GABA agonist, THIP, potentiates a benzodiazepine cue even though, by itself, the agonist does not generalize. It is quite clear from the representation of experiments presented in Table 3, however, that exclusive of the benzodiazepines and barbiturates, compounds other than these do not show reliable and consistent generalization to benzodiazepines.

Therefore, the data generated with the classical anxiolytics do not support a unitary underlying basis of the discriminative stimuli. As reported by others (see "Introduction"), the training doses of the drugs used may be a critical variable in determining the actual basis for the discriminative cue.

3 Novel Nonbenzodiazepine Compounds and Their Interactions with Classical Benzodiazepine Drug Discriminations

Experiments with new non-benzodiazepine chemical entities that bind to the benzodiazepine receptor but do not produce the side effects known to be associated with the benzodiazepines, provide additional information that sheds light on benzodiazepine-mediated discriminative stimuli.

CL 218,872, initially reported to be an anxiolytic with relatively weak side effects (Lippa et al. 1979), showed dose-dependent generalization to both a chlordiazepoxide cue (McElroy and Feldman 1982) and a diazepam cue (Bennett, unpublished data) (Table 3). Initial findings suggested that this drug did not have the muscle relaxant/sedative side effect properties of the benzodiazepines. These benzodiazepine cues, therefore, may have been mediated by a cue other than muscle relaxation or sedation. The case could have been made that the chlordiazepoxide and diazepam cues were related to the anxiolytic property of the training drugs. It has since been reported, however, that CL 218,872 produces a behavioral sedative/muscle relaxant profile similar to the benzodiazepines (Corradi et al. 1985). Therefore, the above interpretation is not necessarily justified. This compound may generalize to a sedative or muscle relaxant effect of the benzodiazepines.

CGS 9896 (Bennett and Petrack 1984; Bernard et al. 1985; Sanger et al. 1985), CGS 9895 (Bennett et al. 1985; Katzman and Shannon 1985), and ZK 91296 (Peterson et al. 1984) are partial benzodiazepine agonists with no pronounced sedation or muscle relaxation properties. CGS 9896 does not generalize to either diazepam (Bennett 1983; Shannon and Herling 1983 b) or lorazepam (Ator and Griffiths 1984) discriminative stimuli. In a similar fashion, CGS 9895 does not generalize to diazepam discriminative stimuli (Bennett et al. 1985; Shannon and Herling 1983 b). In fact, CGS 9895 dose-dependently antagonizes the discriminative stimuli produced by diazepam (Bennett et al. 1985; Shannon and Herling 1983 b).

These results suggest that the discriminative stimuli produced by the classical benzodiazepine compounds are not primarily related to the anxiolytic component of the drug, as these partial benzodiazepine agonists which are known to produce

Table 3. Generalization experiments in animals trained to discriminate a benzodiazepine drug from vehicle

Drug	Generalization	Ref.
Benzodiazepines	Yes	Colpaert et al. (1976), Nielsen et al. (1983), Peterson et al. (1984), Spealman (1985)
Barbiturates	Yes	Colpaert et al. (1976), Sanger et al. (1985), Haug (1983), Haug and Gotestam (1982a), Schechter (1984), Shannon and Herling (1983a)
	No	Spealman (1985), Ator and Griffiths (1983)
Meprobamate	Yes	Sanger et al. (1985)
Chlorpromazine	No	Colpaert et al. (1976)
	No	Haug and Gotestam (1982a)
Haloperidol	No	Sanger et al. (1985), Colpaert et al. (1976)
Clozapine	No	Sanger et al. (1985)
Balcofen	Yes	Bennett and Amrick (1986)
	No	Colpaert et al. (1976)
Glutethimide	Yes	Colpaert et al. (1976)
	No	Haug and Gotestam (1982a)
Cyproheptadine	Partial	Spealman (1985)
	No	Shannon and Herling (1983a)
	No	Sanger et al. (1985)
Cinanserin	Partial	Spealman (1985)
	No	Haug (1983)
THIP	Potentiation	Nielsen et al. (1983)
	No	Nielsen et al. (1983)
	No	Haug (1983)
Muscimol	No	Haug (1983)
	No	Spealman (1985)
AOAA	No	Haug (1983)
Diphenylhydantoin	No	Haug and Gotestam (1982a)
Metaqualone	No	Haug and Gotestam (1982a)
Propranolol	No	Haug and Gotestam (1982a)
Salbutamol	No	Haug and Gotestam (1982a)
PCP	No	Shannon and Herling (1983a)
Ketazocine	No	Shannon and Herling (1983a)
D-Amphetamine	No	Sanger et al. (1985)
Thalidomide	No	Colpaert et al. (1976)
Clozapine	No	Spealman (1985)
Buspirone	No	Spealman (1985)
Diphenhydramine	No	Spealman (1985)
Tripelennamine	No	Spealman (1985)
Caffeine	No	Spealman (1985)

anxiolytic activity in other preclinical test systems do not generalize. The antagonism of the diazepam cue by CGS 9895 further supports the suggestion that the diazepam stimulus is related to a muscle relaxant effect, since CGS 9895 has been shown to selectively antagonize the muscle relaxant effect of diazepam without antagonizing the anxiolytic effects of the drug (Bennett et al. 1985).

This interpretation is not entirely consistent with data reported by Sanger et al. (1985) and Petersen et al. (1984). Some generalization of CGS 9896 to a chlordiazepoxide discriminative cue was noted, although the dose-response curve is shallow and complete generalization is never attained even at a high dose (Sanger et al. 1985). In addition, ZK 91296 generalizes to both a chlordiazepoxide and a diazepam discriminative stimulus (Peterson et al. 1984; Stephens et al. 1984).

These divergent results make it difficult to interpret the generalization results obtained in drug discrimination studies with partial agonists. Why the same drug should generalize in one benzodiazepine discrimination and not generalize to the cue in another study leads one to believe that procedural differences may account for the discrepant results. To assist in characterizing the underlying mechanism of benzodiazepine-mediated stimuli, it was decided to establish, within the same laboratory, a drug discrimination on a classical anxiolytic (i.e., diazepam) and a discrimination on one of the novel non-benzodiazepine partial agonists (i.e., CGS 9896). A direct comparison could then be achieved which would, at least, be interpretable within the context of the experimental situation employed.

4 Comparison of Discriminative Stimuli Established on a Full and Partial Benzodiazepine Agonist

CGS 9896 has been shown to produce a reliable discriminative stimulus when the compound is administered by either the intraperitoneal (Bennett 1985) or the oral (Bennett and Amrick, unpublished data) routes of administration. The CGS 9896 stimuli, like the benzodiazepine cues, appear specifically mediated through the benzodiazepine receptor as evidenced by the dose-dependent antagonism of these stimuli by CGS 8216 (Bennett 1985). Diazepam and other reference anxiolytics (i.e., meprobamate) show dose-dependent generalization to these stimuli (Bennett 1985).

For comparison with the CGS 9896 discrimination, the dose of diazepam that produces 100% generalization to the CGS 9896 (30 mg/kg p.o.) stimulus was also used as the training dose of diazepam (10 mg/kg p.o.) in a separate group of animals. As reported earlier, while diazepam generalizes to CGS 9896 stimuli (Bennett 1985), CGS 9896 does not generalize to diazepam discriminative stimuli (Bennett 1983). This asymmetrical generalization suggests a difference between the discriminative effects produced by the full agonist and the partial agonist, even though both discriminative stimuli are mediated through the benzodiazepine receptor. While diazepam produces an effect inherent in CGS 9896, CGS 9896 does not produce an effect that is characteristic of diazepam. CGS 9895 produces dose-dependent generalization to CGS 9896 stimuli (Bennett et al. 1985) while this compound does not generalize to diazepam (Bennett et al. 1985; Shannon and Herling 1983b). Again, an asymmetry comparable to that noted above with CGS 9896 is evident.

To assess further the muscle relaxant component of each of these cues, the muscle relaxants baclofen and methocarbamol were tested in each discrimination. Dose-dependent generalization of these compounds is noted in the diazepam discriminating animals, with complete generalization seen with 10 mg/kg p.o. of baclofen and 1513 mg/kg p.o. of methocarbamol (Bennett and Amrick, unpublished data). No dose-dependent effects are noted in the CGS 9896 discriminating animals. A maximum of 40% generalization is seen with a 5.4-mg/kg dose of baclofen (10 mg/kg produces complete response suppression) and 60% generalization is noted with the 300- and 1730-mg/kg doses of methocarbamol (doses between these produce less than 50% generalization) (Bennett and Amrick, unpublished data). There is some possibility that methocarbamol may have a slight anxiolytic component associated with it, as this compound has shown effects indicative of preclinical anxiolytic potential in a conflict procedure (Bennett and Amrick 1986).

These results suggest that the diazepam stimulus is associated with a muscle relaxant component of the drug, while the CGS 9896 stimulus is mediated via an effect associated with the anxiolytic component of the drug. The actual effect associated with the anxiolytic component is not clear. Since these animals are not in an anxious state, one cannot attribute the cue to an amelioration of anxiety. Rather, the stimulus is mediated by an interoceptive effect associated with the anxioselective nature of the partial agonist.

To follow the muscle relaxant (diazepam)/anxioselective (CGS 9896) interpretation one step further, an additional study with CGS 9895 was conducted. CGS 9895 has been shown to selectively antagonize the muscle relaxant component of diazepam (Bennett et al. 1985). As previously noted, CGS 9895 dose-dependently blocks diazepam stimuli (Bennett et al. 1985) but dose-dependently generalizes to CGS 9896 stimuli (Bennett and Amrick, unpublished data). CGS 9895 does not produce dose-dependent antagonism of CGS 9896 stimuli, but shows a flat 40% antagonism from 30 to 300 mg/kg (Bennett and Amrick, unpublished data). If the diazepam cue is different from the CGS 9896 cue, then the generalization of diazepam to CGS 9896 should not be dose-dependently blocked by CGS 9895 (since the generalization of diazepam to CGS 9896 would not be dependent upon a muscle relaxant effect). When CGS 9895 was given in combination with diazepam (the dose that produced 100% generalization), a maximum of 50% antagonism was seen at the 30-mg/kg dose with higher doses producing less antagonism (Bennett and Amrick, unpublished data). The muscle relaxant properties of diazepam are not crucial to the underlying basis of the CGS 9896 discriminative stimuli. These results further support the premise that discriminative stimuli produced by the full agonist may be mediated by a muscle relaxant property, while the partial agonist discriminative stimuli are qualitatively different.

Although the discriminative stimuli produced by CGS 9896 appear related to the anxioselective nature of the compound, this cue appears pharmacologically specific. Serotonin agonists which produce preclinical anxiolytic profiles (i.e., buspirone, gepirone and 8 OH-DPAT) do not generalize to CGS 9896 discriminative stimuli (Bennett, unpublished data).

5 Conclusions

Classical benzodiazepine discriminative stimuli are mediated via an interaction with the benzodiazepine macromolecular complex. An underlying basis for these stimuli is not clear and may vary depending upon the training doses and types of training drugs used. Testing with the partial benzodiazepine agonists suggests that the classical stimuli may be mediated by the muscle relaxant properties of the benzodiazepines. When the data from a diazepam discrimination (full agonist) are compared with data from a CGS 9896 discrimination (partial agonist), both of which were conducted in the same laboratory with special consideration of the training doses, the discriminative cues could be differentiated. In this case, the diazepam cue appears mediated via a muscle relaxant effect and the CGS 9896 cue may be attributed to an effect associated with the anxioselective nature of the compound.

References

Ator NA, Griffiths RR (1983) Lorazepam and pentobarbital drug discrimination in baboons: cross-drug generalization and interaction with Ro 15-1788. J Pharmacol Exp Ther 226:776–782

Ator NA, Griffiths RR (1984) Discriminative stimulus properties of novel anxiolytics. Fed Proc 43: 930

Ator NA, Griffiths RR (1985) Lorazepam and pentobarbital discriminations: interactions with CGS 8216 and caffeine. Eur J Pharmacol 107:169–181

Barry III H, Krimmer EC (1978a) Pharmacology of discriminative drug stimuli. In: Ho BT, Richard III BW, Chute DL (eds) Drug discrimination and state dependent learning. Academic, New York, pp 3–32

Barry III H, Krimmer EC (1978b) Similarities and differences in discriminative effects of chlordiazepoxide, pentobarbital, ethanol, and other sedatives. In: Colpaert FC, Rosecrans JA (eds) Stimulus properties of drugs: ten years of progress. Elsevier, Amsterdam, pp 243–252

Bennett DA (1983) The nonsedating potential anxiolytic, CGS 9896 (2-(*p*-chlorophenyl)-pyrazolo [3,4-*c*]quinoline-3(5H)-one, in drug discrimination models. Pharmacologist 25:155

Bennett DA (1985) The non-sedating anxiolytic CGS 9896 produces discriminative stimuli that may be related to an anxioselective effect. Life Sci 37:703–709

Bennett DA, Amrick CL (1986) 2-Amino-7-phosphonoheptanoic acid produces discriminative stimuli and anticonflict effects similar to diazepam. Life Sci 39:2455–2461

Bennett DA, Petrack B (1984) CGS 9896: a nonbenzodiazepine, nonsedating potential anxiolytic. Drug Dev Res 4:75–82

Bennett DA, Amrick CL, Wilson DE, Bernard PS, Yokoyama N, Liebman JM (1985) Behavioral pharmacological profile of CGS 9895: a novel anxiomodulator with selective benzodiazepine agonist and antagonist properties. Drug Dev Res 6:313–325

Bernard PS, Bennett DA, Pastor G, Yokoyama N, Liebman JM (1985) CGS 9896: agonist-antagonist benzodiazepine receptor activity revealed by anxiolytic, anticonvulsant and muscle relaxation in rodents. J Pharmacol Exp Ther 235:98–104

Boast CA, Bernard PS, Barbaz BS, Bergen KM (1983) The neuropharmacology of various benzodiazepine antagonists. Neuropharmacology 22:1511–1521

Bonetti EP, Pieri L, Cumin R, Schaffner R, Pieri M, Gamzu ER, Muller RKM, Haefely W (1982) Benzodiazepine antagonist Ro 15-1788: neurological and behavioral effects. Psychopharmacology 78:8–18

Braestrup C, Nielsen M (1983) Benzodiazepine receptors. In: Iversen LL, Iversen SD, Snyder SH (eds) Handbook of psychopharmacology, vol 17. Plenum, New York, pp 285–384

Braestrup C, Squires RF (1977) Specific benzodiazepine receptors in rat brain characterized by high-affinity 3H-diazepam binding. Proc Natl Acad Sci USA 74:3805–3809

Colpaert FC (1977) Discriminative stimulus properties of benzodiazepines and barbiturates. In: Lal H (ed) Stimulus properties of drugs. Plenum, New York, pp 93–106

Colpaert FC, Desmedt LKC, Janssen PAJ (1976) Discriminative stimulus properties of benzodiazepines, barbiturates and pharmacologically related drugs: relation to some intrinsic and anticonvulsant effects. Eur J Pharmacol 37:113–123

Corradi CL, Bennett DA, Diverio J, Wilson DE (1985) The differential sedative/muscle relaxant profiles of several classical and novel anxiomodulators in four behavioral paradigms. Neurosci Abst 11:425

Dantzer R, Perio A (1982) Behavioural evidence for partial agonist properties of Ro 15-1788, a benzodiazepine receptor antagonist. Eur J Pharmacol 81:655–658

De La Garza R, Johanson CE, Schuster CR (1983) Discriminative stimulus effects of oxazepam and pentobarbital. Fed Proc 42:620
Garcha HS, Rose IC, Stolerman IP (1985) Midazolam cue in rats: generalization tests with anxiolytic and other drugs. Psychopharmacology 87:233–237
Haug T (1983) Neuropharmacological specificity of the diazepam complex: effects of agonists and antagonists. Eur J Pharmacol 93:221–227
Haug T, Gotestam KG (1982a) The diazepam stimulus complex: specificity in a rat model. Eur J Pharmacol 80:225–230
Haug T, Gotestam KG (1982b) Onset and offset of the diazepam stimulus complex. Pharmacol Biochem Behav 17:1171–1174
Herling S, Shannon HE (1982) Ro-15-1788 antagonizes the discriminative stimulus effects of diazepam in rats but not similar effects of pentobarbital. Life Sci 31:2105–2112
Johansson JO, Järbe TUC (1975) Diazepam as a discriminative cue: its antagonism by bemegride. Eur J Pharmacol 30:372–375
Katzman NJ, Shannon HE (1985) Differential diazepam-antagonist effects of the benzodiazepine receptor ligand CGS 9895 in rodents. J Pharmacol Exp Ther 235:589–595
Lippa A, Coupet J, Greenblatt EN, Klepner CA, Beer B (1979) A synthetic non-benzodiazepine ligand for benzodiazepine receptors: a probe for investigating neuronal substrates of anxiety. Pharmacol Biochem Behav 11:99–106
McElroy JF, Feldman RS (1982) Generalization between benzodiazepine- and triazolopyridazine-elicited discriminative cues. Pharmacol Biochem Behav 17:709–713
Mohler H, Okada T (1977a) Properties of ^{3}H-diazepam binding to benzodiazepine receptors in rat cerebral cortex. Life Sci 20:2102–2110
Mohler H, Okada T (1977b) Benzodiazepine receptor: demonstration in the central nervous system. Science 198:849–851
Mohler H, Okada T (1978) Biochemical identification of the site of action of benzodiazepines in human brain by ^{3}H-diazepam binding. Life Sci 22:985–996
Nielsen EB, Valentine JD, Holohean AM, Appel JB (1983) Benzodiazepine receptor mediated discriminative cues: effects of GABA-ergic drugs and inverse agonists. Life Sci 33:2213–2220
Overton DA (1966) State-dependent learning produced by depressant and atropine-like drugs. Psychopharmacologia 10:6–31
Overton DA (1971) Discriminative control of behavior by drug states. In: Thompson T, Pickens R (eds) Stimulus properties of drugs. Appleton Century Crofts, New York, pp 87–100
Overton DA (1976) Discriminable effects of benzodiazepines. Psychopharmacol Commun 2:339–343
Petersen EN, Jensen LH, Honore T, Braestrup C, Kehr W, Stephens DN, Wachtel H, Seidelman D, Schmiechen R (1984) ZK 91296, a partial agonist at benzodiazepine receptors. Psychopharmacology 83:240–248
Sanger DJ, Joly D, Zivkovic B (1985) Behavioral effects of nonbenzodiazepine anxiolytic drugs: a comparison of CGS 9896 and zopiclone with chlordiazepoxide. J Pharmacol Exp Ther 232:831–837
Schechter MD (1984) Specific antagonism of the behavioral effects of chlordiazepoxide and pentobarbital in the rat. Prog Neuro-Psychopharmacol Biol Psychiatry 8:359–364
Shannon HE, Herling S (1983a) Discriminative stimulus effects of diazepam in rats: evidence for a maximal effect. J Pharmacol Exp Ther 227:160–166
Shannon HE, Herling S (1983b) Antagonism of the discriminative effects of diazepam by pyrazoloquinolines in rats. Eur J Pharmacol 92:155–157
Shannon HE, Holtzman SG (1979) Morphine training dose: a determinant of stimulus generalization to narcotic antagonists in the rat. Psychopharmacology 61:238–244
Spealman RD (1985) Discriminative-stimulus effects of midazolam in squirrel monkeys: a comparison with other drugs and antagonism by Ro 15-1788. J Pharmacol Exp Ther 235:456–462
Stephens DN, Shearman GT, Kehr W (1984) Discriminative stimulus properties of B-carbolines characterized as agonists and inverse agonists at central benzodiazepine receptors. Psychopharmacology 83:233–239
Stolerman IP, D'Mello GD (1981) Role of training conditions in discrimination of central nervous system stimulants by rats. Psychopharmacology 73:295–303

Receptor Mechanisms of Opioid Drug Discrimination

J. H. Woods [1,2], A. J. Bertalmio [2], A. M. Young [3], W. D. Essman [2], and G. Winger [1]

Contents

Abstract

Receptor theory of opioid action has provided an extremely useful interpretive framework for the discriminative stimulus effects of opioids. By and large, receptor theory has been applied to opioid actions as they are measured in in vitro and reflex systems. It is clear, however, that it can also assist in interpreting data from experiments addressing operantly conditioned behavior, and provide a link between these data and those obtained using other procedures. The current paper describes the criteria that can be used to determine whether a drug effect is receptor mediated and applies these criteria to the effects of mu and kappa opioids in drug-discrimination studies. Criteria for distinguishing between drug effects occurring through one, as opposed to two, receptor systems are described and again applied to the discriminative stimulus effects of mu and kappa opioids. The potential difficulties that can be caused by postreceptor variability and the presence of multiple receptor systems are noted, since they can modify the effects predicted from simple receptor theory, and are likely to play an important role when studies of opioid action are made in the whole animal. In discrimination studies, complicating variables include dose of the training drug, subject species, nature of the training drug, and context of the discrimination. Finally, the ability of receptor theory to guide future investigation of the phenomenon of partial generalization is explored.

[1] Dept. of Pharmacology, University of Michigan, Ann Arbor, MI 48109, USA.
[2] Dept. of Psychology, University of Michigan, Ann Arbor, MI 48109, USA.
[3] Dept. of Psychology, Wayne State University, Detroit, MI 48202, USA.

Psychopharmacology Series 4
Transduction Mechanisms of Drug Stimuli
Editors: Colpaert, Balster

1 Introduction

Receptor theories of drug action attempt to relate the effects of drugs in an overall system to the actions of these drugs at limited or restricted parts of that system – the drug receptors. Many investigators in the field of drug discrimination have been interested in applying receptor theory to the design and interpretation of studies on the interoceptive properties of drugs. In the following chapter, opioid drug discrimination work will be reviewed selectively in the light of receptor theory, with the hope that the applicability of the theory, the economy it provides in organizing and analyzing the data, and, most of all, its utility in guiding research will be demonstrated.

An initial, brief description of occupation theories of drug-receptor interaction is included here. Occupation theories are perhaps the oldest, most frequently referenced, and most completely developed theories of drug-action. Although there are some differences in various formulations of occupation theory, there are also broad areas of general agreement. Following Clark (1926), these theories assume that the fraction (or number) of receptors occupied by a drug is primarily a function of the experimenter-controlled variable of dose or concentration, and a fixed property of the drug, affinity. The affinity of a drug is determined by the specific structure of the drug molecule. Differences in affinity are required to explain why, at equal doses, some drugs occupy a greater fraction of the receptors than do others. Also, following Stephenson (1956), Ariens (1954), and Furchgott (1966), the theories further assume that mere occupation of receptors will not always lead to the production of an effect. Each drug can be assumed to possess a second property, also determined by its molecular structure, and variously referred to as efficacy, intrinsic activity, or intrinsic efficacy. Differences in efficacy across drugs must be postulated to explain why two drugs that occupy an equal fraction of receptors in a pharmacological system can produce very different levels of effect. If two drugs occupy an equal fraction of the receptors in a preparation, the drug with higher efficacy can be assumed to put a greater fraction of the receptors it occupies into an active state (or the same fraction of receptors for a greater percentage of the time).

The most recent versions of receptor theory assume as well that the test system itself, i.e., postreceptor characteristics, will play a role in determining the effects of the drug. This means that there is no specific relation between the percentage of receptors activated and the resulting level of effect measured across various pharmacological systems. Activation of 50% of the receptors in one system may produce a response that is 50% of maximum, but in another system produce no response or the maximum response. Receptor theory does not specify a necessary relationship between effect and dose, receptor occupation, or receptor activation since levels of effect are subject to a whole host of postreceptor influences. For example, the increase from a 0% to a 100% level of response could occur as receptor activation increased from 0% to 100%, from 17% to 28%, or, if the response is quantal in nature, from 35% to 36%. Finally, application of receptor theories must take into account the fact that drugs often have affinity for and efficacy at more than one type of receptor in a preparation and that the effects they produce can, therefore, be a complex function of their actions at each of these different receptor systems.

Application of receptor theory to drug discrimination has focused on determining whether the discriminative effects of drugs are receptor mediated and, if so, which receptors are involved. This question has at least two logically separable, although related, parts: first, how can it be determined that a drug is producing a discriminative effect through an action on a receptor; and, second, how can it be determined whether the discriminative effects of two different drugs are mediated through a single or through different receptor types?

This chapter is intended to give general guidelines derived from receptor theory, followed by more specific examples from empirical research, of how the two parts of this question have been answered with respect to the discriminative effects of opioid drugs.

2 Determining Receptor Mediation

2.1 General Case

Basic evidence for receptor mediation of effects can be provided by several standard lines of argument which have been developed in studies conducted in a broad range of pharmacological systems.

Potency

If the actions of a particular drug class are receptor mediated, at least some drugs should have the capacity to produce these actions at low doses.

Structure-Activity Relations

Precise structure-activity relations should exist among drugs that act on receptors. Relatively small changes in the part of a drug's chemical structure necessary for receptor attachment or activation – changes that may leave the physiochemical properties largely unchanged – may produce relatively large changes in activity.

Stereospecificity

Stereospecificity, or at least stereoselectivity, should occur with optically active drugs that act at receptors. If two drugs that differ only in the arrangement of groups around an optically active carbon show differences in activity, precise structural requirements suggestive of receptor mediation are indicated.

Antagonism

Receptor-mediated effects can be antagonized by drugs that also interact with the receptor but have low efficacy. As with the agonists, these antagonists should be susceptible to large changes in activity with small changes in chemical structure.

2.2 Receptor Mediation of Discriminative Properties of Opioids

Classical Opioid Drugs

Opioid effects have been clearly shown to be receptor mediated in a number of assay systems. This appears to be no less true when the assay system is one of drug discrimination. The principles described above to define receptor mediation can be easily applied to the situation of opioid drug discrimination.

Potency. Many opioids are extremely potent. Sufentanil, for example, is capable of producing discriminative effects at doses as low as 0.001 mg/kg in codeine-trained rhesus monkeys (Herling and Woods, unpublished observations).

Structure-Activity Relations. Structure-activity relations exist within opioid drugs that produce discriminative effects. For example, addition of only a single ethano group ($-CH=CH_2$) to the methyl nitrogen of morphine converts it to nalorphine, abolishing its capacity to induce generalization in codeine-trained monkeys, while simultaneously conferring the capacity to antagonize codeine's discriminative effects (Winger et al. 1987).

Stereospecificity. For opioid drugs with chiral centers, only one isomer, usually the *l*-isomer, produces morphine, codeine, or fentanyl-like stimulus effects (e.g., Schaefer and Holtzman 1977; Jarbe 1978; Herling and Woods 1981).

Antagonism. The discriminative effects of morphine-like drugs, insofar as they have been tested, can be antagonized by prior administration of the opioid antagonists naloxone, naltrexone, or quadazocine (e.g., Schaefer and Holtzman 1977; Shannon and Holtzman 1976; Jarbe 1978; Valentino et al. 1981; Bertalmio et al. 1984; Dykstra et al. 1987 a).

Kappa Opioid Drugs

This analysis clearly demonstrates that the discriminative effects of morphine-like drugs are mediated through receptors. However, there appear to be opioid drugs whose effects are not mediated through the same receptor that is responsible for the effects of morphine and morphine-like opioids. These drugs, labeled kappa opioids, were first described by Martin et al. (1976) as distinct from morphine or mu opioids. Drug discrimination procedures, particularly in the rhesus monkey, have been extremely useful in the categorization and classification of these more recently described opioids. Mu agonists do not generalize in rhesus monkeys trained to discriminate kappa agonists such as ethylketocyclazocine (EKC) (Hein et al. 1981; Katz et al. 1982; Tang and Collins 1985) or nalorphine (Tang and Code 1983) and EKC does not generalize to either codeine or etorphine in monkeys trained to discriminate these mu agonists. Thus, in this preparation, these two drug classes produce distinct, nonoverlapping effects.

Martin et al. (1976) indicated that the effects of kappa opioids, like those of mu opioids, were receptor mediated. In drug discrimination procedures, the criteria described above can be applied to results obtained with kappa opioids.

Potency. Bremazocine is a kappa agonist that is extremely potent, having discriminative effects at a dose of 0.001 mg/kg in EKC-trained rhesus monkeys (Dykstra et al. 1987 a).

Structure-Activity Relations. Kappa agonists have been evaluated much less rigorously for structure-activity relations than have mu agonists. The kappa agonists ketazocine and EKC represent a slight modification of cyclazocine, a compound that has kappa activity, but considerable opioid antagonist activity as well.

Stereospecificity. As with the mu agonists, the *l*-isomer of kappa agonists with chiral centers is typically more effective in drug discrimination assays of kappa activity. The *d*-isomer of EKC is not active as a discriminative stimulus in monkeys trained to discriminate EKC, and the EKC-like discriminative effects of the racemate UM 1072 appear to reside exclusively in one isomer (Hein et al. 1981).

Antagonism. The discriminative effects of kappa agonists can be shifted to the right by administration of the opioid antagonist quadazocine (Bertalmio et al. 1984; Dykstra et al. 1987a).

These data, taken as a whole, indicate that the discriminative effects of both mu and kappa agonists are mediated through receptors.

3 Receptor Mediation Through One Receptor or Two

3.1 General Case

Although data obtained on the discriminative effects of opioids in the rhesus monkey indicate no overlap between the effects of mu and kappa opioids, data in other assay systems, as discussed in more detail below, are often more ambiguous. It is not always clear that the effects of drugs that have been labeled as mu or kappa opioids are acting on separate receptors. It is thus important to establish ways of determining whether one or more receptors are mediating a common effect in a given assay system. In the general case, several rules can be applied.

Specificity of Agonists

When two agonists produce entirely different, nonoverlapping effects, their actions are likely to be mediated through two different receptors. The discriminative effects of mu and kappa opioids in the rhesus monkey are a demonstration of the specific actions of these two types of opioid agonists.

Selectivity of Agonists

For any group of drugs, if each compound produces a pattern of effects identical to the others, and all display consistent potencies relative to each other, it can generally be assumed that all the drugs in this set are acting at the same receptor. When differences in rank order of potency cannot be simply explained, e.g., by differences in pharmacokinetics, the possibility exists that more than one type of receptor is mediating the effects of one or more of the drugs in this set.

Specificity of Antagonists

Perhaps the most powerful method of differentiating receptor types is through the use of antagonists. If a specific antagonist blocks one action of an agonist, and leaves another action intact, then the agonist is producing its effects through two receptor types. Likewise, if the effects of one agonist are blocked and the effects of another are not modified by a selective antagonist, these drugs are acting through two different receptor types.

Selectivity of Antagonists

Antagonists that are not specific may also be used in differentiating the types of receptors at which an agonist acts. If the antagonist is selective, i.e., possesses greater affinity for one receptor than it does for a second receptor, the dose or concentration of the antagonist required to produce an equal antagonism at the second receptor will be greater. Thus, if a higher dose of a selective antagonist is required to block one effect of an agonist than is required to block a second effect of this agonist, it suggests that the second effect of the agonist may be mediated by a different receptor than the first. Similarly, if two different agonists produce similar effects, but a higher dose of the selective antagonist is required to block the effects of one of these agonists, it suggests that the agonists are acting through different receptors in producing these effects. Such potency measures for selective antagonists are normally expressed as pA_2 values. Equal pA_2 values suggest that the same receptor types are involved. A pA_2 analysis therefore constitutes one of the best tests for the presence of common or multiple receptors (Dykstra et al., this volume).

3.2 Two Receptors Mediating Discriminative Effects of Mu and Kappa Opioids

As described above, the discriminative effects of mu and kappa opioid drugs do not overlap in studies in the rhesus monkey. Although this is, in itself, strong evidence that the drugs do not act on the same receptor systems, it is possible to confirm this initial conclusion by tests of the specificity and selectivity of antagonist action. Beta-funaltrexamine (B-FNA) is an antagonist that is specific for the mu receptor type (Portoghese et al. 1980). B-FNA produced a long-acting antagonism of the discriminative effects of codeine in the rhesus monkey, but did not antagonize the effects of EKC in this species (Dykstra et al. 1987b). Nalorphine, a mixed agonist-antagonist with kappa-like agonist properties and mu antagonist properties, was likewise able to reverse the discriminative effects of codeine at doses that either did not alter or decreased the dose of EKC required for stimulus control (Winger et al. 1987).

Evaluation of the specificity of antagonism of mu and kappa agonists with the broad-spectrum antagonist quadazocine demonstrated further the distinction between the receptor types on which these two compounds act. The pA_2 for quadazocine against a series of mu agonists ranged between 7.7 and 7.9 in drug discrimination tests (Bertalmio et al. 1987). In contrast, the pA_2 of quadazocine against three kappa agonists ranged between 6.1 and 6.4 for both analgesia and drug discrimination (Dykstra et al. 1987b).

This evaluation demonstrates that there can be no doubt that these two drug classes are producing their discriminative effects through two different opioid receptor types in the rhesus monkey. The distinction between mu and kappa drugs in other test systems is less clear, as discussed below, and in order to clarify the more ambiguous results it is important that the tests that have been applied to the rhesus monkey system be applied to other test systems as well.

4 Factors Affecting the Differentiation of Mu and Kappa Agonists in Drug Discrimination Procedures

As discussed in the "Introduction," not only receptor occupancy and receptor activation, but also postreceptor characteristics of individual assay systems can influence the effects of a drug. This is true in studies of drug discrimination just as in other pharmacological systems. There are at least four factors that have been shown to play a critical role in determining results obtained from studies of opioid drug discrimination, and thus may represent alterations in the postreceptor characteristics of this assay system: training dose, subject species, training drug, and pharmacological context of the training stimulus.

4.1 Training Dose

As the dose of the drug used to establish the discrimination increases, the number of receptors activated will most likely increase as well. This implies that with the changes in training dose, the ED_{50} or other measures of potency of drugs that produce full generalization to the training drug will be shifted by an amount comparable to the change in dose of the training drug. This has been demonstrated in opioid systems (Colpaert et al. 1980; Holtzman 1982).

Changes in effects of drugs that produce less than full generalization to the training drug can also be expected to occur with changes in training dose. To the extent that efficacy is a major determinant of the discriminative effects of drugs, there should be greater generalization of drugs with less efficacy when the dose of the training drug is low (Holtzman 1982). This greater generalization with a lower training dose can also occur by other mechanisms, as discussed below, and is not a foolproof method of distinguishing among receptor agonists.

4.2 Species

The ability of various animal species to discriminate between the stimulus properties of mu and kappa agonists is quite different. The rhesus monkey, as described earlier, appears most able to differentiate between these two opioid types. This distinction is less clear in the squirrel monkey and rat, and the pigeon is even less able to distinguish between mu and kappa opioids. For example, morphine-trained squirrel monkeys produce less than 70% drug-appropriate responding when tested with EKC (Teal and Holtzman 1980b) and squirrel monkeys trained to discriminate cyclazocine from saline respond less than 50% of the time on the drug-appropriate lever when tested with morphine (Schaefer and Holtzman 1978). In the rat, the degree of substitution reported for cyclazocine in fentanyl-trained rats varies between 0% (Shearman and Herz 1982c) and approximately 40% (Colpaert and Janssen 1984). Most studies using either morphine- or fentanyl-trained rats report levels of generalization between 20% and 70% for at least one of the purported kappa agonists, ketazocine, EKC, cyclazocine, or nalorphine (e.g., Colpaert and Janssen 1984; Shannon and Holtzman 1977, 1979; Shearman and Herz 1982c). In cyclazocine-trained rats, morphine engenders between 5% and 100% stimulus substitution (McCarten and Lal 1980; Teal and Holtzman 1980a), whereas in rats trained to discriminate between saline and either bremazocine or EKC, the mu agonists morphine, fentanyl, and etorphine produce between 0% and 35% generalization (Shearman and Herz 1981, 1982a, b). The pigeon is the least sensitive to the stimulus properties of kappa agonists of all species tested so far. Initial results indicated that the pigeon cannot discriminate between mu and kappa agonists, i.e., pigeons trained to discriminate between morphine and saline generalize completely to a variety of kappa agonists, including ketazocine and EKC (Herling et al. 1980). The converse study produced similar results: pigeons trained to discriminate EKC generalize to both mu and kappa agonists (Herling and Woods 1981). More recent data indicate that pigeons trained to the more selective kappa agonist, U-50,488, show greater differentiation between mu and kappa agonists. Morphine and EKC produce 30.8% and 66.2% drug-appropriate responding, respectively, in U-50,488-trained pigeons. Bremazocine is the only opioid that produces complete generalization in U-50,488-trained pigeons (Essman and Woods 1986).

There is no a priori assumption that can be made about species differences that would allow a prediction to be made about one species's ability to categorize mu and kappa agonists. Each new animal preparation will have to be measured with a full set of pharmacological tools to determine in each case how that species responds to drugs that have been found to have mu or kappa actions in other preparations.

Data obtained in humans concerning the receptor-mediated effects of opioids, such as their interoceptive and analgesic effects, are important, since opioid effects in this species are of considerable interest, both in psychology and medical practice. It is not clear which of the various nonhuman species most closely reflect the discriminative effects of opioids in humans.

4.3 Training Drug

As is indicated by the effects of opioids in different species, the opioid used to establish a discrimination can greatly affect the capacity of the preparation to distinguish between mu and kappa agonists. It is important that the investigator have a good understanding of the properties of the various opioids available so that he or she can select one that most closely meets the objectives of the study at hand. If the objective is to identify and differentiate opioid receptor types, the most selective agonist available should be used. The peptide Tyr-D-Ala-Gly-(Me)Phe-Gly-ol is the most selective in vitro agonist available at the present time for the mu system, but its effects in vivo have not been determined (Miller et al. 1986). The most selective agonist currently in the kappa system appears be U-50,488 (Von Voightlander et al. 1982). With this drug as a training compound, some differentiation between mu and kappa systems is possible even in the pigeon (Essman and Woods 1986). The choice of selective mu or kappa agonist may vary depending on the species used as subjects in a drug discrimination study. Although EKC is sufficiently selective in the rhesus monkey, it is useless in differentiating mu and kappa agonists in the pigeon. If the objective of the study is to understand the characteristics of a particular drug, the drug should be evaluated against the most selective agonists available and in a number of animal species.

4.4 Context

The pharmacological distinction between mu and kappa opioids can be made by using selective agonists and sensitive subjects. Another way to sharpen the ability of the assay system to differentiate mu and kappa agonists is to alter the context of the discrimination. Establishing a discrimination among three rather than two drug states is one way to effect such a change. For example, establishing a discrimination among cyclazocine, morphine, and saline reduces the cross-generalization between these drugs in the three-key procedure below that observed in two-key procedures (White and Holtzman 1981, 1983). This three-way discrimination procedure may be particularly helpful in a situation where a very selective agonist is not available. Evaluation of both full and partial agonists in a three-way discrimination may allow predictions to be made of effects in a two-way drug discrimination procedure.

These four factors should be taken into careful consideration in the design of drug discrimination experiments. Changes in training dose can result in changes in the apparent spectrum of action of the selected drug. The selected species can also greatly affect the way in which opioid drugs are categorized. Neither is the contribution of a single set of generalization dose-effect curves with a single training drug very helpful in categorizing drugs. This is true, not only of drugs categorized through drug discrimination studies, but also of other effects that are mediated through receptors.

5 Application of Receptor Theory to Partial Generalization

Data described in the preceding discussion include instances of partial generalization, where one drug produces some drug-appropriate responding, but does not produce a full generalization. The analysis of partial generalization data is currently a topic of much interest in opioid drug discrimination research; it is a focus of interest in a number of prominent laboratories (see, e.g., Colpaert and Janssen 1984). The question thus naturally emerges: does receptor theory have anything specific to offer in guiding theory formation, and especially in guiding experimental design, in the analysis of partial generalization data?

Partial generalization curves can be thought of as having two distinct processes: one that leads to increases in training drug-appropriate responding with increasing dose and one that limits or reverses this rise at some point in the dose-response curve.

A sound starting place in the analysis of partial generalization data is to attempt to characterize the receptors responsible for the rising phase of the dose-response curve for those drugs which produce some training drug-appropriate responding, but which do not induce full generalization at any dose. Whereas receptor theory, as outlined in the "Introduction," suggests that such drugs might be true partial agonists, it is clear that the same end point may be reached by interaction with more than one type of receptor or even by nonreceptor-mediated actions. It is helpful to assume initially that drugs producing partial generalization in the opioid system are true partial agonists and thus try to determine whether they are acting through a specific receptor type. The same procedures used in other branches of pharmacology, as discussed above for analysis of full generalization data, can be applied: determine the rank order of potency for a series of agonists in producing partial generalization and study the selectivity of antagonists in blocking it. If the agonist in question is producing partial generalization by activating mu receptors, the pA_2 for an equilibrium competitive receptor-blocking agent in antagonizing this effect should equal the pA_2 for this agent in antagonizing the effect of a drug producing full generalization.

A few words of caution are in order regarding the use of rank order potency analysis to determine receptor mediation of partial agonist data. Since potency is a function of both affinity and efficacy, the rank order of potency may change from one preparation to another, if different levels of sensitivity (fractional receptor activation) are demonstrated across the different preparations (e.g., Miller et al. 1986). Thus, attempts should be made to use a reference procedure with a sensitivity similar to that of the drug discrimination procedure. This is especially important when dealing with partial agonists, which may, in a less-sensitive procedure, show no agonist activity whatsoever, even though they are capable of producing mu receptor-mediated effects in more sensitive systems such as drug discrimination.

In general, the lesson is to approach the characterization of the receptors mediating partial generalization as one would approach the characterization of the receptors mediating an effect in any other pharmacological system. It is not necessary to assume that attainment of equal endpoints necessarily implies equal mechanisms in drug discrimination any more than it does in any other pharmacological system; such findings are quite common in pharmacology. Neither does the attainment of unequal endpoints necessarily imply different mechanisms. Receptor theory and past practice have provided the tools to begin an analysis of the molecular mechanisms underlying partial generalization in opioid drug discrimination, and the explanatory power of receptor theory should be pushed to the limit before it is abandoned for more esoteric strategies.

Acknowledgments. The authors would like to thank Dr. W. Koek for his critical reading of earlier drafts of this manuscript and R. McLaughlin for manuscript preparation.

References

Ariens EJ (1954) Affinity and intrinsic activity in the theory of competitive inhibition. I. Problems and theory. Arch Int Pharmacodyn 99:32–49

Bertalmio AJ, Woods JH (1987) Differentiation between mu and kappa receptor-mediated effects in opioid drug discrimination: apparent pA_2 analysis. J Pharmacol Exp Ther 243:591–597

Clark AJ (1926) The antagonism of acetylcholine by atropine. J Physiol 61:547–556

Colpaert FC, Janssen PAJ (1984) Agonist and antagonist effects of prototype opioid drugs in rats discriminating fentanyl from saline: characteristics of partial generalization. J Pharmacol Exp Ther 230:193–199

Colpaert FC, Neimegeers CJE, Janssen PAJ (1980) Factors regulating drug cue sensitivity: limits of discriminability and the role of a progressively decreasing training dose in fentanyl-saline discrimination. J Pharmacol Exp Ther 212:474–480

Dykstra LA, Gmerek DE, Winger G, Woods JH (1987a) Kappa opioids in rhesus monkeys. I. Diuresis, sedation, analgesia and discriminative stimulus effects. J Pharmacol Exp Ther 242:413–420

Dykstra LA, Gmerek DE, Winger G, Woods JH (1987b) Kappa opioids in rhesus monkeys. II. In vivo pA_2 analysis and specificity of antagonist action. J Pharmacol Exp Ther 242:421–427

Essman WD, Woods JH (1986) Discriminative stimulus and receptor binding properties of kappa agonists in the pigeon. Pharmacol Biochem Behav 25:305 (abstract)

Furchgott RF (1966) The use of beta-haloalkylamines in the differentiation of receptors and in the determination of dissociation constants of receptor-agonist complexes. In: Harper NJ, Simmonds AB (eds) Advances in drug research, vol 3. Academic, New York, pp 21–55

Hein DW, Young AM, Herling S, Woods JH (1981) Pharmacological analysis of the discriminative stimulus characteristics of ethylketazocine in the rhesus monkey. J Pharmacol Exp Ther 218:7–15

Herling S, Woods JH (1981) Discriminative stimulus effects of narcotics: evidence for multiple receptor-mediated actions. Life Sci 28:1571–1584

Herling S, Coale EH, Valentino RJ, Hein DW, Woods JH (1980) Narcotic discrimination in pigeons. J Pharmacol Exp Ther 214:139–146

Holtzman S (1982) Discriminative stimulus properties of opioids in the rat and squirrel monkey. In: Colpaert FC, Slangen JL (eds) Drug discrimination: applications in CNS pharmacology. Elsevier Biomedical, Amsterdam, pp 17–34

Jarbe TUC (1978) Discriminative effects of morphine in the pigeon. Pharmacol Biochem Behav 9:411–416

Katz J, Woods JH, Winger G, Jacobson AE (1982) Compounds of novel structure having kappa-agonist behavioral effects in rhesus monkeys. Life Sci 31:2375–2378

Martin WD, Eades CG, Thompson JA, Huppler RE, Gilbert PE (1976) The effects of morphine- and nalorphine-like drugs in the nondependent and morphine-dependent chronic spinal dog. J Pharmacol Exp Ther 197:517–532

McCarten MD, Lal H (1980) Generalization study with some narcotic and nonnarcotic psychoactive drugs in rats trained to discriminate between cyclazocine and saline. In: Way EL (ed) Endogenous and exogenous opioid agonists and antagonists. Pergamon, New York, pp 439–442

Miller L, Shaw JS, Watkins EM (1986) The contribution of intrinsic activity to the action of opioids in vitro. Br J Pharmacol 87:595–601

Portoghese PS, Larson DL, Sayre LM, Fries DS, Takemori AE (1980) A novel opioid receptor site directed alkylating agent with irreversible narcotic antagonistic and reversible agonist activities. J Med Chem 23:233–234

Schaefer GJ, Holtzman SG (1977) Discriminative effects of morphine in the squirrel monkey. J Pharmacol Exp Ther 201:67–75

Schaefer GJ, Holtzman SG (1978) Discriminative effects of cyclazocine in the squirrel monkey. J Pharmacol Exp Ther 205:291–302

Shannon HE, Holtzman SG (1976) Evaluation of the discriminative effects of morphine in the rat. J Pharmacol Exp Ther 198:54–65

Shannon HE, Holtzman SG (1977) Further evaluation of the discriminative effects of morphine in the rat. J Pharmacol Exp Ther 201:55–66

Shannon HE, Holtzman SG (1979) Morphine training dose: a determinant of stimulus generalization to narcotic antagonists in the rat. Psychopharmacology 61:239–244

Shearman GT, Herz A (1981) Discriminative stimulus properties of bremazocine in the rat. Neuropharmacology 20:1209–1213

Shearman GT, Herz A (1982a) Generalization and antagonism studies with rats trained to discriminate an effect of the proposed sigma opioid receptor agonist *N*-allylnormetazocine (SKF 10,047). In: Colpaert FC, Slangen JL (eds) Drug discrimination: applications in CNS pharmacology. Elsevier Biomedical, Amsterdam, pp 37–48

Shearman GT, Herz A (1982b) Discriminative stimulus properties of narcotic and non-narcotic drugs in rats trained to discriminate opioid kappa receptor agonists. Psychopharmacology 78:63–66

Shearman GT, Herz A (1982c) Evidence that the discriminative stimulus properties of fentanyl and ethylketocyclazocine are mediated by an interaction with different opioid receptors. J Pharmacol Exp Ther 221:735–739

Stephenson RP (1956) A modification of receptor theory. Br J Pharmacol 11:379–393

Tang AH, Code RA (1983) Discriminative stimulus properties of nalorphine in the rhesus monkey. J Pharmacol Exp Ther 227:563–569

Tang AH, Collins RJ (1985) Behavioral effects of a novel kappa opioid analgesic, U-50,488, in rats and rhesus monkeys. Psychopharmacology 85:309–314

Teal JJ, Holtzman SG (1980a) Discriminative stimulus effects of cyclazocine in the rat. J Pharmacol Exp Ther 212:368–376

Teal JJ, Holtzman SG (1980b) Discriminative stimulus effects of prototype opioid receptor agonists in monkeys. Eur J Pharmacol 68:1–10

Valentino RJ, Herling S, Woods JH, Medzihradsky F, Merz H (1981) Quaternary naltrexone: evidence for the central mediation of discriminative stimulus effects of narcotic agonists and antagonists. J Pharmacol Exp Ther 217:652–659

Von Voightlander PF, Lahti RA, Ludens JH (1982) U-50,488: a selective and structurally novel non-mu (kappa) opioid agonist. J Pharmacol Exp Ther 224:7–12

White JM, Holtzman SG (1981) Three-choice drug discrimination in the rat: morphine, cyclazocine and saline. J Pharmacol Exp Ther 217:254–262

White JM, Holtzman SG (1983) Further characterization of the three-choice morphine, cyclazocine and saline discrimination paradigm: opioids with agonist and antagonist properties. J Pharmacol Exp Ther 224:95–99

Winger G, Dykstra LA, Gmerek DE, Woods JH (1987) Behavioral effects of kappa-type agonist-antagonist nalorphine, cyclazocine, and bremazocine in rhesus monkeys. Fed Proc 46:382

Wood PL, Charleson SE, Lane D, Hudgin RL (1981) Multiple opioid receptors: differential binding of mu, kappa and delta agonists. Neuropharmacology 20:1215–1220

Discriminative and Analgesic Effects of Mu and Kappa Opioids: In Vivo pA_2 Analysis

L. A. Dykstra [1], A. J. Bertalmio [2], and J. H. Woods [2]

Contents

Abstract

This paper illustrates the use of antagonists to study receptor mediation of the discriminative stimulus effects of opioids and to determine if their analgesic effects are mediated in a similar manner. Analysis by pA_2 was used to quantify interactions between the opioid antagonist quadazocine and several opioid agonists. Interactions were examined under a drug discrimination procedure and under a tail withdrawal analgesia procedure. Values of pA_2 for quadazocine in combination with the kappa agonists bremazocine, ethylketazocine, and U50,488 ranged between 6.1 and 6.4 under the drug discrimination and tail withdrawal procedures. In contrast, the pA_2 values for the mu agonists etorphine, fentanyl, and morphine under these procedures ranged between 7.6 and 8.2. The difference between these pA_2 values indicates that the affinity of quadazocine for the receptors mediating the effects of the kappa agonists is over one log unit lower than its affinity for the receptors mediating the effects of the mu agonists. This suggests that the discriminative stimulus and analgesic effects of kappa opioid agonists are mediated at the same opioid receptor type, which is different from the type of opioid receptor at which mu agonists produce their discriminative stimulus and analgesic effects.

1 Introduction

In 1976 Martin and his colleagues utilized the concept of multiple opioid receptors to explain the different patterns of activity observed for morphine, ketocyclazocine, and *N*-allylnormetazocine in the dog. Briefly, they proposed that the diverse behavioral and physiological effects observed following administration of

[1] Departments of Psychology and Pharmacology, University of North Carolina, Chapel Hill, NC 27599, USA.
[2] Departments of Pharmacology and Psychology, University of Michigan, Ann Arbor, MI 48109, USA.

Psychopharmacology Series 4
Transduction Mechanisms of Drug Stimuli
Editors: Colpaert, Balster

each of these compounds might be due to differences in activity at three opiate receptor types. The three receptor types proposed were the mu receptor type, which was thought to mediate morphine-like analgesia and euphoria; the kappa receptor type, which was thought to mediate ketocyclazocine-like sedation and analgesia; and the sigma receptor type, which was thought to mediate *N*-allylnormetazocine-like psychotomimetic effects.

Subsequent investigations of the behavioral effects of these classes of opioids have provided additional evidence for multiple opioid receptor types – particularly those of the mu and kappa type. Among the numerous behavioral characteristics that can be used to distinguish between various opioid agonists, those related to their discriminative stimulus properties have been well characterized. For example, the discriminative stimulus effects of the kappa agonists can be clearly distinguished from those of the mu agonists in both rats and rhesus monkeys. Moreover, the discriminative stimulus effects of both mu and kappa agonists appear to be receptor mediated in that they display precise structure activity relationships and stereoselectivity, and are antagonized by opioid antagonists (Herling and Woods 1981 b; Holtzman 1985). Distinctions between mu and kappa opioids can also be made by examining the conditions under which their analgesic effects are apparent. For example, whereas both kappa and mu agonists are active in the stretching test (Schmauss and Yaksh 1984), only mu agonists tend to be active in rodent tests employing thermal stimuli such as the hot plate or tail flick (see Dykstra 1985 for a review of these data).

Although unique behavioral profiles among different opioids suggest that these compounds are acting at different receptor types, more convincing evidence can be obtained with drug antagonism studies. That is, if an antagonist such as naloxone differentially blocks the effects of mu and kappa opioid agonists, it might be inferred that the effects of these agonists are mediated at different receptor types. In the present chapter, we will discuss the use of pA_2 analysis to examine these types of interactions. Briefly, this analysis provides a way to quantify the potency of a given antagonist to attenuate the effects of different agonists. It is assumed that if an antagonist is equally potent in attenuating the effects of two different agonists, then the agonist-antagonist interaction is mediated by the same receptor population. Conversely, if an antagonist is more potent in antagonizing the effects of one agonist over another, it is assumed that different receptor populations mediate the effects of the two agonists. Thus, pA_2 analysis provides a parsimonious way to illustrate similarities and differences between diverse compounds and relate them to receptor-mediated activity.

2 Determination of an Apparent pA_2

Analysis utilizing the concept of pA_2 (Schild 1947) was originally developed to quantify interactions between agonists and antagonists in vitro. More recently, this analysis has been used successfully to characterize interactions between opioid agonists and antagonists in vivo (Takemori 1974). It is conventional to refer to an in vivo pA_2 value as an apparent pA_2.

2.1 Method of Calculation

In vivo, the apparent pA_2 is defined as the negative log of the dose of antagonist which when combined with an agonist doubles the dose of agonist necessary to produce a particular level of response. In most studies the "particular level of response" is expressed as an A50, i.e., the dose of agonist which produces 50% of the maximal response. In drug discrimination experiments, A50 refers to the dose of agonist that produces 50% responding on the drug-appropriate response alternative. In order to compare the effects of an agonist alone with its effects in the presence of an antagonist, a dose ratio is determined. The determination of the dose ratio is illustrated in Fig. 1. Briefly, the left portion of Fig. 1 shows four dose-effect curves, one for an agonist alone (broken lines), the other three for the agonist in the presence of increasing doses of an antagonist (solid lines). It can be seen that the agonist alone produced dose-dependent increases in the percentage effect with 0.3 mg/kg of the agonist alone producing 50% of the maximal effect (i.e., A50). Increasing doses of the antagonist shifted the agonist dose-effect curve to the right in a dose-dependent fashion, yielding A50 doses for the agonist in the presence of 0.01, 0.1, and 1.0 mg/kg of the antagonist of approximately 0.6, 3.0, and 30.0 mg/kg, respectively. The dose ratio is determined by dividing the A50 dose of the agonist in the presence of each antagonist dose by the A50 dose of the agonist alone. It can be seen from Fig. 1 that these ratios are 2, 10, and 100 for the 0.01-, 0.1-, and 1.0-mg/kg dose of antagonist, respectively.

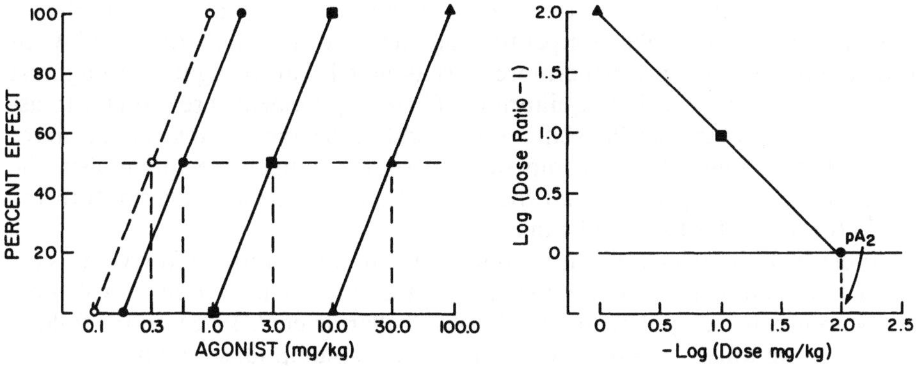

	A_{50}	Ratio	Log (Ratio-I)	-Log (Antag Dose)
Agonist Alone	0.3			
plus Antag (mg/kg)				
● 0.01	0.6	2	0	2
■ 0.10	3.0	10	0.95	1
▲ 1.00	30.0	100	2.00	0

Fig. 1. *Left* Dose-effect curves for agonist alone (*broken line*) and in combination with three doses (0.01, 0.1, and 1.0 mg/kg) of an antagonist (*solid line*). *Ordinate,* percentage of maximal effect. *Abscissa,* agonist dose in milligrams per kilogram, log scale. *Right* Schild plot for data shown *on left*. *Ordinate,* log (dose ratio minus 1). *Abscissa,* negative log of the antagonist dose shown as milligrams per kilogram. Dose ratios were obtained by dividing the A50 dose of agonist in the presence of a given dose of antagonist by the A50 dose of the agonist alone. The log (dose ratio minus 1) was then plotted as a function of the negative log of the antagonist dose. The pA_2 is the point *on the abscissa* where the regression line crosses the x-axis. In the present case the pA_2 is 2.0

Once the dose ratio has been determined, this ratio is plotted as shown on the right side of Fig. 1. Each dose ratio is represented on the ordinate as log (dose ratio −1) and expressed as a function of the negative logarithm of the dose of antagonist which produced that ratio. A plot of this kind is often referred to as a Schild plot. It is conventional to present the antagonist dose as a molar concentration of the base, but for purposes of the present description we present it as the negative log of the mg/kg dose. The dose ratios obtained for each dose of antagonist are shown on the right portion of Fig. 1, along with the line of best fit drawn through these points. In the present case, this line yields a slope of −1.

Since the pA_2 is defined as the dose of antagonist which doubles the dose of agonist necessary to produce a given effect, it is derived by determining the dose of antagonist which yields a dose ratio of 2. We can see that the point on the ordinate corresponding to a dose ratio of 2 is zero [i.e., log (2-1)]. The pA_2 value is the point at which the regression line intercepts the x-axis. In the present case, the pA_2 is 2.0, which corresponds to the 0.01-mg/kg dose of antagonist. A quick check of the left portion of Fig. 1 shows clearly that the 0.01-mg/kg dose is the dose of the antagonist which doubles the dose of agonist necessary to produce a 50% effect.

2.2 Theoretical Assumptions

In deriving a pA_2, several assumptions are made. For example, it is assumed that the antagonist used is a pure antagonist with no agonist effects. Moreover, derivation of the pA_2 requires that both the agonist and antagonist interact with the receptor in a fully reversible, competitive manner, and that they obey the law of mass action in so doing. Furthermore, it is assumed that the agonist-antagonist interactions are examined at equilibrium. These requirements are difficult to assess in vivo; however, parallel shifts to the right in the agonist dose-effect curves suggest that the interaction is competitive. Also, for competitive antagonists, the mathematical model predicts that the slope of the regression line used to fit the data in the Schild Plot should be minus one (−1).

Similarly, although equilibrium conditions cannot be verified in vivo, it is assumed that examination of interactions at the time of peak effects will most closely approximate equilibrium conditions (Tallarida et al. 1978). In the following section, we will discuss the way in which these assumptions were met in a specific series of studies in which pA_2 analysis was used to examine the potency of quadazocine as an antagonist of the discriminative and analgesic effects of mu and kappa agonists in the rhesus monkey.

3 Use of In Vivo pA_2 Analysis to Examine the Discriminative and Analgesic Effects of Mu and Kappa Opioids in Rhesus Monkeys

The utility of pA_2 analysis is best illustrated by examining data to which this analysis has been applied. In this section, we present data from a series of experiments (Bertalmio et al. unpublished observations; Dykstra et al. 1987a, b) in which this analysis was used to quantify the dose-related antagonism of the effects of several different opioid agonists. Data are presented from two different behavioral procedures: a drug discrimination procedure and a procedure designed to measure analgesia. The effects of six different opioid agonists were examined under these two procedures. These included three compounds classified as kappa agonists and three compounds classified as mu agonists. The effects of these compounds were examined alone and in combination with the opioid antagonist quadazocine.

3.1 Methods

Animals

Fourteen individually housed male rhesus monkeys weighing between 4 and 11 kg were used. Monkeys used in the tail withdrawal procedure were allowed free access to food. Monkeys used in the drug discrimination procedure were fed restricted amounts of food.

Pharmacological Procedure

Three different kappa and three different mu agonists were administered subcutaneously. The kappa agonists included bremazocine, a 6,7-benzomorphan with both kappa agonist (Romer et al. 1980) and mu antagonist activity (Von Voigtlander and Lewis 1982; Gmerek et al. 1987); ethylketazocine, another benzomorphan, which has been used traditionally as a prototype kappa agonist, although it is now known to have mu agonist properties in some preparations (e.g., Porreca et al. 1982), and the benzeneacetamide U-50,488 [*trans*-3,4-dichloro-*N*-methyl-*N*[2-(1-pyrrolidinyl)cyclohexyl] benzeneacetamide methanesulfonate hydrate], a structurally novel opioid (Von Voigtlander et al. 1983) which is thought to have the most selective kappa agonist profile in rodents (Lahti et al. 1982). The mu agonists included morphine as the prototypic agonist along with etorphine and fentanyl, both of which closely parallel morphine in their effects. Etorphine is a highly potent agonist from the oripavine series whereas fentanyl is a very potent phenylpiperidine (Casy and Parfitt 1986).

The effects of these agonists were examined in combination with the opioid antagonist quadazocine (Win 44, 441; Michne et al. 1978). As an antagonist, quadazocine is more potent than naloxone and has a much longer duration of action (Ward et al. 1983; Wood et al. 1984). Evidence suggests that quadazocine has antagonist activity in assays for mu, kappa, and delta receptor agonists (Ward et al. 1983; Wood et al. 1984; Richards and Sadee 1985).

Behavioral Procedures

The drug discrimination procedure has been described in detail previously (Bertalmio et al. 1982). Briefly, two groups of monkeys were placed in chambers equipped with two response levers and trained to respond differentially on the two levers depending on whether they received an injection of a drug or of saline. One group was trained to discriminate between the presence and absence of ethylketazocine (0.0032 mg/kg or 0.0056 mg/kg) and the other group was trained to discriminate between the presence and absence of codeine (1.8 mg/kg). Each experimental session was divided into several 15-min trials. Each trial started with a 10-min blackout period which was followed by the onset of a light in the chamber. In the presence of the stimulus light, completion of 100 consecutive responses on the appropriate lever in 5-min or less was followed by delivery of 10 food pellets. Each trial was followed by as many as five additional 15-min trials to comprise a daily session. In the monkeys trained to discriminate ethylketazocine, dose-effect curves were obtained by administering bremazocine, ethylketazocine, or U-50,488 alone and in combination with quadazocine. In the monkeys trained to discriminate codeine, dose-effect curves were obtained by administering morphine, etorphine, or fentanyl alone or in combination with quadazocine.

The tail withdrawal procedure has also been described previously (Dykstra and Woods 1986). Briefly, monkeys were placed in chairs which loosely restrained them at the neck and positioned them in a seated position such that their shaved tails hung free. The lower 10–12 cm of each monkey's tail was immersed into a thermos of water maintained at 50 °C and the time it took to withdraw its tail from the water was determined. If a monkey did not remove its tail from the water within 20 s, the tail was removed by the experimenter and the latency assigned a value of 20 s. Tail-withdrawal latencies were measured once on each of several daily trials and trials were spaced 30 min apart. Dose-effect curves were obtained by administering doses of morphine, bremazocine, ethylketazocine, or U-50,488 alone, and in combination with quadazocine.

Data Analysis

All data are expressed as the percentage of effect as a function of dose of agonist. Under the tail withdrawal procedure, percentage of effect refers to percentage of the maximal tail withdrawal latency (i.e., 20 s). Under the drug discrimination procedure, percentage of effect refers to percentage of total responses made on the drug-appropriate lever. Data are presented as the mean of all monkeys examined under each procedure. Dose-effect curves were determined for each agonist alone and in the presence of graded doses of the antagonist. For each of these dose-effect curves, the dose required to produce 50% effect (A50) was estimated by least square linear regression analysis. Using these A50 values, dose ratios were calculated as described above. Then, in a Schild plot, the log of the (dose ratio minus one) was plotted as a function of the negative logarithm of the corresponding dose of quadazocine, presented here as moles per kilogram using 480 as the mol. wt. of the quadazocine salt. Regression lines were determined for these data; slope and x-axis intercepts (apparent pA_2 values) were determined.

3.2 Results

Discriminative Effects

Figure 2 shows the effects of one mu agonist (etorphine) and one kappa agonist (ethylketazocine) alone and in combination with doses of quadazocine under the drug discrimination procedure. The effects of etorphine are shown in monkeys trained to discriminate between codeine and saline; the effects of ethylketazocine are shown in monkeys trained to discriminate between ethylketazocine and saline.

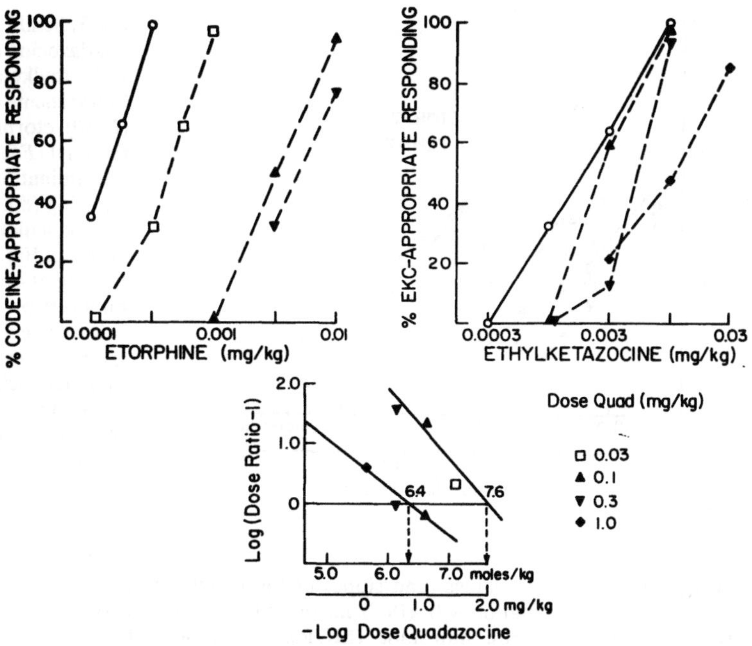

Fig. 2. *Top* Percentage of codeine-appropriate responding following doses of etorphine (*left panel*) or of ethylketazocine-appropriate responding following doses of ethylketazocine (*right panel*), alone (*solid line*) and in combination with doses of quadazocine ranging from 0.03 to 1.0 mg/kg (*broken lines*). *Ordinate*, percentage of total responses on the codeine-appropriate (*left panel*) or ethylketazocine-appropriate (*right panel*) lever. *Abscissa*, dose of etorphine or ethylketazocine. Each point is the mean of one determination in at least three monkeys. *Bottom* Schild plot for data shown above. *Ordinate*, log of (dose ratio minus one). *Abscissa*, negative log of the dose of quadazocine, based on moles per kilogram (*top abscissa*) or milligrams per kilogram (*bottom abscissa*). Dose ratios were obtained by dividing the A50 dose of agonist in the presence of a given dose of quadazocine by the A50 dose of the agonist alone. The pA_2 values shown here were 7.6 for quadazocine/etorphine and 6.4 for quadazocine/ethylketazocine combinations.

Etorphine produced dose-dependent increases in codeine-appropriate responding in the codeine-trained monkeys and ethylketazocine produced dose-dependent increases in ethylketazocine-appropriate responding in the ethylketazocine-trained monkeys. When quadazocine was administered prior to each agonist, the dose-effect curves were shifted to the right in a dose-dependent manner. It can be seen that a dose of 0.1 mg/kg quadazocine produced over a ten fold shift in the etorphine dose-effect curve, whereas this dose of quadazocine produced a very small rightward shift in the ethylketazocine dose-effect curve. A larger dose of quadazocine (1.0 mg/kg) was required to shift the ethylketazocine dose-effect curve approximately ten fold to the right.

Figure 2 also shows the Schild plot from which the pA_2 value was determined. The Schild plot was constructed using dose-ratio values obtained from regression analyses of the dose-effect curves shown in the upper portion of Fig. 2. The apparent pA_2 value is the point at which the regression line intercepts the x-axis and the dose ratio equals two. The apparent pA_2 for etorphine was 7.6 with a slope of -1.2 and the apparent pA_2 for ethylketazocine was 6.4 with a slope of -0.8.

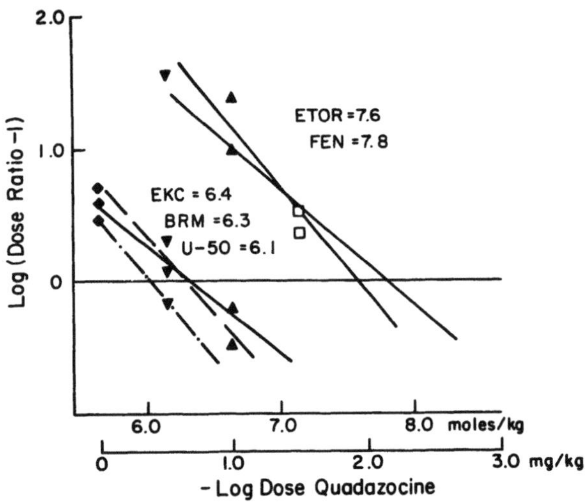

Fig. 3. Schild plot obtained for quadazocine in combination with ethylketazocine (*EKC*), bremazocine (*BRM*), U-50,488 (*U-50*), etorphine (*ETOR*), or fentanyl (*FEN*) under the drug discrimination procedure. *Ordinate,* log (dose ratio minus one), calculated as described above in Fig. 2. *Abscissa,* negative log of quadazocine based on moles per kilogram (*top abscissa*) or milligrams per kilogram (*bottom abscissa*). Doses of quadazocine in milligrams per kilogram correspond to the following symbols: □, 0.03 mg/kg; ▲, 0.1 mg/kg; ▼, 0.3 mg/kg; ◆, 1.0 mg/kg

Table 1. Apparent pA_2 values (and slopes of the regression line) for quadazocine in combination with ethylketazocine (EKC), bremazocine (BRM), U-50,488 (U50), morphine (MOR), etorphine (ET), and fentanyl (FEN) under the tail withdrawal and drug discrimination procedures

	pA_2	
	Tail withdrawal	Drug discrimination
EKC	6.4 (−1.3)	6.4 (−0.8)
BRM	6.1 (−1.0)	6.3 (−1.0)
U50	6.4 (−1.4)	6.1 (−1.2)
MOR	8.2 (−0.6)	7.7 (−1.0)
ET		7.6 (−1.2)
FEN		7.8 (−0.9)

Figure 3 summarizes the pA_2 values obtained for quadazocine in combination with three kappa and two mu agonists under the drug discrimination procedure. It can be seen that when doses of quadazocine were combined with the mu agonists, morphine, etorphine and fentanyl, the pA_2 values ranged between 7.6 and 7.8, respectively. In contrast, the pA_2 values obtained when doses of quadazocine were combined with the kappa agonists, ethylketazocine, bremazocine and U50,488, ranged between 6.1 and 6.4. See Table 1 for corresponding slopes.

Comparisons Between Discriminative and Analgesic Effects

Figure 4 compares the effects of the kappa agonist bremazocine alone and in combination with doses of quadazocine under the tail withdrawal (top, left panel) and the drug discrimination procedure in which monkeys were trained to discriminate between ethylketazocine and saline (top, right panel). Bremazocine alone pro-

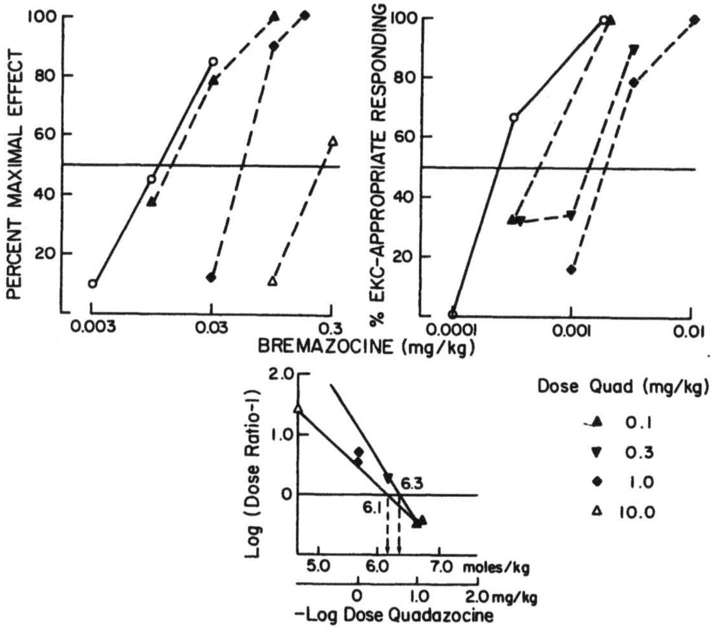

Fig. 4. *Top* Percentage of maximal tail withdrawal latency (*left panel*) or percentage of ethylketazocine (*EKC*)-appropriate responding (*right panel*) following bremazocine alone (*solid line*), or in combination with doses of quadazocine ranging from 0.1 to 10.0 mg/kg (*broken lines*). Each point is the mean of one determination in at least three monkeys. *Bottom* Schild plot obtained for quadazocine in combination with bremazocine under the drug discrimination and tail withdrawal procedures. Data presented as in Fig. 2. The pA_2 value was 6.3 under the drug discrimination procedure and 6.1 under the tail withdrawal procedure

duced dose-dependent increases in tail withdrawal latencies and in responding on the ethylketazocine-appropriate level. Doses of quadazocine produced approximately equivalent shifts to the right under both the tail withdrawal and drug discrimination procedures. The Schild plot shown in the bottom portion of Fig. 4 indicates that the pA_2 values for bremazocine/quadazocine combinations under the two procedures are similar, yielding a value of 6.1 (with a slope of -1.0) under the tail withdrawal procedure and 6.3 (with a slope of -1.2) under the discrimination procedure. Table 1 compares the pA_2 values obtained for several mu and kappa agonists under the drug discrimination and tail withdrawal procedures. It can be seen that the pA_2 values obtained for the mu agonists were similar under both the tail withdrawal and discrimination procedures whereas the values obtained for the kappa agonists, although similar to each other under both procedures, were different from those of the mu agonists.

3.3 Discussion

The data presented here indicate that the pA_2 values obtained for quadazocine in combination with the kappa agonists ranged between 6.1 and 6.4 under both the drug discrimination and tail withdrawal procedures. In contrast, the pA_2

Table 2. pA_2 values for naloxone in combination with mu and kappa agonists

pA_2		Species	Ref.
Mu	Kappa		
Analgesia			
7.2	6.3	Rhesus monkey	Yaksh and Rudy (1977)
7.0	5.9–6.6	Rats	Schmauss and Yaksh (1984)
6.8	5.2–6.0	Mice	Von Voigtlander et al. (1983)
7.1	6.7	Mice	Ward and Takemori (1983)
Discriminative stimulus effects			
7.6		Rats	Krimmer and Barry (1982)
	6.8	Rats	Herling and Shannon (1982)
7.8		Rats	Shannon et al. (1984)
Guinea pig ileum			
8.5	7.8		Ward et al. (1983)
8.6	7.8–7.9		Kosterlitz et al. (1974)

values for the mu agonists under these procedures ranged between 7.6 and 8.2. The difference between these two sets of values indicates that quadazocine has substantially lower affinity for the receptors at which the kappa agonists act than it has for the receptors at which the mu agonists act.

Previous investigations in which pA_2 values for naloxone have been compared for mu and kappa agonists have also found lower pA_2 values for naloxone in combination with kappa agonists than with mu agonists. Table 2 summarizes data from a few studies in which pA_2 values were obtained for naloxone in combination with both mu and kappa agonists in analgesia, drug discrimination, and smooth muscle assays. It can be seen that the pA_2 for naloxone in combination with morphine was approximately 7 in both the primate shock titration procedure and the rodent writhing procedure; whereas it was closer to 6 when the kappa agonists were examined under similar procedures. Different pA_2 values for naloxone in combination with mu or kappa agonists were also found in smooth muscle assays (see Table 2).

A pA_2 value of 7 for naloxone/morphine combinations is consistent with data obtained for a variety of mu agonists under a number of different analgesia assays (Takemori et al. 1972; Hayashi and Takemori 1971). Interestingly, different pA_2 values have been obtained for naloxone antagonism of the thermoregulatory, respiratory, and gastrointestinal effects of morphine (Takemori et al. 1969; McGilliard and Takenori, 1978). Although direct comparisons cannot be made between the pA_2 values obtained for naloxone/morphine interactions and those obtained in the present study in which quadazocine was used as an antagonist, the important point is that quadazocine, like naloxone, yielded different pA_2 values when combined with mu agonists than when combined with kappa agonists. (See also Bertalmio et al. 1984; Bertalmio and Woods 1987.) Ward et al. (1983) have shown similar differences for quadazocine in combination with mu and kappa agonists in the guinea pig ileum.

Although the apparent pA_2 values obtained here for quadazocine when it was combined with the kappa agonists were clearly quite different from those obtained when combined with morphine and other mu agonists, it is important to consider whether such quantitative differences truly indicate underlying receptor differences or whether they could have arisen from errors in the way the values were determined. Thus, it is important to examine how well the present analysis met the assumptions upon which determination of apparent pA_2 values rest.

We noted that the pA_2 analysis assumes that observations are made at equilibrium, i.e., the rate at which the agonist and antagonist are continually associating with the receptors must be equal to the rate of breakdown of these drug-receptor complexes. Although the attainment of equilibrium conditions cannot be verified in vivo, it is generally assumed (Takemori 1974; Tallarida et al. 1978) that equilibrium conditions are best approximated if observations are made when both the antagonist and agonist are at the peak of their effects. On the basis of studies in our laboratory indicating that quadazocine has a very long duration of action, we assumed that the effects of quadazocine were relatively constant at, or near, peak effect throughout our 2- to 3-h experimental sessions. On the other hand, although we did not directly determine time course and peak effects for all of our mu and kappa agonists, we did examine agonists with remarkably different time parameters and found that they yielded consistent pA_2 values. For example, despite the fact that morphine has a very long duration of action and fentanyl a short one, the pA_2 values for morphine and fentanyl were similar to each other, i.e., 7.7 and 7.8, respectively, and different from the kappa agonists.

Moreover, quadazocine produced parallel shifts in the agonists' dose-response curves, and generally yielded regression lines in the Schild plot analysis which were close to minus one. This pattern is what one would expect to see for a pure, equilibrium, competitive, receptor antagonism – the type required for the derivation of accurate, meaningful pA_2 values. Thus, the data presented here are consistent with the pattern one would expect to see if all conditions had been met successfully. It is therefore likely that the difference in apparent pA values found here between mu and kappa agonists reflects activity at different receptor types.

4 Practical Considerations Related to the Use of pA_2 Analysis

In the present section we would like to address some of the practical considerations which might be taken into account when using in vivo pA_2 analyses, particularly in drug discrimination procedures. We have already discussed the pharmacological assumptions of equilibrium and competition which underlie the determination of an apparent pA_2. We have also shown how these assumptions were met (or approximated) within the studies reviewed here. Meeting these assumptions, however, is not the only important consideration in using pA_2 analysis. Indeed, there are a number of other pharmacological considerations including the selection of appropriate agonists and antagonists, and of appropriate dose ranges of these compounds.

In designing the present studies, we made use of evidence from other experiments to select appropriate mu and kappa agonists. Indeed, a number of studies have shown that the mu and kappa agonists examined here are clearly differentiated by rhesus monkeys under drug discrimination procedures (Dykstra et al. 1987a; Hein et al. 1981; Herling and Woods 1981a) and that they exhibit unique profiles in analgesic, smooth muscle, and receptor-binding assays (James and Goldstein 1984; Hayes and Kelly 1985; Lahti et al. 1982; Schmauss and Yaksh 1984).

Moreover, it is desirable to use agonists with a high degree of potency so that rightward shifts in the dose-effect curve are not limited by the solubility of the compound. For example, in drug discrimination procedures, a 2 log unit shift in the morphine dose-effect curve would require the administration of an impractical volume of approximately 250 ml of a maximally saturated solution in a 10-kg monkey. Moreover, very large doses of a nonselective agonist might have opioid activity at other receptors or nonopioid effects which would interfere with reliable estimates of pA_2 values.

Similar considerations need to be made in the selection of an antagonist. We have already discussed the importance of selecting an antagonist with a relatively long duration of action so that it exerts comparable pharmacological effects throughout an experimental session. We show here that quadazocine is an especially good compound in this respect whereas compounds even purported to have long durations, e.g., naltrexone, have been shown to be variable in their antagonist effects because of their duration of action (Teal and Holtzman 1980).

Another important consideration relates to the dose range over which the antagonist can be examined. Since reliable pA_2 estimates rest on examining a range of antagonist doses, it is important that the antagonist be examined over a range of doses that shift the agonist dose-effect curves severalfold without producing disruptive effects of its own. Indeed, it has been shown that certain opioid antagonists have direct effects on behavior which interfere with observation of their antagonist action (e.g., Young and Woods 1982; Dykstra 1983). In the present studies quadazocine was effective as an antagonist over a large dose range (0.001–10.0 mg/kg in the tail withdrawal studies; 0.01–3.2 mg/kg in the discrimination studies) and produced only very limited direct effects on behavior. Moreover, the effects of quadazocine did not appear to change as the result of repeated administration, which is in contrast to our own observations with antagonists such as naloxone and naltrexone to which monkeys have been shown to develop sensitivity upon repeated administration (Dykstra 1983; Winger, personal communication), thereby limiting their use in experiments such as these.

Finally, one might ask whether this analysis can be used to examine data from individual animals as is often done in behavioral studies. In this context, it should be noted that pA_2 values in individual monkeys were similar to those obtained by averaging across monkeys (Bertalmio and Woods 1987; Dykstra, unpublished observations).

Thus, it appears that in vivo pA_2 analysis can be used successfully to differentiate mu and kappa opioid agonists in the rhesus monkey. Although the apparent pA_2 values obtained here may be specific to the species and antagonist examined as well as to parameters related to time and route of drug administration, it is important to note that the differences described herein are similar in magnitude to those observed with other antagonists and agonists, with other assays, and in other species. Thus, we can conclude that this analysis is useful for characterizing receptor-mediated behavioral effects of diverse classes of opioids, and, more specifically, that it may be applicable to the examination of a variety of behaviors.

In summary, the different pA_2 values obtained for quadazocine in antagonizing both the discriminative and analgesic effects of kappa and mu agonists indicate that the affinity of quadazocine for the receptors mediating the effects of the kappa agonists is over tenfold less than its affinity for the receptors mediating the effects of the mu agonists. Hence one can conclude that mu agonists act at the same type of opioid receptor to produce their discriminative and analgesic effects, but that these are different from the type of opioid receptor at which kappa agonists produce their discriminative and analgesic effects. This is in keeping with previous reports indicating that the discriminative stimulus effects of opioid agonists are highly correlated with their analgesic effects (Colpaert et al. 1976a, b, 1978a, b). These data, along with those from other studies using pA_2 analysis (Krimmer and Barry 1982; Herling and Shannon 1982; Shannon et al. 1984; Bertalmio et al. 1984; Dykstra et al. 1987b; Bertalmio and Woods 1987), demonstrate the feasibility of this approach to the study of transduction mechanisms of drug stimuli.

Acknowledgments. This work was supported by U.S. Public Service Grants DA00154, 00254, and 02749 from the National Institute on Drug Abuse. L.A.D. is a recipient of Research Scientist Award DA00033.

References

Bertalmio AJ, Woods JH (1987) Differentiation between mu and kappa receptor-mediated effects in opioid drug discrimination: apparent pA_2 analysis. J Pharmacol Exp Ther 243:591–597

Bertalmio AJ, Herling S, Hampton RY, Winger G, Woods JH (1982) A procedure for rapid evaluation of the discriminative stimulus effects of drugs. J Pharmacol Methods 7:289–299

Bertalmio AJ, Winger GD, Woods JH (1984) Mu and kappa receptor-mediated opioid discriminative effects in rhesus monkeys: pA analysis. Fed Proc 43:966

Casy AF, Parfitt RT (1986) Opioid analgesics: chemistry and receptors. Plenum, New York

Colpaert FC, Niemegeers CJE, Janssen PAJ (1976a) On the ability of narcotic antagonists to produce the narcotic cue. J Pharmacol Exp Ther 197:180–187

Colpaert FC, Niemegeers CJE, Janssen PAJ (1976b) The narcotic discriminative stimulus complex: relation to analgesic activity. J Pharm Pharmacol 28:183–187

Colpaert FC, Niemegeers CJE, Janssen PAJ (1978a) Narcotic cueing and analgesic activity of narcotic analgesics: associative and dissociative characteristics. Psychopharmacology 57:21–26

Colpaert FC, Niemegeers CJE, Janssen PAJ, Van Ree JM (1978b) Narcotic cueing properties of intraventricularly administered sufentanil, fentanyl, morphine and met-enkephalin. Eur J Pharmacol 47:115–119

Dykstra LA (1983) Development of enhanced sensitivity to naloxone. Life Sci 33:2079–2089

Dykstra LA (1985) Behavioral and pharmacological factors in opioid analgesia. In: Seiden LS, Balster RL (eds) Behavioral pharmacology: the current status. Liss, New York, pp 111–129

Dykstra LA, Woods JH (1986) A tail withdrawal procedure for assessing analgesic activity in rhesus monkeys. J Pharmacol Methods 15:263–269

Dykstra LA, Gmerek DE, Winger G, Woods JH (1987a) Kappa opioids in rhesus monkeys. I. Diuresis, sedation, analgesia and discriminative stimulus effects. J Pharmacol Exp Ther 242:413–420
Dykstra LA, Gmerek DE, Winger G, Woods JH (1987b) Kappa opioid agonists in rhesus monkeys. II. Analysis of the antagonist actions of quadazocine and β-funaltrexamine. J Pharmacol Exp Ther 242:421–427
Gmerek DE, Dykstra LA, Woods JH (1987) Kappa opioids in rhesus monkeys: III. Dependence associated with chronic administration. J Pharmacol Exp Ther 242:428–436
Hayashi G, Takemori AE (1971) The type of analgesic-receptor interaction involved in certain analgesic assays. Eur J Pharmacol 16:63–66
Hayes A, Kelly A (1985) Profile of activity of *k* receptor agonists in the rabbit vas deferens. Eur J Pharmacol 110:317–322
Hein DW, Young AM, Herling S, Woods JH (1981) Pharmacological analysis of the discriminative stimulus characteristics of ethylketazocine in the rhesus monkey. J Pharmacol Exp Ther 218:7–15
Herling S, Shannon HE (1982) Discriminative effects of ethylketazocine in the rat: stereospecificity and antagonism by naloxone. Life Sci 31:2371–2374
Herling S, Woods JH (1981a) Discriminative stimulus effects of etorphine in rhesus monkeys. Psychopharmacology 72:265–267
Herling S, Woods JH (1981b) Discriminative stimulus effects of narcotics: evidence for multiple receptor-mediated actions. Life Sci 28:1571–1584
Holtzman SG (1985) Discriminative stimulus properties of opioids that interact with mu, kappa, and PCP/sigma receptors. In: Seiden LS, Balster RL (eds) Behavioral pharmacology: the current status. Liss, New York, pp 131–147
James IF, Goldstein A (1984) Site-directed alkylation of multiple opioid receptors 1. Binding selectivity. Mol Pharmacol 25:337–342
Kosterlitz HW, Waterfield AA, Berthoud V (1974) Assessment of the agonist and antagonist properties of narcotic analgesic drugs by their actions on the morphine receptor in the guinea pig ileum. In: Braude MC, Harris LS, May EL, Smith JP, Villarreal JE (eds) Advances in biochemical psychopharmacology, vol 8. Raven, New York, pp 319–334
Krimmer EC, Barry H III (1982) Measuring naloxone antagonism of discriminative opioid stimulus. Fed Proc 41:2319–2322
Lahti RA, Von Voigtlander PF, Barsuhn C (1982) Properties of a selective kappa agonist, U-50,488H. Life Sci 31:2257–2260
Martin WR, Eades CG, Thompson JA, Huppler RE, Gilbert PE (1976) The effects of morphine- and nalorphine-like drugs in the nondependent and morphine-dependent chronic spinal dog. J Pharmacol Exp Ther 197:517–532
McGilliard KL, Takemori AE (1978) Antagonism by naloxone of narcotic-induced respiratory depression and analgesia. J Pharmacol Exp Ther 207:494–503
Michne WF, Lewis TR, Michalec SJ, Pierson AK, Gillan MGC, Paterson SJ, Robson LE, Kosterlitz HW (1978) Novel developments of *n*-methylbenzomorphan narcotic antagonists. In: van Ree JM, Terenius L (eds) Characteristics and function of opioids. Elsevier, Amsterdam, pp 197–206
Porreca F, Cowan A, Raffa RB, Tallarida RJ (1982) On the criteria for classifying opiate agonists in rats. J Pharm Pharmacol 34:525–526
Richards ML, Sadée W (1985) In vivo binding of benzomorphans to mu, *delta* and *kappa* opioid receptors: Comparison with urine output in the rat. J Pharmacol Exp Ther 233:425–432
Römer D, Büscher H, Hill RC, Maurer R, Petcher TJ, Welle HBA, Bakel HCCK, Akkerman AM (1980) Bremazocine: a potent, long-acting opiate kappa-agonist. Life Sci 27:971–978
Schild HO (1947) Drug antagonism. Br J Pharmacol Chemother 2:189–206
Schmauss C, Yaksh TL (1984) In vivo studies on spinal opiate receptor systems mediating antinociception. II. Pharmacological profiles suggesting a differential association of *mu, delta* and *kappa* receptors with visceral chemical and cutaneous thermal stimuli in the rat. J Pharmacol Exp Ther 228:1–12
Shannon HE, Cone EJ, Gorodetzky CW (1984) Morphine-like discriminative stimulus effects of buprenorphine and demethoxybuprenorphine in rats: quantitative antagonism by naloxone. J Pharmacol Exp Ther 229:768–774

Takemori AE (1974) Determination of pharmacological constants: use of narcotic antagonists to characterize analgesic receptors. In: Braude MC, Harris LS, May EL, Smith JP, Villarreal JE (eds) Advances in biochemical psychopharmacology, vol 8. Raven, New York, pp 335–344

Takemori AE, Kupferberg HJ, Miller JW (1969) Quantitative studies of the antagonism of morphine by nalorphine and naloxone. J Pharmacol Exp Ther 169:39–45

Takemori AE, Hayashi G, Smits SE (1972) Studies on the quantitative antagonism of analgesics by naloxone and diprenorphine. Eur J Pharmacol 20:85–92

Tallarida RJ, Harakal C, Maslow J, Geller EB, Adler MW (1978) The relationship between pharmacokinetics and pharmacodynamic action as applied to in vivo pA_2: application to the analgesic effect of morphine. J Pharmacol Exp Ther 206:38–45

Teal JJ, Holtzman SG (1980) Stimulus effects of morphine in the monkey: quantitative analysis of antagonism. Pharmacol Biochem Behav 12:587–593

Von Voigtlander PF, Lewis RA (1982) U-50,488, a selective kappa opioid agonist: comparison to other reputed kappa agonists. Prog Neuro-Psychopharmacol Biol Psychiatry 6:467–470

Von Voigtlander PF, Lahti RA, Ludens JH (1983) U-50,488: a selective and structurally novel non-*mu* (*kappa*) opioid agonist. J Pharmacol Exp Ther 224:7–12

Ward SJ, Takemori AE (1983) Relative involvement of *mu, kappa* and *delta* receptor mechanisms in opiate-mediated antinociception in mice. J Pharmacol Exp Ther 224:525–530

Ward SJ, Pierson AK, Michne WF (1983) Multiple opioid receptor profile in vitro and activity in vivo of the potent opioid antagonist WIN 44,441-3. Life Sci 33, Sup 1:303–306

Wood PL, Pilapil C, Thakur M, Richard JW (1984) WIN 44,441: a stereospecific and long-acting narcotic antagonist. Pharm Res 1:46–48

Yaksh TL, Rudy TA (1977) A dose ratio comparison of the interaction between morphine and cyclazocine with naloxone in rhesus monkeys on the shock titration task. Eur J Pharmacol 46:83–92

Young AM, Woods JH (1982) Limitations on the antagonistic actions of opioid antagonists. Fed Proc 41:2333–2338

Receptor Mediation of the Discriminative Stimulus Properties of Phencyclidine and Sigma-Opioid Agonists

R. L. BALSTER and J. WILLETTS

Contents

Abstract

Evidence has accumulated that the discriminative stimulus effects of phencyclidine (PCP) may be transduced by a specific receptor in mammalian brain. Two major lines of evidence support this hypothesis. One is the structure-activity correlation that arylcyclohexylamine analogs of PCP have for PCP-like discriminative effects and displacement of [^{3}H]PCP binding. The other, even stronger, evidence is that representatives of some nonarylcyclohexylamine classes of drugs are both generalized from PCP and have activity at PCP-binding sites. These include the psychotomimetic sigma-agonist opioids such as (+)-*N*-allylnormetazocine (NANM), the 1,3-substituted dioxolanes dexoxadrol and etoxadrol and some benz(f)isoquinolines. Early evidence suggested that there may be a complete overlap in the discriminative effects and receptor systems for PCP and sigma-agonists, and there continues to be evidence to support the commonality of these drug groups. On the other hand, binding studies with radiolabeled sigma agonists have revealed a non-PCP site. The role of this site in the behavioral actions of sigma-agonists is at present unknown. PCP analogs and other PCP-like drugs also can function as reinforcers for self-administration behavior, suggesting that the same cellular mechanisms may be responsible for both discriminative and reinforcing stimulus effects of these drugs. PCP has been shown to block many of the in vitro and in vivo effects of N-methyl-D-aspartate (NMDA), a putative specific agonist for a subtype of excitatory amino acid receptor. Recent evidence that NMDA antagonists are generalized from PCP in rats and pigeons provides evidence that modification of excitatory amino neurotransmission may be a physiological function of the PCP receptor and that this receptor complex may be involved in PCP's discriminative stimulus effects.

Department of Pharmacology and Toxicology, Medical College of Virginia, Virginia Commonwealth University, Richmond, VA 23298, USA.

Psychopharmacology Series 4
Transduction Mechanisms of Drug Stimuli
Editors: Colpaert, Balster

1 Introduction

Phencyclidine (PCP) is a major representative of a class of arylcyclohexylamine drugs with a unique spectrum of behavioral pharmacological effects (Domino 1964; Balster 1986a). Initially developed as an injectable anesthetic, PCP gained notoriety with an epidemic of abuse emerging in the late 1960s and continuing until today. Because of this epidemic of abuse and because PCP has such an interesting pharmacological profile, drug discrimination studies with PCP began to appear in 1979 and in the intervening years a great deal has been learned about its discriminative stimulus effects.

One of the important areas of drug discrimination research with PCP has been concerned with the neural substrate for its discriminative stimulus properties. Sufficient data have accumulated to hypothesize that these properties of PCP are mediated by a specific receptor located in the mammalian central nervous system. This receptor has been quite well characterized and progress is rapidly being made in determining its physiological function in the brain. There is also some evidence that a number of other drugs may exert their discriminative stimulus effects at least partly through this same receptor system. The purpose of this paper is to review evidence for receptor mediation of the discriminative stimulus effects of PCP as well as newer developments in studies of transduction of PCP stimuli.

2 Neuropharmacology of PCP

Phencyclidine has a broad range of behavioral effects and interacts potently with a number of neurochemical systems (Balster 1986a; Johnson and Snell 1986). A postulated biochemical mechanism by which PCP exerts these behavioral effects is that of direct interaction with receptors for one or more classical neurotransmitter systems. However, in binding studies, PCP has no affinity for sites labeled by dopamine, serotonin, various amino acids, and other putative neurotransmitters (Vincent et al. 1981) and only low affinity for muscarinic and opiate binding sites (Vincent et al. 1978). It is, therefore, unlikely that direct actions at these sites are primarily responsible for PCP's pharmacological effects.

Evidence that PCP acts through its own receptor first came from Zukin and Zukin and Vincent et al. in 1979 when a specific, relatively high affinity site for [^{3}H]PCP binding was discovered. This site has biochemical characteristics of a receptor since binding is saturable, reversible, and selective. No other known neurotransmitters or neuromodulators tested to date have a high affinity for the site. Additionally, proteolytic enzymes and heat decrease binding, indicating the proteinaceous nature of the site. The distribution of PCP binding throughout the brain is heterogeneous, with highest density in the cortex and hippocampus and lowest in the brain stem and spinal cord (Quirion et al. 1981; Zukin and Zukin 1983). Thus, this binding site fulfills many of the criteria for being a receptor. The rest of this paper will address another, and most important, criterion for identification of a receptor, the correlation of binding activity with a pharmacological function.

3 Evidence for Receptor Mediation of the Discriminative Stimulus Effects of PCP

Important evidence that PCP might have a specific neural transduction mechanism for its discriminative stimulus effects is the specificity of these effects. Even though PCP shares some effects with drugs from differing pharmacological categories, for example, anticholinergics, sympathomimetic stimulants, CNS depressants, and hallucinogens, these drugs do not share PCP's discriminative stimulus properties (Poling et al. 1979; Shannon 1983; Browne et al. 1983; Browne 1986) nor do they displace PCP from its binding site (Vincent et al. 1979). This rules out many possible transduction mechanisms for PCP stimuli; for example, if PCP's catecholaminergic actions were responsible for its discriminative stimulus effects, then PCP should be generalized to sympathomimetics such as amphetamine. In addition, common neurotransmitter receptor blocking drugs including opiate, cholinergic, dopaminergic, serotonergic, adrenergic, and nicotinic antagonists fail to antagonize PCP's discriminative stimulus effects (Poling et al. 1979; Browne et al. 1983; Browne 1986) and likewise do not displace PCP from its binding site (Vincent et al. 1979). While it was reported that PCP's discriminative stimulus effects could be blocked by adenosine agonists such as 1-phenylisopropyladenosine (Browne and Welch 1982), subsequent studies showed that PCP did not interfere with adenosine binding and a biodispositional rather than pharmacological interaction was implicated (Browne et al. 1983).

Thus, because PCP's discriminative effects do not appear to be mediated through commonly studied neurotransmitter systems, it could be concluded that these effects might be mediated by a unique transduction mechanism. A logical candidate to be involved in PCP's cellular mechanism of action is the PCP binding site. One of the strongest pieces of evidence in support of this site being the receptor mediating PCP's discriminative effects is the failure of commonly studied agonists and antagonists for other receptor systems to either potently displace PCP binding or mimic or block its discriminative stimulus effects. The next sections will consider those compounds which do interact with both PCP binding and discrimination.

3.1 Generalization Studies with Arylcyclohexylamines

Phencyclidine is an arylcyclohexylamine. A number of arylcyclohexylamine analogs of PCP have been studied by ourselves (Brady and Balster 1981, 1982a) and others (Shannon 1981a; Browne et al. 1983; Browne 1986) for their discriminative stimulus effects. Many of these analogs are generalized from PCP and also have affinity for the PCP receptor (Table 1). Indeed, there is no case of an arylcyclohexylamine which has PCP-like discriminative stimulus effects which does not have a reasonable affinity for the receptor. There is even a good correlation ($r=0.83$) between binding and discrimination potencies (Fig. 1). For the results shown in Table 1 and Fig. 1, only studies in rats were considered. The potency of each of the drugs relative to PCP in each study was calculated. When the same compound was evaluated in more than one study, the relative potencies from these studies were averaged. For Table 1, drugs were considered to have a reason-

Table 1. Comparison of PCP-like discriminative stimulus effects and binding properties of arylcyclohexylamines in rats

Drug	Generalization from PCP	Displacement of [^{3}H]PCP binding (relative potency to PCP >0.1)
N-Substituted analogs of PCP		
(±)-Ketamine	Yes[a,b]	Yes[b]
(+)-Ketamine	Yes[b]	Yes[b]
(−)-Ketamine	Yes[b]	Weak[b]
PCE	Yes[a]	Yes[c,d]
PCPY	Yes[a]	Yes[c,d]
PCM	Yes[a]	Yes[c,d]
NPPCA	Yes[a]	Yes[c]
NIPPCA	Yes[a]	Yes[e]
PCDEA	Yes[a]	Yes[c,d]
(+)-PCMP	Yes[g]	Yes[e]
(−)-PCMP	No[g]	No[e]
Aromatic-Ring Analogs of PCP		
TCP	Yes[a]	Yes[c−f]
TCPY	Yes[a]	Yes[c,d]
TCM	Yes[a]	Yes[c]/No[d]
PCC	No[a]	No[c,d]
Cycloalkyl-Ring Analog of PCP		
P(3M)CP	No[a]	Yes[f]

Abbreviations defined in Shannon (1983)
[a] Shannon (1983).
[b] Zukin et al. (1984).
[c] Zukin and Zukin (1979).
[d] Quirion et al. (1983).
[e] Contreras et al. (1986).
[f] Vincent et al. (1979).
[g] Browne et al. (1983).

able affinity for [^{3}H]PCP binding if they were at least one-tenth as potent as PCP.

One of the requirements in showing receptor mediation of drug action is stereospecificity. While PCP does not have a chiral center, several arylcyclohexylamine analogs are optically active. The most studied of these are ketamine and the 3-methyl piperidine derivative of PCP, PCMP. In both discrimination and binding studies, the (+)-isomers of these compounds are more potent than their (−)-isomers (Brady and Balster 1982a; Browne et al. 1983; Zukin et al. 1984; Contreras et al. 1986).

3.2 Generalization Studies with PCP Biologs

Opioids

Even more important in establishing the role of the PCP receptor in PCP's discriminative stimulus effects is the finding that a number of drugs which are not

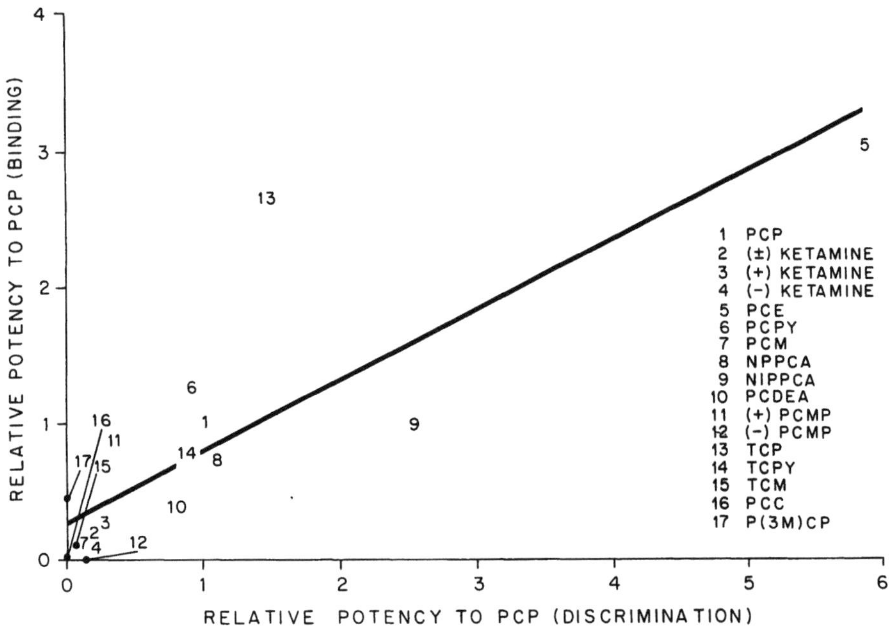

Fig. 1. Correlation between the relative potency of arylcyclohexylamine analogs of PCP for displacement of [^{3}H]PCP binding and for generalization from PCP in rats. Abbreviations are defined in Shannon (1983)

arylcyclohexylamine analogs of PCP also have affinity for the PCP receptor and are generalized from PCP (Fig. 2, Table 2). These drugs could be referred to as PCP "biologs." They include psychotomimetic opioids, with the most studied example being the prototypic sigma-agonist *N*-allylnormetazocine (NANM, SKF 10,047). Using the chronic spinal dog preparation in which Martin et al. (1976) arrived at their model of opiate receptor subtypes, Vaupel and Jasinski (1979) found NANM, PCP, and, subsequently, various arylcyclohexylamine analogs of PCP (Vaupel 1983) to produce similar profiles of effects. In drug discrimination studies, NANM was generalized from PCP (Shannon 1981 b) and had reasonable affinity for the PCP-binding site (Zukin and Zukin 1979). This lent important early support for the hypothesis that the site was involved in mediating PCP's discriminative stimulus effects. As shown in Table 2, other opioids have also subsequently been shown to have PCP-like discriminative stimulus effects and affinity for PCP-binding sites in rats.

The enantiomeric selectivity of PCP-like opioids is a somewhat complicated issue. Some studies with stereoisomers of benzomorphans have indicated that only the (+)-isomers are generalized from PCP (Brady et al. 1982; Brady and Balster 1982 b; Holtzman 1982). For example, with cyclazocine, studies in both rats and squirrel monkeys have shown that only the (+)-isomer, not the (−)-isomer nor racemate, was generalized from PCP (Brady and Balster 1982 b; Holtzman 1982; Slifer and Balster 1985 b; Zukin et al. 1984). However, (−)-cyclazocine is not inactive and causes response rate decreases at lower doses than is necessary

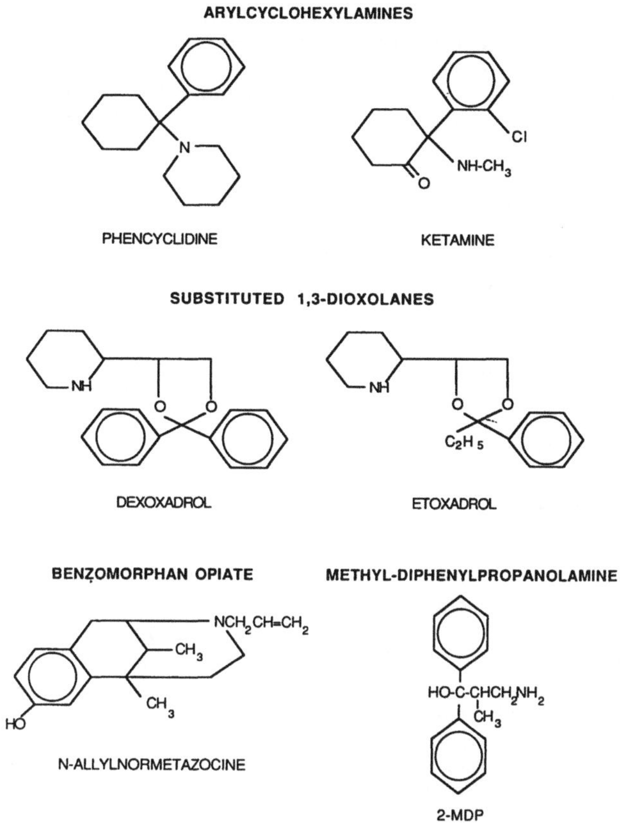

Fig. 2. Structures of representative PCP analogs and biologs which have PCP-like discriminative stimulus effects

for the (+)-isomer to produce PCP-lever responding. This response rate decrease can be attributed to kappa-agonist effects which predominate in the (−)-isomer (S. R. Zukin and Zukin 1981). When naltrexone is used to block the kappa-agonist effects of (−)-cyclazocine, response rate decreases are reduced, and at higher doses of (−)-cyclazocine PCP-like discriminative stimulus effects are observed (Holtzman 1982; Slifer and Balster 1985 b). Thus, with opioid isomers the (+)-enantiomers are more selective for PCP-like effects, generally lacking the mu- and kappa-agonist and/or antagonist activity of the (−)-enantiomers and racemates (Zukin et al. 1984).

1,3-Substituted Dioxolanes

The drugs which have shown the greatest stereospecificity for PCP-like effects are the 1,3-substituted dioxolanes. Etoxadrol and its diphenyl analog, dexoxadrol [alpha (+)-dioxadrol], are both generalized from PCP (Shannon 1983; Slifer and Balster 1985 a; Tang et al. 1984) and have affinity for the PCP receptor (Mendelsohn et al. 1984). On the other hand, the alpha (−)-enantiomer of dioxadrol, levoxadrol, is neither generalized from PCP (Shannon 1983; Slifer and Balster 1985 a) nor binds to the PCP receptor (Mendelsohn et al. 1984).

Tabele 2. Comparison of the PCP-like discriminative stimulus effects and binding properties of selected PCP biologs in rats

Drug	Generalization from PCP	Displacement of [^{3}H]PCP binding (relative potency to PCP >0.1)
Opioids		
(±)-NANM	Yes[c]/No[a]	Yes[a]
(+)-NANM	Yes[a, c]	Yes[a]
(−)-NANM	Yes[c]/No[a]	Yes[a]
(±)-Cyclazocine	Yes[c]/No[a]	Yes[a]
(+)-Cyclazocine	Yes[a, c]	Weak[a]
(−)-Cyclazocine	Yes[c]/No[a]	Yes[a]
(±)-Pentazocine	No[c]	No[b]
(±)-Ketocyclazocine	No[c]	No[b]
Dextrorphan	Yes[c]	Yes[e]
Levallorphan	No[b]	No[d]
1,3-Substituted dioxolanes		
Dexoxadrol	Yes[c, f]	Yes[g]
Levoxadrol	No[c, f]	No[g]
Etoxadrol	Yes[h]	Yes[g]
Benz-(f)-isoquinolines		
LY154045	Yes[i]	Yes[g]
LY154005	No[i]	No[g]

[a] Zukin et al. (1984).
[b] Quirion et al. (1983).
[c] Shannon (1983).
[d] Zukin and Zukin (1979).
[e] Murray and Leid (1984).
[f] Slifer and Balster (1985a).
[g] Mendelsohn et al. (1984).
[h] Tang et al. (1984).
[i] In pigeons and monkeys: Zimmerman et al. (1983).

The beta-series enantiomers of dioxadrol have recently been synthesized and neither were found to have PCP-like effects (Zenk, May and Balster, unpublished data). Currently, synthesis and testing of the phenyl-ethyl analog of dioxadrol, etoxadrol, which has three chiral centers and thus eight possible isomers, is underway. Early evidence confirms that there are strict structural requirements for PCP-like effects in this series, consistent with a receptor-mediated action (Zenk et al. 1986).

Benz-(f)-isoquinolines

Another distinct chemical class of drugs producing PCP-like effects are the bridged and unbridged benz-(f)-isoquinoline derivatives described by Zimmerman et al. (1983). Members of this class produce a distinctive PCP-like catalepsy (Zimmerman et al. 1983; Mendelsohn et al. 1984). PCP-like discriminative effects (Zimmerman et al. 1983) and binding to the PCP receptor (Mendelsohn et al. 1984) have been demonstrated for one of one of the series, LY154045. On the other hand, the analog LY154005 is neither generalized (Zimmerman et al. 1983) nor binds to the receptor (Mendelsohn et al. 1984).

Diphenylpropanolamines

Another chemical group with PCP-like effects are the diphenylpropanolamines, of which the (+)-isomer of 2-methyl-3,3-diphenyl-3-propanolamine (2-MDP) has been shown to have PCP-like discriminative effects with evidence for stereospecificity of action (Tang et al. 1984). Binding data on this compound are not yet available.

MK-801

MK-801, the (+)-isomer of 5-methyl-10,11-dihydro-5*H*-dibenzo[*a,d*] cyclohepten-5,10-imine has potent PCP-like discriminative stimulus effects (Willetts and Balster 1987) and the binding of [^{3}H]MK-801 is potently displaced by PCP (Wong et al. 1986). The potency and rigid chemical structure of MK-801 should make it an important probe for PCP-like effects.

Thus, there is generally a very good correspondence between drugs with affinity for PCP-binding sites and with PCP-like discriminative stimulus effects. This is true for chemical analogs as well as chemically unrelated biologs of PCP. These findings are important evidence for the physiological relevance of the PCP-binding site and support a role for this receptor in mediating the discriminative stimulus effects of PCP and related compounds.

4 Overlap Between PCP and Sigma Receptors

The data reviewed above showing that sigma-agonist opioids have both affinity for [^{3}H]-PCP-binding sites and PCP-like behavioral effects have led many to believe that the PCP and sigma opiate receptor are the same, and, indeed, it has been referred to as the "PCP/sigma" receptor (Quirion et al. 1981). Evidence supporting the coidentity of PCP and sigma receptors has been presented by Zukin and Zukin (1983), Mendelsohn et al. (1985), and Sircar et al. (1986). Behavioral data are consistent in showing common effects of PCP and representative sigma opioids such as (+)-NANM, and rats cannot be trained to discriminate between the two (Balster and Brady 1982). However, a more complicated picture has arisen from biochemical studies comparing binding of radiolabeled PCP and sigma agonists. Initial binding studies used racemic mixtures of various benzomorphans. Therefore, when Pasternak et al. (1981) and Johnson (1983) found evidence for two binding sites for [^{3}H]($\pm$)-NANM, this could be explained by one being a more classical opiate site for which the (−)-isomer had selective affinity, the other being the sigma site. Consequently, in order to study the sigma site, investigators added high concentrations of drugs (e.g., naloxone or etorphine) which would block the classical opiate sites (R. S. Zukin and Zukin 1981; Su 1982; Tam 1983) or, more recently, used the more selective sigma agonist (+)-NANM as the ligand (Martin et al. 1984; Tam and Cook 1984; Itzhak et al. 1985; Sircar et al. 1986). Results from these studies indicated that this (+)-NANM-binding site has many dissimilarities from the PCP receptor. For example, (+)-NANM binding is displaceable by haloperidol and other dopamine receptor blocking drugs, whereas PCP binding is not. Other differences include displacement by the (+)-isomers of a number of benzomorphans including (+)-ketocyclazocine and (+)-ethylketocyclazocine (Martin et al. 1984; Tam 1985), neither of which interact with PCP binding. In addition, PCP has a lower affinity for this site than does (+)-NANM, whereas PCP is more potent than (+)-NANM in drug discrimination studies (Balster and Brady 1982). While these data clearly indicate different binding sites for PCP and (+)-NANM, Sircar et al. (1986) have recently proposed that (+)-NANM itself binds to two distinct sites, one being a high-affinity, haloperidol-sensitive site, while the other is a low-affinity site corresponding to the PCP receptor.

To assess the involvement of the high-affinity (+)-NANM-binding site in the transduction of (+)-NANM's discriminative stimulus effects, drugs with high affinity for the (+)-NANM-binding site have been tested in rats trained to discriminate between (+)-NANM and saline. Haloperidol and (+)-ketocyclazocine are not generalized from (+)-NANM, neither do they block the discriminative stimulus effects of (+)-NANM (Balster 1987). Although there was initially some evidence that the dopamine analog (+)-3-PPP which potently displaces (+)-NANM binding (Largent et al. 1984; Gundlach et al. 1986), while failing to be generalized from (+)-NANM, blocked (+)-NANM's discriminative stimulus properties (Balster 1986b), subsequent studies have failed to confirm this (Balster 1987). Based on these results, there is little evidence that the high-affinity (+)-NANM-binding site is involved in the transduction of the discriminative stimulus effects of (+)-NANM. The role for this binding site in the pharmacology of sigma-agonist drugs is presently unknown. Thus, support remains for the hypothesis that PCP and sigma receptors are the same or closely related.

5 Receptor Mediation of Other Effects of PCP

5.1 Motor Performance

The evidence reviewed above provides strong support that PCP's discriminative stimulus effects are mediated by a specific PCP receptor. It is not unreasonable to hypothesize that this receptor also mediates other behavioral or pharmacological effects of PCP. Evidence in support of this hypothesis comes from Lazdunski et al. (1983), who used the rotarod test to measure disruption of motor performance by PCP and a number of related arylcyclohexylamines. A good correlation ($r = 0.95$) was found between potency for motor disruption and PCP receptor binding affinity. A similarly good correlation ($r = 0.81$) exists between the rotarod data of Su et al. (1980) and the binding data of S. R. Zukin and Zukin (1981). Thus, there is some evidence that, at least for the arylcyclohexylamines, the PCP receptor is also likely to be involved in mediating their motor performance effects (but see Contreras et al. 1986).

5.2 Reinforcing Effects

There is also correspondence between the discriminative stimulus and other properties of PCP, providing further evidence for PCP receptor mediation of these effects. One particularly important property of PCP in the context of its abuse is its reinforcing effect for self-administration (Balster et al. 1973). Many drugs which share PCP's discriminative stimulus properties, including a number of arylcyclohexylamines, etoxadrol and dexoxadrol, and the opioids dextrorphan, (+)-NANM, and (+)-cyclazocine, also have reinforcing properties (Young and Woods 1981; Risner 1982; Brady et al. 1982; Slifer and Balster 1983; Lukas et al. 1984). Drugs such as the (−)-enantiomers and racemates of the sigma-agonist benzomorphans are not self-administered. This is not surprising in view of the previously discussed observation that (−)-cyclazocine was only generalized from PCP under conditions of opioid receptor blockade. It could be hypothesized that the opioid actions of this drug, such as its kappa-agonist effects, would interfere with any reinforcing capabilities. Thus, the good correspondence between the discriminative and reinforcing properties of PCP-like drugs leads to the conclusion that these two stimulus properties of PCP may be mediated by a common receptor mechanism.

The role of the PCP receptor in mediating other effects of PCP is less clear. It may play a role in the anticonvulsant activity of PCP since many of the PCP biologs, including (+)-NANM, dexoxadrol, and (+)-2-MDP, are also anticonvulsants (Hayes and Balster 1985).

6 Physiological Role of the PCP Receptor

The physiological role of the PCP receptor is unknown, nor has its endogenous ligand been clearly identified. However, electrophysiological, biochemical, and, more recently, behavioral evidence suggests that PCP modifies excitatory amino acid neurotransmission, specifically, that class mediated by *N*-methyl-D-aspartate (NMDA)-sensitive glutamate or aspartate receptors. PCP selectively antagonizes NMDA-induced excitation of spinal neurons (Anis et al. 1983) and prevents long-term potentiation in the hippocampus (Stringer et al. 1983), an event which may be initiated by NMDA receptors. NMDA-induced release of acetylcholine from superfused striatal slices is also specifically antagonized by PCP (Snell and Johnson 1985). While NMDA does not displace [^{3}H]PCP from its binding site (Johnson and Snell 1986), Maragos et al. (1986) showed a high correlation in the distribution of NMDA receptors and the binding of the PCP-analog [^{3}H]TCP in rat brain. These results suggest that PCP and NMDA receptors are distinct but intimately associated, perhaps similar to the relationship between benzodiazepine and GABA receptors in the CNS.

Recently, Leander et al. (1986) found that PCP antagonized NMDA-induced response rate decreases in pigeons. We assessed the involvement of NMDA receptors in the discriminative stimulus properties of PCP using the NMDA receptor blocker 2-amino-7-phosphonoheptanoate (APH) in rats trained to discriminate PCP from saline (Willetts et al. 1986). There was evidence that intracerebral administration of APH was generalized from PCP. Similar findings have been reported by Koek et al. (1986), who showed that the NMDA receptor blocker 2-amino-5-phosphonovalerate (APV) was generalized from PCP in pigeons. These results provide support for the involvement of NMDA receptors in transducing PCP's discriminative stimulus effects and suggest a physiological role for the PCP receptor in modulating excitatory amino acid neurotransmission.

7 Summary

Evidence has been presented that the discriminative stimulus properties of PCP are produced via its interaction with a specific receptor. A physiological function of this receptor may be the regulation of excitatory amino acid neurotransmission. The behavioral function subserved by these biochemical systems is, however, a matter of speculation. By continuing to investigate the behavioral effects of PCP, this speculation may be resolved. PCP produces a diverse and interesting array of effects, including production of schizophreniform behaviors, reinforcing effects, specific effects on learning, disruption of sensory processing, and unique subjective effects. It is possible that the PCP receptor and its endogenous ligands may play a role in these and other behavioral processes.

Acknowledgments. The preparation of this review and the research from our laboratory described herein were supported by National Institute on Drug Abuse Grant DA-01442.

References

Anis NA, Berry SC, Burton NR, Lodge D (1983) The dissociative anaesthetics, ketamine and phencyclidine, selectively reduce excitation of central mammalian neurones by *N*-methyl-aspartate. Br J Pharmacol 79:565–575

Balster RL (1986a) Clinical implications of behavioral pharmacology research on phencyclidine. In: Clouet DH (ed) Phencyclidine: an update. Natl Inst Drug Abuse Res Monogr 64:148–162 [DHHS publication (ADM) 86–1443]

Balster RL (1986b) (+)-3-PPP antagonizes the discriminative stimulus effects of (+)-*N*-allylnormetazocine. Eur J Pharmacol 127:283–286

Balster RL (1987) Drugs with a high affinity for ^{3}H-(+)-*N*-allylnormetazocine (NANM) binding do not have activity in (+)-NANM discrimination in rats. Fed Proc 46:(to be published)

Balster RL, Brady KT (1982) Similarities in the discriminative stimulus properties of (+)-*N*-allylnormetazocine and phencyclidine in rats. In: Colpaert FC, Slangen JL (eds) Drug discrimination: applications in CNS pharmacology. Elsevier Biomedical, Amsterdam, pp 99–108

Balster RL, Johanson CE, Harris RT, Schuster CR (1973) Phencyclidine self administration in the rhesus monkey. Pharmacol Biochem Behav 1:167–172

Brady KT, Balster RL (1981) Discriminative stimulus properties of phencyclidine and five analogues in the squirrel monkey. Pharmacol Biochem Behav 14:213–218

Brady KT, Balster RL (1982a) Discriminative stimulus properties of ketamine stereoisomers in phencyclidine-trained rats. Pharmacol Biochem Behav 17:291–295

Brady KT, Balster RL (1982b) Discriminative stimulus properties of stereoisomers of cyclazocine in phencyclidine-trained squirrel monkeys. Life Sci 31:541–549

Brady KT, Balster RL, May EL (1982) Stereoisomers of *N*-allylnormetazocine: phencyclidine-like behavioral effects in squirrel monkeys and rats. Science 215:178–180

Browne RG (1986) Discriminative stimulus properties of PCP mimetics. In: Clouet DH (ed) Phencyclidine: an update. Natl Inst Drug Abuse Res Monogr 64:134–147 [DHHS publication (ADM) 86–1443]

Browne RG, Welch WM (1982) Stereoselective antagonism of phencyclidine's discriminative properties by adenosine receptor agonists. Science 217:1157–1159

Browne RG, Welch WM, Kozlowski MR, Duthu G (1983) Antagonism of PCP discrimination by adenosine analogs. In: Kamenka JM, Domino EF, Geneste P (eds) Phencyclidine and related arylcyclohexylamines. NPP, Ann Arbor, pp 639–666

Contreras PC, Rice KC, Jacobson AE, O'Donohue TL (1986) Stereotyped behavior correlates better than ataxia with phencyclidine-receptor interactions. Eur J Pharmacol 121:9–18

Domino EF (1964) Neurobiology of phencyclidine (Sernyl), a drug with an unusual spectrum of pharmacological activity. Int Rev Neurobiol 6:303–347

Gundlach AL, Largent BL, Snyder SH (1986) Characterization of phencyclidine and sigma receptor-binding sites in brain. In: Clouet DH (ed) Phencyclidine: an update. Natl Inst Drug Abuse Res Monogr 64:1–13 [DHHS publication (ADM) 86–1443]

Hayes BA, Balster RL (1985) Anticonvulsant properties of phencyclidine-like drugs in mice. Eur J Pharmacol 117:121–125

Holtzman SG (1982) Phencyclidine-like discriminative stimulus properties of opioids in the squirrel monkey. Psychopharmacology 77:295–300

Itzhak Y, Hiller JM, Simon EJ (1985) Characterization of specific binding sites for [^{3}H](*d*)-*N*-allylnormetazocine in rat brain membranes. Mol Pharmacol 27:46–52

Johnson KM (1983) Partial characterization of the specific binding of ^{3}H-*N*-allylnormetazocine (SKF 10047) to rat brain membranes. Res Commun Subst Abuse 4:33–47

Johnson KM, Snell LD (1986) Involvement of dopaminergic, cholinergic, and glutaminergic mechanisms in the actions of phencyclidine-like drugs. In: Clouet DH (ed) Phencyclidine: an update. Natl Inst Drug Abuse Res Monogr 64:52–66 [DHHS publication (ADM) 86–1443]

Koek W, Woods JH, Ornstein P (1986) Phencyclidine like behavioral effects induced by systemic administration of the excitatory amino acid antagonist, 2-amino-5-phosphonovalerate. Life Sci 39:973–978

Largent BL, Gundlach AL, Snyder SH (1984) Psychotomimetic opiate receptors labeled and visualized with (+)-[^{3}H]3-(3-hydroxyphenyl)-*N*-(1-propyl)piperidine. Proc Natl Acad Sci USA 81:4983–4987

Lazdunski M, Bidard JN, Romey G, Tourneur Y, Vignon J, Vincent JP (1983) The different sites of action of phencyclidine and its analogues in nervous tissue. In: Kamenka JM, Domino EF, Geneste P (eds) Phencyclidine and related arylcyclohexylamines. NPP, Ann Arbor, pp 83–106

Leander JD, Zimmerman DM, Ornstein PL (1986) Antagonism of the behavioral effects of *N*-methyl-*d*-aspartic acid by phencyclidine-like drugs. Fed Proc 45:430 (abstract)

Lukas SE, Griffiths RR, Brady JV, Wurster RM (1984) Phencyclidine-analogue self-injection by the baboon. Psychopharmacology 83:316–320

Maragos WF, Chu DCM, Greenamyre JT, Penny JB, Young AB (1986) High correlation between the localization of [^{3}H]TCP binding and NMDA receptors. Eur J Pharmacol 123:173–174

Martin BR, Katzen JS, Woods JA, Tripathi HL, Harris LS, May EL (1984) Stereoisomers of [^{3}H]-*N*-allylnormetazocine bind to different sites in mouse brain. J Pharmacol Exp Ther 231:539–544

Martin WR, Eades CG, Thompson JA, Huppler RE, Gilbert PE (1976) The effects of morphine- and nalorphine-like drugs on the nondependent and morphine dependent chronic spinal dog. J Pharmacol Exp Ther 197:517–531

Mendelsohn LG, Kerchner GA, Kalra V, Zimmerman DM, Leander JD (1984) Phencyclidine receptors in rat brain cortex. Biochem Pharmacol 33:3529–3535

Mendelsohn LG, Kalra V, Johnson BG, Kerchner GA (1985) *Sigma* opioid receptor: characterization and co-identity with the phencyclidine receptor. J Pharmacol Exp Ther 233:597–602

Murray TF, Leid ME (1984) Interaction of dextrorotatory opioids with phencyclidine recognition sites in rat brain membranes. Life Sci 34:1899–1911

Pasternak GW, Carroll-Buatti M, Spiegel K (1981) The binding and analgesic properties of a *sigma* opiate, SKF 10,047. J Pharmacol Exp Ther 219:192–198
Poling AD, White FJ, Appel JB (1979) Discriminative stimulus properties of phencyclidine. Neuropharmacology 18:459–463
Quirion R, Hammer RP, Herkenham M, Pert CB (1981) Phencyclidine (angel dust)/σ "opiate" receptor: visualization by tritium-sensitive film. Proc Natl Acad Sci USA 78:5881–5885
Quirion R, O'Donohue TL, Everist H, Pert A, Pert CB (1983) Phencyclidine receptors and possible existence of an endogenous ligand. In: Kamenka JM, Domino EF, Geneste P (eds) Phencyclidine and related arylcyclohexylamines. NPP, Ann Arbor, pp 667–684
Risner ME (1982) Intravenous self administration of phencyclidine and related compounds in the dog. J Pharmacol Exp Ther 221:637–644
Shannon HE (1981 a) Evaluation of phencyclidine analogs on the basis of their discriminative stimulus properties in the rat. J Pharmacol Exp Ther 216:543–551
Shannon HE (1981 b) Pharmacological evaluation of *N*-allylnormetazocine (SKF 10,047) on the basis of its discriminative stimulus properties in the rat. J Pharmacol Exp Ther 225:144–152
Shannon HE (1983) Discriminative stimulus effects of phencyclidine: structure-activity relationships. In: Kamenka JM, Domino EF, Geneste P (eds) Phencyclidine and related arylcyclohexylamines. NPP, Ann Arbor, pp 311–336
Sircar R, Nichtenhauser R, Ieni JR, Zukin SR (1986) Characterization and autoradiographic visualization of (+)-[^{3}H]SKF10,047 binding in rat and mouse brain: further evidence for phencyclidine/"*sigma* opiate" receptor commonality. J Pharmacol Exp Ther 237:681–688
Slifer BL, Balster RL (1983) Reinforcing properties of stereoisomers of the putative *sigma* agonists *N*-allylnormetazocine and cyclazocine in rhesus monkeys. J Pharmacol Exp Ther 225:522–528
Slifer BL, Balster RL (1985 a) Phencyclidine-like discriminative stimulus properties of the stereoisomers of dioxadrol. Subst Alcohol Actions Misuse 5:273–280
Slifer BL, Balster RL (1985 b) Phencyclidine-like behavioral effects of stereoisomers of *beta*-cyclazocine. Paper presented at the American College of Neuropsychopharmacology, Maui
Snell LD, Johnson KM (1985) Antagonism of *N*-methyl-D-aspartate-induced transmitter release in the rat striatum by phencyclidine-like drugs and its relationship to turning behavior. J Pharmacol Exp Ther 235:50–57
Stringer JL, Greenfield LJ, Hackett JT, Guyenet PG (1983) Blockade of long-term potentiation by phencyclidine and opiates in the hippocampus in vivo and in vitro. Brain Res 280:127–138
Su TP (1982) Evidence for *sigma* opioid receptor: binding of [^{3}H]SKF-10047 to etorphine-inaccessible sites in guinea-pig brain. J Pharmacol Exp Ther 223:284–290
Su TP, Cone EJ, Shannon H, Vaupel DB (1980) Relative potencies of phencyclidine and analogs in the opiate receptor binding assay: correlation with relative potencies determined in vivo in mouse and rat. Res Commun Subst Abuse 1:85–87
Tam SW (1983) Naloxone-inaccessible sigma receptor in rat central nervous system. Proc Natl Acad Sci USA 80:6703–6707
Tam SW (1985) (+)-[^{3}H]SKF 10,047, (+)-[^{3}H]ethylketocyclazocine, μ, κ, σ and phencyclidine binding sites in guinea pig brain membranes. Eur J Pharmacol 109:33–41
Tam SW, Cook L (1984) Sigma opiates and certain antipsychotic drugs mutually inhibit (+)[^{3}H]SKF-10,047 and [^{3}H]-haloperidol binding in guinea pig brain membranes. Proc Natl Acad Sci USA 81:5618–5621
Tang AH, Cangelosi AA, Code RA, Franklin SR (1984) Phencyclidine-like behavioral effects of 2-methyl-3,3-diphenyl-3-propanolamine (2-MDP). Pharmacol Biochem Behav 20:209–213
Vaupel DB (1983) Phencyclidine studies in the dog: pharmacology of PCP, profiles of eight analogues abused by man and the similarity of PCP to the opioid NANM (*N*-allylnormetazocine or SKF 10,047). In: Kamenka JM, Domino EF, Geneste P (eds) Phencyclidine and related arylcyclohexylamines. NPP, Ann Arbor, pp 347–368
Vaupel DB, Jasinski DR (1979) Acute single dose effects of phencyclidine (PCP) in the dog. Fed Proc 38:435

Vincent JP, Cavey D, Kamenka JM, Geneste P, Lazdunski M (1978) Interaction of phencyclidine with the muscarinic and opiate receptors in the central nervous system. Brain Res 152:176–182

Vincent JP, Kartalovski B, Geneste P, Kamenka JM, Lazdunski M (1979) Interaction of phencyclidine ("angel dust") with a specific receptor in rat brain membranes. Proc Natl Acad Sci USA 76:4678–4682

Vincent JP, Vignon J, Kartalovski B, Lazdunski M (1981) Receptor sites for phencyclidine in mammalian brain and peripheral organs. In: Domino EF (ed) PCP (phencyclidine): historical and current perspectives. NPP, Ann Arbor, pp 83–103

Willetts J, Balster RL (1987) MK-801 has potent phencyclidine-like discriminative stimulus effects in rats. Fed Proc 46:(to be published)

Willetts J, Chapman MY, Balster RL (1986) Discriminative stimulus effects of 2-amino-7-phosphonoheptanoic acid (APH) in phencyclidine-trained rats. Soc Neurosci Abs 12:909

Wong EHF, Kemp JA, Priestley T, Knight AR, Woodruff GN, Iverson LL (1986) The anticonvulsant MK-801 is a potent *N*-methyl-D-aspartate antagonist. Proc Natl Acad Sci USA 83:7104–7108

Young AM, Woods JH (1981) Maintenance of behavior by ketamine and related compounds in rhesus monkeys with different self administration histories. J Pharmacol Exp Ther 218:720–727

Zenk P, May EL, Balster RL (1986) Synthesis of the diastereoisomers of 2-ethyl-2-phenyl-4-(2-piperidyl)-1,3-dioxolane (etoxadrol) as probes of the phencyclidine receptor. Paper presented at the annual meeting of the Virginia Academy of Sciences, Harrisonburg

Zimmerman DM, Woods JH, Hynes MD, Cantrell BE, Reamer M, Leander JD (1983) Discovery and characterization of the phencyclidine-like actions of a new series of benz(f)isoquinoline derivatives. In: Kamenka JM, Domino EF, Geneste P (eds) Phencyclidine and related arylcyclohexylamines. NPP, Ann Arbor, pp 59–70

Zukin RS, Zukin SR (1981) Demonstration of [^{3}H]cyclazocine binding to multiple opiate receptor types. Mol Pharmacol 20:246–254

Zukin RS, Zukin SR (1983) A common receptor for phencyclidine and the *sigma* opiates. In: Kamenka JM, Domino EF, Geneste P (eds) Phencyclidine and related arylcyclohexylamines: present and future applications. NPP, Ann Arbor, pp 107–124

Zukin SR, Zukin RS (1979) Specific [^{3}H]phencyclidine binding in rat central nervous system. Proc Natl Acad Sci USA 76:5372–5376

Zukin SR, Zukin RS (1981) Identification and characterization of [^{3}H]phencyclidine binding to specific brain receptor sites. In: Domino EF (ed) PCP (phencyclidine): historical and current perspectives. NPP, Ann Arbor, pp 105–130

Theory, Concepts, and Research Strategies

Neural Mechanisms of Drug Stimuli: Experimental Approaches

S. G. HOLTZMAN [1] and K. W. LOCKE [2]

Contents

Abstract

Drug discrimination methodology affords a behavioral end point that reflects actions of a drug at the cellular level. The stimulus control of behavior by drugs of many pharmacological classes satisfies criteria for a receptor-mediated phenomenon. Within many classes of drugs, order of relative of potency for discriminative effects correlates highly with the order of potency of the drugs in other procedures that involve interactions with a defined population of receptors. Examples include the morphine-like opioids and the benzodiazepines. Specific competitive antagonists are of paramount importance for defining neuronal substrates that subserve the discriminative effects of a drug. Quantitative pharmacological approaches can be used not only to define the receptor mediating the discriminative effects of a drug but also to assess the intrinsic activity of the drug at the receptor. The primary site of stimulus control of behavior by most drugs can be shown by indirect and direct means to be of central rather than peripheral origin; however, neuroanatomical localization of specific sites of drug action remains elusive. Attention is starting to be focused on endogenous neuropeptides, their discriminative stimulus properties and potential role in modulating discriminative effects of drugs. Some of the most important contributions of drug discrimination methodology may arise in the rapidly emerging area of peptide neuropharmacology.

1 Introduction

Drug discrimination methodology affords a behavioral end point that can reflect the actions of a drug at the cellular level. For example, it often has been possible to identify the type of receptor that mediates the discriminative stimulus effects of a drug, and to relate this information to actions of the drug in other contexts, behavioral, physiological, and biochemical. The result has been new insights into

[1] Department of Pharmacology, Emory University School of Medicine, Atlanta, GA 30322, USA.
[2] American Hoechst Corporation, Route 202–206 North Sommerville, NJ 08876, USA.

Psychopharmacology Series 4
Transduction Mechanisms of Drug Stimuli
Editors: Colpaert, Balster

the pharmacological properties of the drug itself as well as into the cellular bases of drug actions.

There exists a particularly large body of literature on the discriminative stimulus properties of the diverse compounds referred to generically as "opioids." This situation arose from a confluence of factors. Included among these were the research interests of many of the early investigators in the field of drug discrimination and an extensive clinical literature on opioids with which findings in animals could be correlated. However, most important was the availability of the pharmacological tools essential for a thorough and systematic study of the system: a variety of agonists and (especially) specific competitive antagonists whose pharmacology had been carefully documented in numerous assay procedures, both in vivo and in vitro. This review of experimental approaches for studying neural mechanisms of drug stimuli will draw heavily from the opioid literature for examples, particularly those concerning the morphine-like compounds. Many points can also be illustrated well by work done on drugs from other pharmacological classes, notably the benzodiazepines and phencyclidine (PCP) and its analogs.

2 Receptor Mediation

A major reason for the current widespread interest in drug discrimination methodology is that, for many classes of drugs, stimulus control of behavior satisfies the usual criteria for a receptor-mediated phenomenon:

1. Similar effects by similar drugs. Drugs with comparable spectra of activity in other assay systems owing to interactions with a defined population of receptors should have qualitatively similar discriminative stimulus properties as reflected in their ability to generalize with one another.
2. Order of relative potency. The order of potency with which drugs of a particular class produce discriminative effects should parallel their order of potency in other procedures involving interactions with the receptor.
3. Steric selectivity. Where pairs of enantiomers exist, the characteristic discriminative effects should be produced exclusively or predominantly by the member of the pair possessing the pharmacological properties representative of the drug class.
4. Specific antagonists. Antagonists that bind reversibly to the receptor should block the discriminative effects of drugs of the class. The antagonism should be surmountable so that increasing the dose of the agonist results in a parallel shift to the right of the stimulus generalization curve with no decrement in peak effect.

2.1 Orders of Relative Potency and Their Correlations

That stimulus control of behavior by morphine and morphine-like drugs, such as fentanyl, meets the foregoing criteria for mediation by the mu-opioid receptor has been demonstrated consistently in several animal species, and has been reviewed in detail (e.g., Lal et al. 1977; Colpaert 1978; Herling and Woods 1981; Holtzman

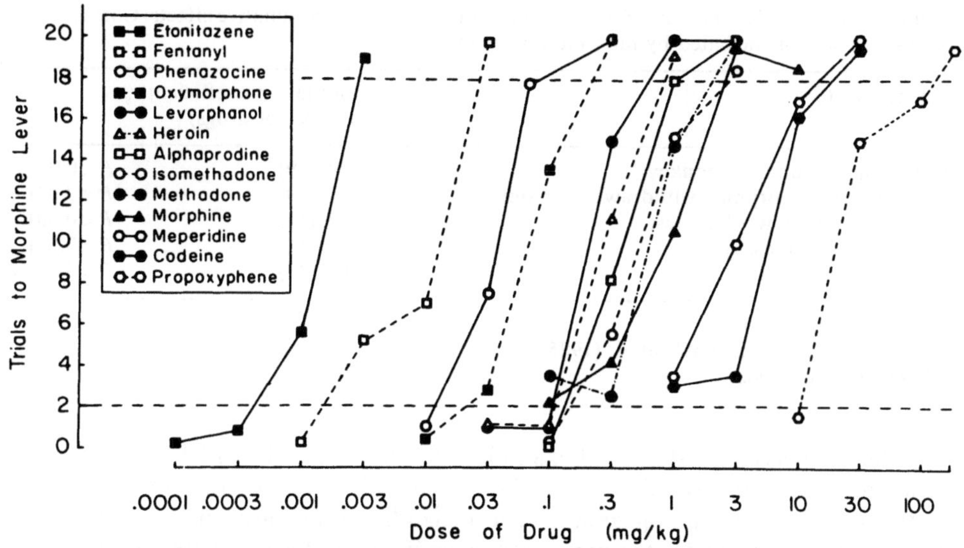

Fig. 1. Drugs classified as morphine-like (mu-receptor) agonists produce discriminative stimulus effects comparable to those of morphine, with the predicted order of potency. Drugs were administered s.c., 30 min before the start of a session, to rats trained to discriminate between saline and 3.0 mg/kg morphine in a two-choice discrete-trial avoidance paradigm. Each point is the mean number of trials completed on the morphine-appropriate lever in a 20-trial session; the remaining trials of the session were completed on the choice lever appropriate for saline. Four or more rats were tested at each dose. The *upper* and *lower horizontal dashed lines* indicate the minimum levels at which the discrimination performance of the animals was maintained during training sessions with 3.0 mg/kg morphine and saline, respectively. Data are adapted from Shannon and Holtzman (1976, 1977a) and reproduced from Holtzman (1983), with permission

1982, 1983). Figure 1 illustrates the order of potency of a series of morphine-like opioids tested for generalization with 3.0 mg/kg morphine in the rat. It was shown previously that order of potency for producing morphine-like discriminative effects correlates closely with analgesic potency (Colpaert et al. 1976). Drawing from the drugs in Fig. 1, Table 1 provides additional correlations between order of potency for morphine-like discriminative effects and relative potency for various effects involving interactions with the mu-opioid receptor. The high correlations for self-administration by rhesus monkeys and for suppression of withdrawal in rhesus monkeys physically dependent upon morphine are noteworthy in view of the implications of these models for abuse potential in humans (Griffiths and Balster 1979; Woolverton and Schuster 1983). Also noteworthy is the fact that the poorest correlation, albeit still significant, is with receptor binding, which presumably affords the most direct measure of the interactions between morphine-like compounds and the mu receptor. The reason for such an outcome is unclear, but may be due to the use of an agonist [D-Ala^{2}NMePhe^4-Gly^5(ol)enkephalin, DAGO] rather than an antagonist as the labeled ligand. Be that as it may, the importance of valid behavioral and physiological correlates of drug action in addition to correlations derived from in vitro binding assays cannot be overstated.

Table 1. Potency order of opiates for producing morphine-like discriminative effects: Correlation with other actions mediated by mu-opioid receptor

Opiate effect	Correlation coefficient r_s (number of drugs)	
Self-administration by monkeys[a]	1.00 (9)	$P<0.001$
Suppression of morphine withdrawal in monkeys[a]	0.98 (9)	$P<0.001$
Analgesia: mouse writhing test[b]	0.97 (11)	$P<0.001$
Guinea pig ileum assay[c]	0.93 (11)	$P<0.001$
Mu-receptor binding[c]	0.68 (10)	$P<0.05$

[a] Young et al. (1981).
[b] Pearl et al. (1968); Schmidt et al. (1985).
[c] Magnan et al. (1982).

There are several additional examples of drug classes in which order of potency for discriminative effects correlates well with potency order in other procedures, including binding assays. The benzodiazepines are one such class of drugs. Table 2 illustrates the correspondence between diazepam-like discriminative effects in the rat, muscle relaxation in the cat, and inhibition of [^{3}H]diazepam binding in rat cerebral cortex. These data expand upon an earlier report (Colpaert 1977) of significant correlations between chlordiazepoxide-like discriminative effects, ataxia, and anticonvulsant activity within a series of benzodiazepines and barbiturates.

In their initial demonstration of the existence of saturable [^{3}H]PCP-binding sites in rat brain homogenates, Zukin and Zukin (1979) showed that a series of arylcyclohexylamines displaced the labeled ligand with an order of potency that correlated highly with the potency order for producing ataxia in the mouse and

Table 2. Potency order of benzodiazepines for producing diazepam-like discriminative effects: correlation with muscle relaxation and receptor binding

Drug	Diazepam-like discriminative effects[a]	Muscle relaxation in cats[b]	Receptor binding: displacement of [^{3}H]diazepam[b]
Flunitrazepam	1	1	2
Clonazepam	2	2	1
Bromazepam	3	4	6
Diazepam	4	5	4
Lorazepam	5	6	3
Nitrazepam	6	3	5
Chlordiazepoxide	7	9	9
Oxazepam	8	7	8
Flurazepam	9	8	7
Medazepam	10	10	10
Correlation coefficient r_s		0.89 ($P<0.001$)	0.85 ($P<0.01$)

[a] Rats were trained to discriminate between saline and 1.0 mg/kg diazepam (Shannon and Herling 1983).
[b] Möhler et al. (1978).

Table 3. Potency order of drugs for producing PCP- and *d,l*-NANM (SKF 10,047)-like discriminative effects in the rat and for displacing binding of [^{3}H]PCP from rat brain membranes

	Discriminative effects Training drug		Binding[c]
	PCP[a] (2.0–3.0 mg/kg)	NANM[b] (3.0–5.0 mg/kg)	
l-Cyclazocine	1	1	3
PCP	2	2	2
Dexoxadrol	3	3	1
d-NANM	4	4	4
d,l-Metazocine	5	8	–[d]
l-NANM	6	5	5
Dextrorphan	7	9	6
d-Cyclazocine	8	7	7
Ketamine	9	10	8
Levallorphan	10	6	–[d]

[a] Holtzman (1980); Shannon (1982, 1983a); PCP, phencyclidine.
[b] Shearman and Herz (1982a); Shannon (1983b); NANM, *N*-allylnormetazocine.
[c] Mendelsohn et al. (1984).
[d] Not tested.
Spearman rank correlation coefficient:
$r_s = 0.81$ for column I vs. II ($P < 0.005$)
$r_s = 0.90$ for column I vs. III ($P < 0.005$)
$r_s = 0.88$ for column II vs. III ($P < 0.005$).

PCP-like discriminative effects in the rat. The correspondence between binding affinity and discriminative stimulus potency was extended subsequently to psychotomimetic opioids of the benzomorphan and morphinan families and to dioxolane dissociative anesthetics (Table 3). Such relationships do not establish causality. However, they go far to buttress the view that the discriminative stimulus effects of the various "PCP/sigma" agonists are mediated at a neuronal site common to all of these drugs (Quirion et al. 1981; Zukin and Zukin 1981; Herling and Woods 1981).

2.2 Defining Neuronal Substrates of Drug Action: Importance of Antagonists

A competitive selective antagonist is probably the single most important pharmacological tool for defining the neuronal substrates of drug action. If a series of generalization curves is constructed by testing graded doses of an agonist with several fixed doses of an antagonist, it is possible to determine an apparent pA_2 value for the agonist-antagonist pair. The apparent pA_2 is the negative logarithm of the molar dose of antagonist that reduces the effect of a double dose of agonist to that of a single dose. The apparent pA_2 reflects the affinity of the antagonist for the receptor mediating the effects of the agonist, and has the advantage of being independent of agonist affinity (i.e., potency). Derived from dose-ratio anal-

Table 4. Apparent pA_2 values in rats trained to discriminate between saline and either morphine (3.0 mg/kg) or ethylketocyclazocine (0.1 mg/kg)[a]

Agonist-antagonist pairing	Apparent pA_2 (95% CL)
Morphine-naloxone	7.65 (7.40–7.90)
Ethylketocyclazocine-naloxone	6.77 (6.59–6.96)

[a] Herling and Shannon (1982a).

yses originally developed for in vitro assays (Schild 1947; Arunlakshana and Schild 1959), the pA_2 determination can be applied in vivo, with certain caveats (e.g., Takemori 1974; Tallarida et al. 1979). Table 4 shows the apparent pA_2 values for pairings of the opiate antagonist, naloxone, with morphine or ethylketocyclazocine, prototypic mu- and kappa-opioid receptor agonists (Martin et al. 1976), respectively. The data are consistent with the view that the discriminative effects of morphine and ethylketocyclazocine are subserved by different populations of receptors (e.g., Herling and Woods 1981; Holtzman 1985). The affinity of naloxone for the receptor mediating the discriminative effects of ethylketocyclazocine is almost ten times lower than its affinity for the receptor mediating the discriminative effects of morphine. This difference in relative affinity is concordant with results from assays of radioligand binding to putative mu and kappa sites in brain (Kosterlitz et al. 1981). The apparent pA_2 for naloxone-morphine shown in Table 4 is the same as the value derived from the pairing of naltrexone with morphine, 7.69 ± 0.07 (Holtzman 1983). It is also similar to the apparent pA_2 of 7.95 ± 0.19 derived for naloxone and morphine in an assay of analgesic activity in the rat (Tallarida et al. 1979).

In the preceding example, the two drugs being compared have discriminative stimulus properties that are readily distinguishable from one another in most animal species (Herling and Woods 1981; Holtzman 1982, 1983). However, antagonists can also be of value for distinguishing between drugs that have similar discriminative stimulus properties, but which are engendered through different neuronal sites of action. Benzodiazepines and barbiturates often share discriminative stimulus properties, cross-generalizing with one another completely. This apparently is a consequence of drugs from the two classes interacting with the same macromolecular receptor complex, but at different recognition sites (e.g., Williams 1984). Antagonists, such as Ro 15-1788, that interact exclusively with the benzodiazepine recognition site, block completely and surmountably the discriminative effects of benzodiazepines without modifying stimulus control by barbiturates, regardless of whether the drug used for training is a benzodiazepine or a barbiturate (Herling and Shannon 1982b; Ator and Griffiths 1983).

Table 5 provides another example of an antagonist modifying differentially apparently similar discriminative stimulus effects. Three drugs that share common PCP/sigma stimulus properties, the optical isomers of *N*-allylnormetazocine and PCP (Table 3), each generalized completely with butorphanol, an opioid mixed agonist/antagonist (unpublished observation). However, equi-discriminable doses of the drugs were not equally sensitive to blockade by naltrexone. Ten

Table 5. Butorphanol-like discriminative effects of PCP/*sigma* agonists: differential sensitivity to antagonism by naltrexone[a]

Naltrexone dose (mg/kg)	Drug tested in combination with naltrexone		
	l-NANM (0.3 mg/kg)	*d*-NANM (0.3 mg/kg)	PCP (1.0 mg/kg)
Percentage trials to butorphanol lever			
0	92	96	92
0.003	89	–[b]	–
0.01	61	–	–
0.03	11	94	–
0.1	6	80	91
0.3	–	37	91
1.0	–	8	90

[a] Rats were trained to discriminate between saline and 0.3 mg/kg butorphanol in a discrete-trial avoidance paradigm. Drugs were injected s.c. 30 min before the start of a 20-trial session. The doses of the three drugs, 0.3 mg/kg *l-N*-allylnormetazocine (*l*-NANM), 0.3 mg/kg *d-N*-allylnormetazocine (*d*-NANM), and 1.0 mg/kg phencyclidine (PCP), were equieffective in producing butorphanol-like stimulus control of behavior, as determined from the complete stimulus generalization curve for each drug. Each value is the average percentage of trials completed on the butorphanol choice lever by a group of five rats; remaining trials were completed on the saline choice lever.
[b] Not tested.

times more naltrexone was required to block the butorphanol-like discriminative effects of *d-N*-allylnormetazocine than was needed to block those of its levorotatory counterpart. The butorphanol-like discriminative effects of PCP were unaffected by naltrexone over the range of doses tested. These results suggest that a comparable behavioral outcome was produced by three drugs, each acting at a separate and distinct neuronal site. A more definitive conclusion awaits further experimental verification.

The lack of a specific antagonist has been a principal obstacle to full characterization of the neuronal substrates that mediate the discriminative effects of many of the major classes of behaviorally active drugs.

3 Intrinsic Activity: Characterization of Partial Agonists

Quantitative pharmacological approaches have been used successfully not only to define the receptor that mediates the discriminative effects of a drug but also to assess the drug's intrinsic activity at the receptor. Colpaert et al. (1982) tested a series of serotonin receptor-blocking drugs in rats trained to discriminate between saline and 0.16 mg/kg of LSD. Most of the drugs generalized partially with LSD when tested alone, and antagonized partially the discriminative effects of LSD when tested in combination. The covariance of LSD-like agonist and LSD antagonist activity is consonant with the conclusion that the compounds are, in fact, partial LSD agonists (Colpaert et al. 1982). A similar experimental approach was used in a study of opioid mixed agonist/antagonists in rats trained to discriminate between saline and 0.04 mg/kg fentanyl (Colpaert and Janssen 1984). Here too, the fentanyl-like agonist and fentanyl antagonist activity of the mixed-acting opioids covaried, and were mutually exclusive. The more that a drug was generalized with fentanyl, the less able it was to block the discriminative effects of fentanyl and vice versa. These results accord well with those of a study (Holtzman 1983) designed to determine if the partial generalization of mixed agonist/antagonists with morphine could be accommodated by the theory of competitive dualism (Ariëns et al. 1964). Rats trained to discriminate between saline and 3.0 mg/kg morphine were tested for stimulus generalization to the mixed agonist/antagonists nalorphine (0.03–30 mg/kg) and cyclazocine (0.03–1.0 mg/kg). The partial generalization of these drugs with morphine is shown in Fig. 2. The curves for nalorphine and cyclazocine were then redetermined in combination with 0.3, 1.0 or 3.0 mg/kg morphine. The family of curves for the combination of nalorphine and morphine and for cyclazocine and morphine converge upon the corresponding curve for nalorphine or cyclazocine alone (Fig. 2). These outcomes are good in vivo approximations of the theoretical log-concentration-response curves for the combination of a partial agonist (nalorphine, cyclazocine) and a full agonist (morphine) acting on a single population of receptors (Ariens et al. 1964).

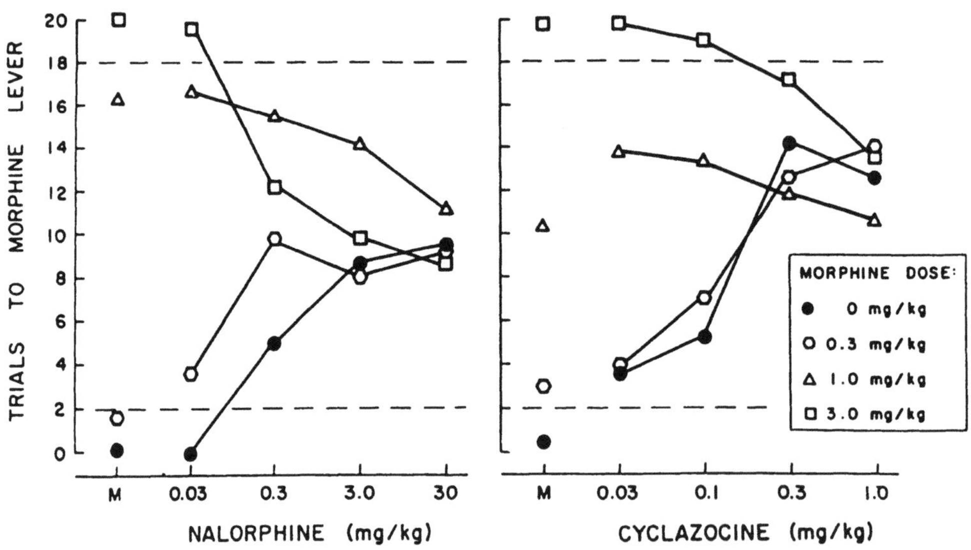

Fig. 2. Nalorphine and cyclazocine appear to be partial morphine agonists in the rat. Stimulus generalization curves are shown for each of the opioid agonist/antagonists tested alone (morphine dose, 0 mg/kg) or in combination with 0.3, 1.0, or 3.0 mg/kg morphine in rats trained to discriminate between saline and 3.0 mg/kg morphine. The *individual points above M* indicate the mean number of trials completed on the morphine-appropriate choice lever in sessions that followed the administration of 0 (saline), 0.3, 1.0, or 3.0 mg/kg morphine alone. Each point is a mean based upon one observation in each of five rats. Other details are the same as in Fig. 1. Modified from Holtzman (1983), with permission

4 Anatomical Site of Stimulus Control

It has been a relatively simple matter to demonstrate that the primary site of stimulus control of behavior by most drugs is central as opposed to peripheral. This is usually accomplished through indirect methods. Polar compounds, which do not readily penetrate the blood-brain barrier, are administered systemically. Discriminative stimulus effects that are mediated centrally should neither be mimicked by polar agonists nor blocked by polar antagonists administered peripherally. For example, the discriminative effects of nicotine are blocked by mecamylamine but not by hexamethonium (e.g., Barry 1974). Both are nicotinic-cholinergic receptor-blocking drugs; however, the former is lipophilic and the latter is not. Similarly, the discriminative effects of atropine are mimicked by scopolamine but not by polar quaternary analogs (Barry 1974), which effectively block muscarinic receptors in the periphery. Other such examples abound.

A more direct approach is to show that animals trained with a drug administered systemically generalize to the drug when it is administered centrally, and that the drug is more potent by the latter route than by the former. This approach is difficult technically for a number of reasons. Highly trained subjects must be fitted surgically with a brain cannula, and patent, healthy, and behaviorally stable preparations must be maintained over long periods of repeated drug testing. In addition, the pharmacokinetic characteristics of drugs injected centrally are often

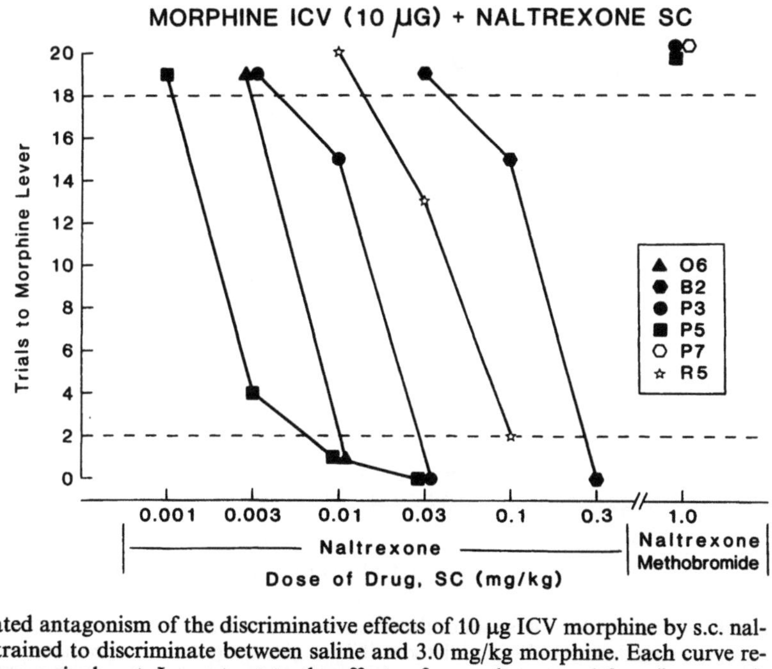

Fig. 3. Dose-related antagonism of the discriminative effects of 10 μg ICV morphine by s.c. naltrexone in rats trained to discriminate between saline and 3.0 mg/kg morphine. Each curve represents data from a single rat. In contrast to the effects of s.c. naltrexone, 1.0 mg/kg s.c. naltrexone methobromide did not block the discriminative effects of 10 μg ICV morphine in the three rats in which it was tested (*unconnected points at upper right*). Other details are the same as in Fig. 1. From Locke and Holtzman (1985), with permission

uncertain at best. Nevertheless, discriminative stimulus properties of drugs as diverse pharmacologically as morphine, nicotine, and LSD have been examined following central administration (e.g., Rosecrans and Glennon 1979). Locke and Holtzman (1985) used a variation of the direct and indirect methods to characterize further the discriminative stimulus properties of morphine in the rat (Locke and Holtzman 1985). It was shown that rats trained to discriminate saline from 3.0 mg/kg morphine s.c. generalize to morphine administered into the lateral cerebral ventricle (ICV) (see also Shannon and Holtzman 1977b). The discriminative effects of ICV morphine (10 μg/rat) were blocked completely by 0.01–0.3 mg/kg s.c. naltrexone but were unaffected by 1.0 mg/kg s.c. naltrexone methobromide, a polar quaternary analog of naltrexone (Fig. 3). Conversely, stimulus control of behavior by the s.c. training dose of morphine was blocked by 0.03–3.0 μg/rat ICV naltrexone methobromide but not by 500 μg/rat (approximately 1.25 mg/kg) of the antagonist injected s.c. (Fig. 4).

Attempts to identify the specific brain areas mediating discriminative drug effects have met with limited success (e.g., Shannon and Holtzman 1977b; for discussion see Rosecrans and Glennon 1979; Holtzman 1983). If the cuing properties of opioids and other drugs represent a compound stimulus derived from the full spectrum of the drug actions, a single discrete neuroanatomical locus of stimulus control may be difficult if not impossible to define (Colpaert 1978).

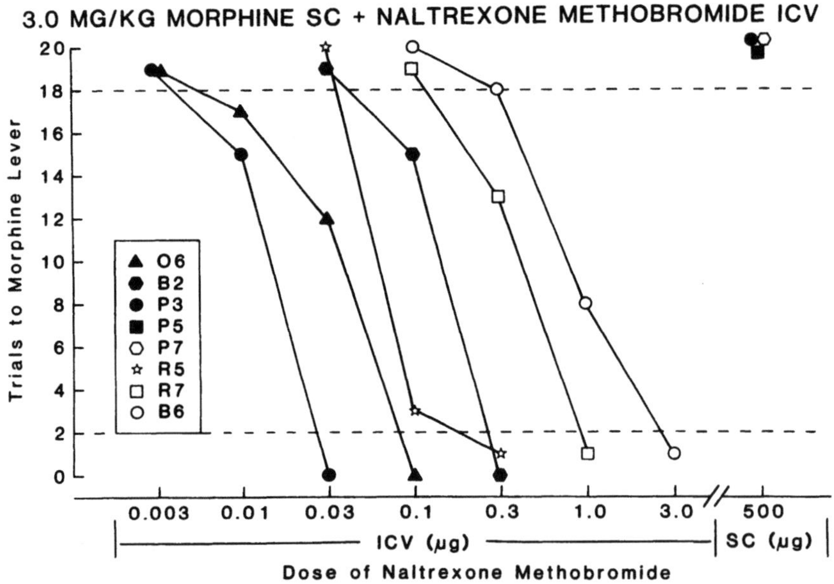

Fig. 4. Dose-related antagonism of the discriminative effects of the training dose of morphine by naltrexone methobromide administered ICV to rats trained to discriminate between saline and 3.0 mg/kg morphine. Each curve represents data from a single animal. When 500 μg naltrexone methobromide was administered s.c. it did not block the discriminative effects of the morphine training dose (*unconnected points at upper right*). From Locke and Holtzman (1985), with permission

5 Endogenous Receptor Ligands: Neuropeptides

The potential role of "traditional" neurotransmitters, such as monoamines, in the genesis of discriminative stimulus effects has been examined widely, most often in connection with psychomotor stimulants (e.g., Ho and Silverman 1978) and LSD-type hallucinogens (e.g., Glennon et al. 1982). Far less investigative attention has been directed toward another important group of endogenous neuroactive substances, the neuropeptides. Issues of basic interest include whether or not the peptides themselves have discriminative stimulus properties and how brain peptides might interact with drugs to modulate the discriminative effects of the latter. Because many neuropeptides do not penetrate the blood-brain barrier from the periphery, they must be administered centrally, with all of the attendant problems of central drug administration discussed previously. In addition, many peptides are degraded rapidly in vivo by peptidases and can be prohibitively expensive to obtain and use in large quantities. In light of the foregoing, it is understandable why discrimination research on neuropeptides has lagged behind studies on other types of compounds.

Thyrotropin-releasing hormone (TRH) is a tripeptide that is behaviorally active following systemic administration. Rats could be trained to discriminate 20 mg/kg TRH from saline, which generalized with 25 μg TRH applied ICV (Jones et al. 1978). In view of reports that TRH reverses behavioral depressant effects of barbiturates and ethanol (Breese et al. 1975), it would be of fundamental interest to determine if the discriminative effects of these drugs are also modified by TRH. A fragment of adrenocorticotrophic hormone, $ACTH_{4-10}$, was found to alter the threshold for stimulus control of behavior by fentanyl in the rat (Colpaert et al. 1978).

As in other areas of drug discrimination research, opioids have been a prominent factor in studies of neuropeptides. To date, almost all opioid peptides examined have been found to share discriminative effects with morphine-like drugs regardless of the relative selectivity of the peptides for mu- versus delta-opioid receptors (e.g., Lord et al. 1977) in binding assays or in bioassays on isolated tissue preparations. The metabolically stable analogs of methionine enkephalin, FK 33,824, which is selective for the mu receptor, and metkephamid, which exhibits selectivity for the delta receptor, generalized with morphine-like agonists when administered systemically to rhesus monkeys (Young et al. 1983). The delta-receptor agonist DADL (see legend to Fig. 5 for explanation of abbreviations), administered to rats ICV, generalized with fentanyl but occasioned only vehicle-appropriate responding in rats trained with the kappa agonist, ethylketocyclazocine (Shearman and Herz 1982b).

We (Locke and Holtzman 1986) recently evaluated a series of opioid peptides having relative selectivity either for the mu- or delta-opioid receptor. The peptides were tested ICV in rats trained to discriminate 3.0 mg/kg morphine (s.c.) from saline. The mu-selective peptides (DAGO, FK 33,824, and morphiceptin), the delta-selective peptides (DADL and metkephamid), and beta-endorphin, which interacts equally well with both receptor types, all produced dose-dependent morphine-like stimulus control of behavior (Fig. 5). On a molar basis, the order of relative potency of these peptides is FK 33,824 $\geqq$ DAGO > beta-endorphin >

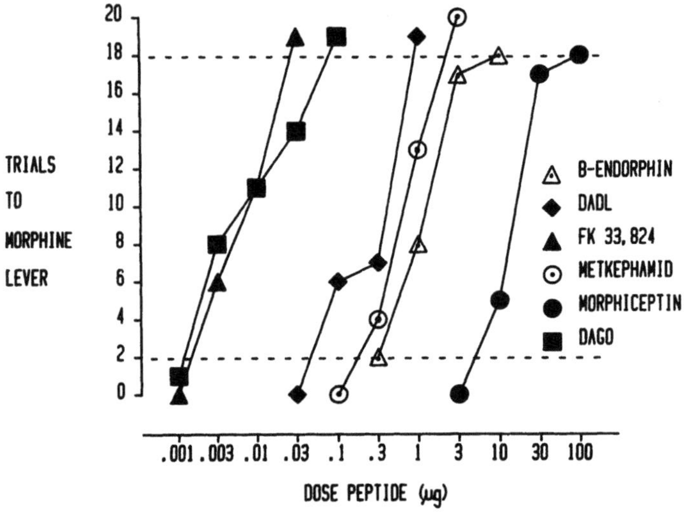

Fig. 6. The morphine-like discriminative effects of approximately equidiscriminable doses of morphine and three opioid peptides are blocked dose dependently and completely over a similar range of naltrexone doses administered s.c. The morphine dose was 10 μg; the dose of each peptide was determined from the results shown in Fig. 5, and varied within each group according to the sensitivity of each subject. Each point is a mean based upon observations in at least five rats. The other details are the same as in Fig. 1. (Locke and Holtzman, unpublished observations)

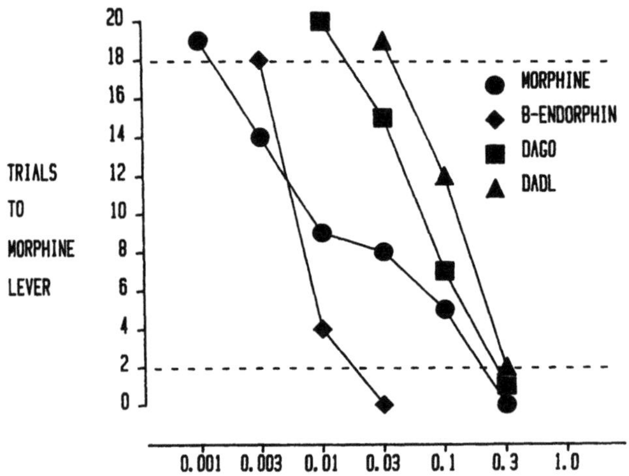

Fig. 5. Regardless of their relative selectivity for mu- or delta-opioid receptors, opioid peptides administered ICV produce dose-dependent morphine-like discriminative effects in rats trained to discriminate between saline and 3.0 mg/kg morphine. Each point is a mean based upon one observation in each of four to seven animals. Note that peptide doses are shown as micrograms per rat. Other details are the same as in Fig. 1. *DADL*, D-Ala^2-D-Leu^{5}enkephalin; *FK 33,824*, D-Ala^2-NMePhe^{4}Met(O)^{5}ol enkephalin; *metkephamid*, D-Ala^2-NMeMet^5-NH_2 enkephalin; *morphiceptin*, Tyr-Pro-Phe-Pro-NH_2; *DAGO*, D-Ala^{2}NMePhe^4-Gly^5(ol) enkephalin. (Locke and Holtzman, unpublished observations)

DADL > metkephamid > morphiceptin. The low potency of morphiceptin is probably due to its rapid degradation in vivo (Kreil et al. 1983). The morphine-like discriminative effects of DAGO, beta-endorphin and DADL were tested for sensitivity to antagonism by naltrexone. Discriminative effects of the peptides were blocked dose-dependently and completely by naltrexone over the same range of doses that blocked the discriminative effects of ICV morphine (Fig. 6). Thus, in producing morphine-like discriminative effects, the mu-selective peptides FK 33,824 and DAGO are approximately two orders of magnitude more potent than the delta-selective peptides DADL and metkephamid. This fact taken with the similar and high sensitivity of opioid peptide effects to antagonism by naltrexone suggests that interactions of opioid peptides with mu receptors in the brain is a critical element in the genesis of their morphine-like discriminative effects.

As we and others have shown, systematic study of the discriminative stimulus properties of neuroactive peptides is clearly possible, albeit sometimes technically difficult. It is in this emerging area of peptide neuropharmacology that drug discrimination methodologies may well make some of their most important contributions in the years ahead.

Acknowledgments. Research performed by the authors at Emory University and preparation of this report were supported, in part, by grant DA 00541 and by Research Scientist Award KO5 DA 00008, both from National Institute on Drug Abuse, U.S. Public Health Service.

References

Ariens EJ, Simonis AM, Van Rossum JM (1964) Drug-receptor interaction: interaction of one or more drugs with one receptor system. In: Ariens EJ (ed) Molecular pharmacology, the mode of action of biologically active compounds, vol 1. Academic, New York, pp 119–286

Arunlakshana O, Schild HO (1959) Some quantitative uses of drug antagonists. Br J Pharmacol 14:48–58

Ator NA, Griffiths RR (1983) Lorazepam and pentobarbital drug discrimination in baboons: cross-drug generalization and interaction with Ro 15-1788. J Pharmacol Exp Ther 226:776–782

Barry H (1974) Classification of drugs according to their discriminable effects in rats. Fed Proc 33:1814–1824

Breese GR, Cott JM, Cooper BR, Prange AJ, Lipton MA, Plotnikoff NP (1975) Effects of thyrotropin releasing hormone (TRH) on the actions of pentobarbital and other centrally acting drugs. J Pharmacol Exp Ther 193:11–22

Colpaert FC (1977) Discriminative stimulus properties of benzodiazepines and barbiturates. In: Lal H (ed) Discriminative stimulus properties of drugs. Plenum, New York, pp 93–106

Colpaert FC (1978) Discriminative stimulus properties of narcotic analgesic drugs. Pharmacol Biochem Behav 9:863–887

Colpaert FC, Janssen PAJ (1984) Agonist and antagonist effects of prototype opiate drugs in rats discriminating fentanyl from saline: characteristics of partial generalization. J Pharmacol Exp Ther 230:193–199

Colpaert FC, Niemegeers CJE, Janssen PAJ (1976) The narcotic discriminative stimulus complex: relation to analgesic activity. J Pharm Pharmacol 28:183–187

Colpaert FC, Niemegeers DJE, Janssen PAJ, Van Ree JM, de Wied (1978) Selective interference of $ACTH_{4-10}$ with discriminative responding based on the narcotic cue. Psychoneuroendocrinology 3:203–221

Colpaert FC, Niemegeers CJE, Janssen PAJ (1982) A drug discrimination analysis of lysergic acid diethylamide (LSD): in vivo agonist and antagonist effects of purported 5-hydroxytryptamine antagonists and of pirenperone, a LSD-antagonist. J Pharmacol Exp Ther 221:206–214

Glennon RA, Rosecrans JA, Young R (1982) The use of the drug discrimination paradigm for studying hallucinogenic agents. A review. In: Colpaert FC, Slangen JL (eds) Drug discrimination: applications in CNS pharmacology. Elsevier Biomedical, Amsterdam, pp 69–96

Griffiths RR, Balster RL (1979) Opioids: similarity between evaluations of subjective effects and animal self-administration results. Clin Pharmacol Ther 25:611–617

Herling S, Shannon HE (1982a) Discriminative effects of ethylketazocine in the rat: stereospecificity and antagonism by naloxone. Life Sci 31:2371–2374

Herling S, Shannon HE (1982b) Ro 15-1788 antagonizes the discriminative stimulus effects of diazepam in rats but not similar effects of pentobarbital. Life Sci 31:2105–2112

Herling S, Woods JH (1981) Discriminative stimulus effects of narcotics: evidence for multiple receptor-mediated actions. Life Sci 28:1571–1584

Ho BT, Silverman PB (1978) Stimulus as discriminative stimuli. In: Colpaert FC, Rosecrans JA (eds) Stimulus properties of drugs: ten years of progress. Elsevier, Amsterdam, pp 53–68

Holtzman SG (1980) Phencyclidine-like discriminative effects of opioids in the rat. J Pharmacol Exp Ther 214:614–619

Holtzman SG (1982) Discriminative stimulus properties of opioids in the rat and squirrel monkey. In: Colpaert FC, Slangen JL (eds) Drug discrimination: applications in CNS pharmacology. Elsevier Biomedical, Amsterdam, pp 17–36

Holtzman SG (1983) Discriminative stimulus properties of opioid agonists and antagonists. In: Cooper SJ (ed) Theory in psychopharmacology, vol 2. Academic, London, pp 1–45

Holtzman SG (1985) Discriminative stimulus properties of opioids that interact with mu, kappa and PCP/sigma receptors. In: Seiden LS, Balster RL (eds) Behavioral pharmacology: the current status. Liss, New York, pp 131–147

Jones CN, Grant LD, Prange AJ Jr (1978) Stimulus properties of thyrotropin-releasing hormone. Psychopharmacology 59:217–224

Kosterlitz HW, Paterson SJ, Robson LE (1981) Characterization of the *kappa*-subtype of the opiate receptor in the guinea pig brain. Br J Pharmacol 73:939–949

Kreil G, Umbach M, Brantl V, Teschemacher H (1983) Studies on the enzymatic degradation of B-casomorphins. Life Sci 33 [Suppl I]:137–140

Lal H, Gianutsos G, Miksic S (1977) Discriminable stimuli produced by analgesics. In: Lal H (ed) Discriminative stimulus properties of drugs. Plenum, New York, pp 23–45

Locke KW, Holtzman SG (1985) Characterization of the discriminative stimulus effects of centrally administered morphine in the rat. Psychopharmacology 87:1–6

Locke KW, Holtzman SG (1986) Behavioral effects of opioid peptides selective for mu or delta receptors. I. Morphine-like discriminative stimulus effects. J Pharmacol Exp Ther 238:990–996

Lord JA, Waterfield AA, Hughes J, Kosterlitz HW (1977) Endogenous opioid peptides: multiple agonists and receptors. Nature 267:495–499

Magnan J, Paterson SJ, Tavani A, Kosterlitz HW (1982) The binding spectrum of narcotic analgesic drugs with different agonist and antagonist properties. Naunyn Schmiedebergh's Arch Pharmacol 319:197–205

Martin WR, Eades CG, Thompson JA, Huppler RE, Gilbert PE (1976) The effects of morphine- and nalorphine-like drugs in the nondependent and morphine-dependent chronic spinal dog. J Pharmacol Exp Ther 197:517–532

Mendelsohn LG, Kerchner GA, Kalra V, Zimmerman DM, Leander JD (1984) Phencyclidine receptors in rat brain cortex. Biochem Pharmacol 33:3529–3535

Möhler H, Okada T, Heitz P, Ulrich J (1978) Biochemical identification of the site of action of benzodiazepines in human brain by ^{3}H-diazepam binding. Life Sci 22:985–995

Pearl J, Aceto MD, Harris LS (1968) Prevention of writhing and other effects of narcotics and narcotic antagonists in mice. J Pharmacol Exp Ther 160:217–230

Quirion R, Hammer RP Jr, Herkenham M, Pert CB (1981) Phencyclidine (angel dust)/sigma "opiate" receptor: visualization by tritium-sensitive film. Proc Natl Acad Sci 78:5881–5885

Rosecrans JA, Glennon RA (1979) Drug-induced cues in studying mechanisms of drug action. Neuropharmacology 18:981–989

Schild HO (1947) pA, A new scale for the measurement of drug antagonism. Br J Pharmacol 2:189–200

Schmidt WK, Tam SW, Shotzberger GS, Smith DH Jr, Clark R, Vernier VG (1985) Nalbuphine. Alcohol Drug Depend 14:339–362

Shannon HE (1982) Pharmacological analysis of the phencyclidine-like discriminative stimulus properties of narcotic derivatives in rats. J Pharmacol Exp Ther 222:146–151

Shannon HE (1983a) Discriminative stimulus effects of phencyclidine. In: Kamenka JM, Domino EF, Geneste P (eds) Phencyclidine and related arylcyclohexylamines: present and future applications. NPP, Ann Arbor, pp 311–335

Shannon HE (1983b) Pharmacological evaluation of *N*-allylnormetazocine (SKF 10,047) on the basis of its discriminative stimulus properties in the rat. J Pharmacol Exp Ther 225:144–152

Shannon HE, Herling S (1983) Discriminative stimulus effects of diazepam in rats: evidence for a maximal effect. J Pharmacol Exp Ther 227:160–166

Shannon HE, Holtzman SG (1976) Evaluation of the discriminative effects of morphine in the rat. J Pharmacol Exp Ther 198:54–65

Shannon HE, Holtzman SG (1977a) Further evaluation of the discriminative effects of morphine in the rat. J Pharmacol Exp Ther 201:55–66

Shannon HE, Holtzman SG (1977b) Discriminative effects of morphine administered intracerebrally in the rat. Life Sci 21:585–594

Shearman GT, Herz A (1982a) Non-opioid psychotomimetic-like discriminative stimulus properties of *N*-allylnormetazocine (SKF 10,047) in the rat. Eur J Pharmacol 82:167–172

Shearman GT, Herz A (1982b) Evidence that the discriminative stimulus properties of fentanyl and ethylketocyclazocine are mediated by an interaction with different opiate receptors. J Pharmacol Exp Ther 221:735–739

Takemori AE (1974) Determination of pharmacological constants: use of narcotic antagonists to characterize analgesic receptors. In: Braude MC, Harris LS, May EL, Smith JP, Villarreal JE (eds) Narcotic antagonists. Raven, New York, pp 335–344

Tallarida RJ, Cowan A, Adler MW (1979) pA_2 and receptor differentiation: a statistical analysis of competitive antagonism. Life Sci 25:637–654

Williams M (1984) Molecular aspects of the action of benzodiazepine and nonbenzodiazepine anxiolytics: a hypothetical allosteric model of the benzodiazepine receptor complex. Prog Neuro-Psychopharmacol Biol Psychiatry 8:209–247

Woolverton WL, Schuster CR (1983) Behavioral and pharmacological aspects of opioid dependence: mixed agonist-antagonists. Pharmacol Rev 35:33–52

Young AM, Swain HH, Woods JH (1981) Comparison of opioid agonists in maintaining responding and in suppressing morphine withdrawal in rhesus monkeys. Psychopharmacology 74:329–335

Young AM, Woods JH, Herling S, Hein DW (1983) Comparison of the reinforcing and discriminative stimulus properties of opioids and opioid peptides. In: Smith JE, Lane JD (eds) The neurobiology of opiate reward processes. Elsevier, New York, pp 147–174

Zukin RS, Zukin SR (1981) Multiple opiate receptors: emerging concepts. Life Sci 29:2681–2690

Zukin SR, Zukin RS (1979) Specific [^{3}H]phencyclidine binding in rat central nervous system. Proc Natl Acad Sci 76:5372–5376

Intrinsic Activity and Discriminative Effects of Drugs

F. C. Colpaert

Contents

Abstract

A conceptual framework has been proposed which coherently specifies the transduction mechanisms of opiate drug stimuli at the behavioral, pharmacological, and molecular levels.

One assertion that is inherent in a one-receptor theory of opiate drug discrimination is that the different magnitudes of intrinsic activity that can be produced at the receptor are associated with discrimination effects that differ qualitatively. A number of theoretical inferences are made here concerning the apparent training drug-like agonist and training-drug antagonist effects of compounds that differ in the maximal intrinsic activity which they produce at the receptor. The predicted patterns of mixed and partial agonist and antagonist effects are consistent with available data on the effects of prototype opiate compounds.

1 Introduction

Detailed analyses (Colpaert and Janssen 1984; Colpaert et al. 1982) of partial generalization have led us to propose a conceptual framework (Colpaert 1986) which attempts coherently to describe the transduction mechanisms of opiate drug stimuli at three levels of analysis, i.e., behavioral, pharmacological, and molecular. The control of discriminative behavior by opiate drugs is presumably contingent upon the pharmacological activity which the training drug produces at a given receptor. The framework implied that the existence of no more than one opiate receptor needed to be assumed in order to account for the available opiate drug discrimination data.

Earlier molecular accounts of opiate drug discrimination (Herling and Woods 1981; Holtzman 1982a, b; Shearman and Herz 1982) supported the multiple opiate receptors theory proposed by Martin (Gilbert and Martin 1976; Martin et al. 1976; Iwamoto and Martin 1981). These accounts proposed that the discriminative effects of opiate drugs are mediated by mu, kappa, and sigma opiate receptors. This multiple opiate receptors interpretation was based on evidence that the

Neurobiology Division, FONDAX-Groupe de Recherches Servier, 7, rue Ampère, 92800 Puteaux, France

Psychopharmacology Series 4
Transduction Mechanisms of Drug Stimuli
Editors: Colpaert, Balster

discriminative effects of different opiate drugs differ qualitatively as well as in terms of their naloxone reversibility.

One assertion that is inherent in our single receptor account of opiate drug discrimination is that the different magnitudes of intrinsic activity that can be generated at the receptor are associated with discriminative effects that differ qualitatively (Colpaert 1986). Note that such a relation between a single quantitative variate (i.e., intrinsic activity) and qualitative variation of sensory effect is not unantecedented; manipulating optic energy along the single quantitative variate of wavelength yields sensory effects (i.e., colors) that differ in a manner which is said to be qualitative. This chapter further specifies in what manner this assertion accounts for differences among opiate drugs in terms of both the quality and the naloxone reversibility of their discriminative effects. Theoretical predictions will also be inferred regarding the training-drug-like agonist and the training-drug-antagonist effects of compounds that differ in the intrinsic activity which they produce. The discussion will adhere to elementary principles of molecular pharmacology (Ariëns 1964). One such principle that has particular significance is that the magnitude of the pharmacological activity which is generated at a given receptor depends on the dose or concentration and, also, on the maximal intrinsic activity of the compound. We will apply this principle to the discriminative effects of a set of drugs that each interact with one single receptor. The receptor can be of an opiate or, for that matter, of any other nature.

2 Partial and Complete Agonist and Antagonist Effects

Let us consider three compounds A, B, and C such that the maximal intrinsic activity which they generate at the receptor differs among them in the manner $A<B<C$ (Fig. 1, upper panel). The compounds may or may not differ in their affinity for the receptor, but there is no need to assume that the relationship between dose and degree of receptor occupation differs among them (Fig. 1, lower panel).

Animals are trained to discriminate the training dose d_t of the stronger agonist C. The activity generated by the training dose varies, if only because of pharmacokinetic variability, both within and among subjects and defines a range a to b of intrinsic activity that is associated with a particular quality of discriminative effect (Fig. 2, upper panel). Tests of stimulus generalization with other than training doses of compound C will yield generalization which occurs from dose d_1 onward and continues to increase at higher doses (Fig. 2, lower panel).

Compound A in these animals will at no dose produce sufficient intrinsic activity for it to be generalized with dose d_t of compound C. Sufficiently large doses of compound A will nonetheless occupy the receptor, and because it can displace compound C, it will antagonize the discriminative effects of dose d_t of compound C in a dose-dependent manner. The antagonism will proceed from 0% to 100%; it will be linear, and will have a relatively steep slope (not shown). Compound B in these animals will produce sufficient intrinsic activity for it to yield generalization in some but not in all animals. The compound will thus generate a generalization gradient the slope of which is shallower than that of compound C and

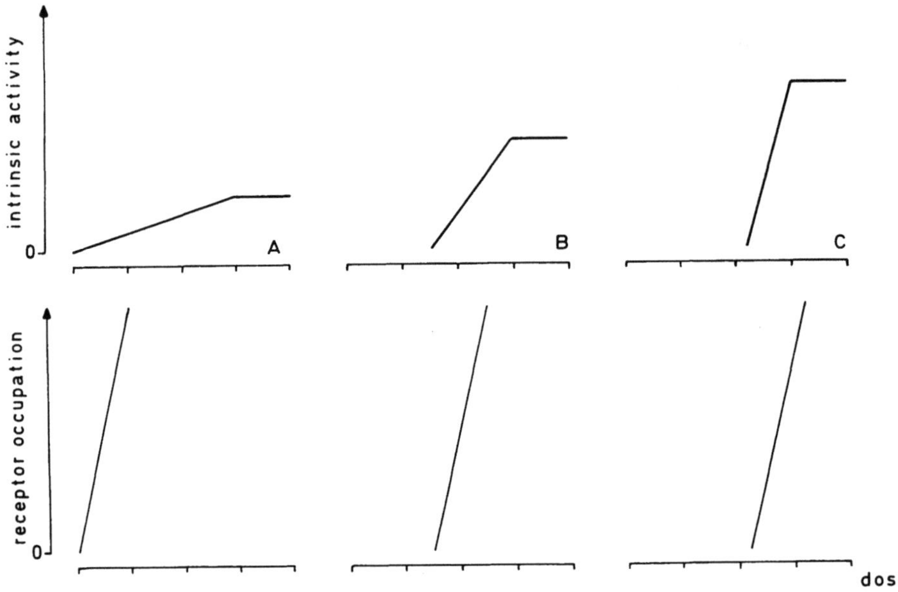

Fig. 1. Schematic representation of the intrinsic activity and of the receptor occupation as a function of the dose of three compounds A, B, and C that differ in the maximal intrinsic activity which they produce at a single receptor

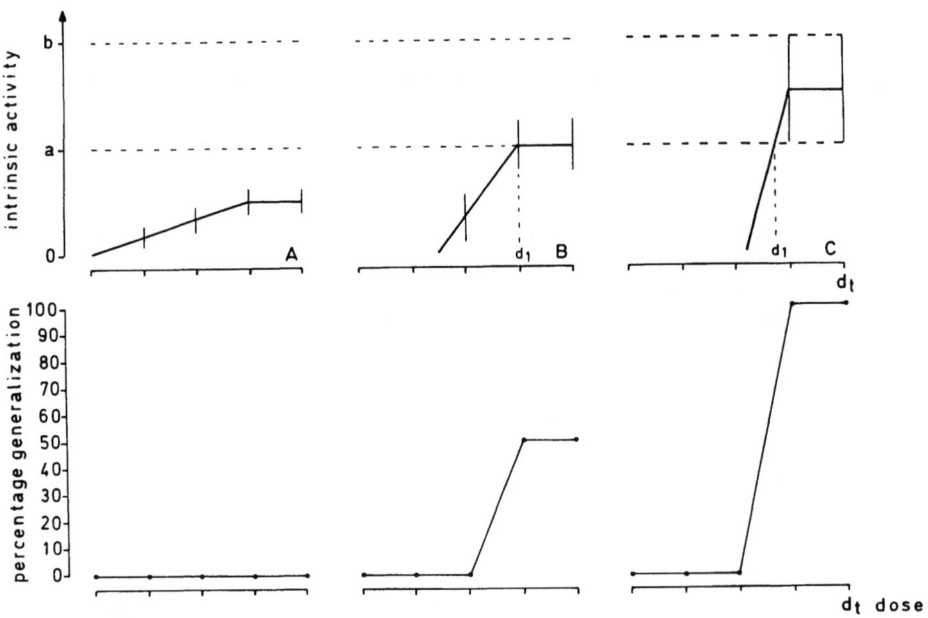

Fig. 2. Schematic representation of generalizations produced by compounds A, B, and C in animals that are trained to discriminate dose d_t of compound C from saline. See text for explanation

which reaches a maximal magnitude of generalization less than 100% (Fig. 2, lower panel). Compound B also competes for the receptor in all animals, and the animals in which doses greater than d_1 do not produce generalization are predicted to be animals in which the compound will antagonize the stimulus effects of dose d_t of compound C. This is unlike animals that generalize compound B and in whom no antagonism of C by B is expected to occur. It is thus predicted that compound B will be effective in all animals, that it will act as a C-like agonist in some and as a C-antagonist in other animals, and that the agonist and antagonist effects will occur at roughly the same doses of compound B. It is also anticipated that other compounds whose maximal intrinsic activity is similar to that of compound B will produce apparent agonist and antagonist effects in the same manner and in the same animals as did compound C itself.

In animals trained on dose d_t of compound C, compound A will thus act as a complete and pure antagonist, and compound B will act as a mixed and partial agonist/antagonist. Let us consider then what the effects of A and B would be in animals trained on a lower dose (e.g., dose d_1) of compound C (not shown). In animals so trained, compound A will still act as a complete or almost complete antagonist. Compound B, however, will now act as an almost complete if not complete agonist. Though compound B may thus produce complete generalization, its effects will still differ from those of compound C in at least two ways. One is that the slope of the generalization gradient of B is predictably shallower than that of C. A second difference is that the slope of, for example, compound A in antagonizing the complete generalization of compound B would be shallower than that in antagonizing C. A case to this effect is that of the effects of naloxone on the generalization of buprenorphine and fentanyl in rats discriminating 0.04 mg/kg fentanyl from saline (Colpaert 1978). Taking this reasoning further, it will become apparent that, as a function of the dose at which a strong agonist is used in training, a weaker agonist may appear to act as a pure antagonist (at high training doses), as a mixed and partial agonist/antagonist (at intermediate training doses), and as a complete agonist (at low training doses).

In animals that discriminate dose d_t of compound B from saline (Fig. 3), compound B itself will yield dose-dependent and complete generalization. Compound A will not produce stimulus generalization and will, in fact, act as a pure antagonist. Compound C will produce generalization in the d_1 to d_2 range of doses; higher doses, however, will generate an intrinsic activity greater than b: the stimulus effects at this greater activity differ qualitatively and will no longer be generalized. The generalization will thus be partial (as in Fig. 3, lower panel) and would assume an inverted u-shape if all intermediate doses were tested. Note that a compound with slightly stronger intrinsic activity than compound A would yield a generalization curve which is similar to that of the much stronger agonist C (not shown).

Animals discriminating dose d_t of compound A will generalize the compound in a dose-dependent manner (Fig. 4). Compound B in those animals will be generalized at doses ranging from d_1 to d_2, but not at higher doses. Higher than d_2 doses of compound B will in fact antagonize the stimulus effects of compound A. Compound B will thus be effective in all animals tested, but will act as a mixed and partial agonist/antagonist. Compound C will produce generalization within

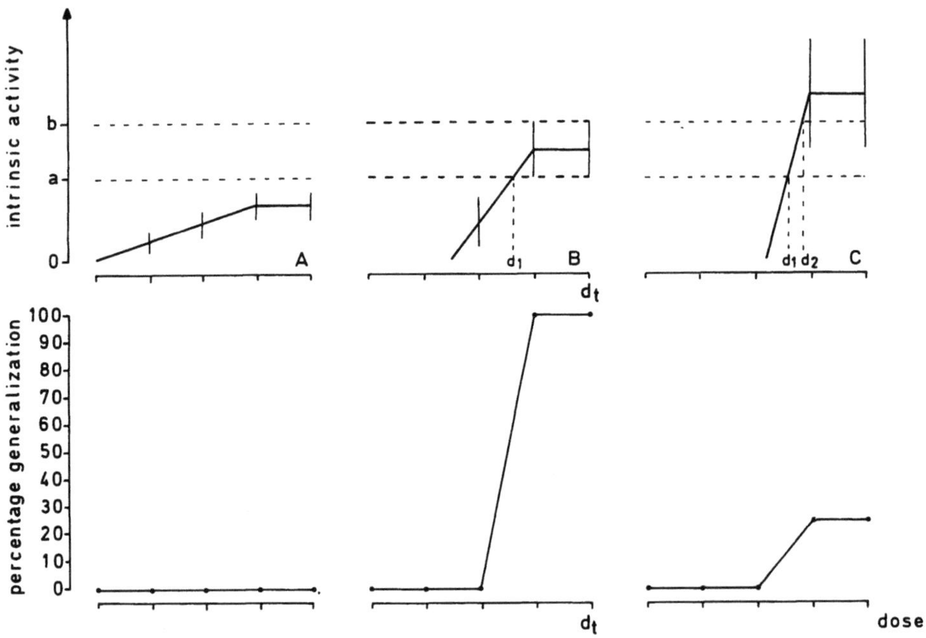

Fig. 3. Schematic representation of generalizations produced by compounds A, B, and C in animals that are trained to discriminate dose d_t of compound B from saline. See text for explanation

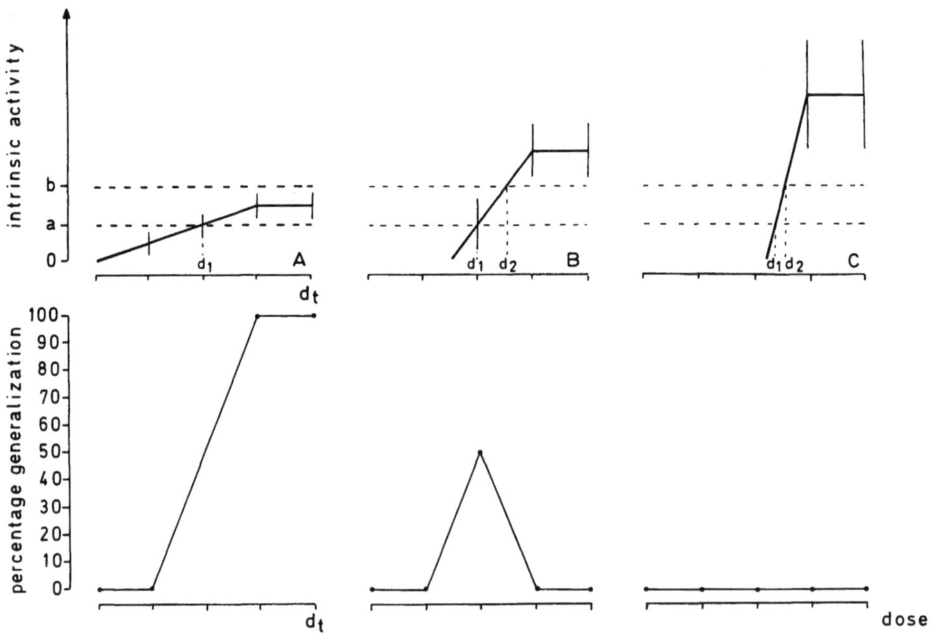

Fig. 4. Schematic representation of generalizations produced by compounds A, B, and C in animals that are trained to discriminate dose d_t of compound A from saline. See text for explanation

the range d_1 to d_2, but the generalization within this narrow range may well go unnoticed because the chance that this critical range is hit by the experimenter is very limited. In addition, the critical dose is likely to vary among subjects. Sufficiently large doses of C should antagonize the stimulus effects of compound A, but it will again be difficult to obtain the antagonism. This is because a high d_t and, thus, a high degree of receptor occupation is required for compound A to produce even the modest magnitude *a–b* of intrinsic activity. Relatively high doses of C will therefore be required to displace A from the receptor. In addition, strong agonists tend to depress behavior, perhaps in part because of the peculiar quality of their stimulus effects.

While compound A antagonizes the d_t effects of compounds B and C, it is also predicted that the ED_{50} doses of A in doing so will vary; the ED_{50} dose is expected to be higher for B than for C.

3 Conclusions

One parsimonious account of available opiate drug discrimination data suggests that the different magnitudes of intrinsic activity that can be generated at a single opiate receptor are associated with discriminative effects that differ in a qualitative manner (Colpaert 1986). In this chapter, some theoretical inferences have been made concerning the apparent training drug-like agonist and training drug-antagonist effects of compounds that differ in their maximal intrinsic activity. These inferences are being made while taking into account elementary principles (Ariëns 1964) of molecular pharmacology. The patterns of mixed and partial agonist and antagonist effects that are predicted to occur are consistent with available data (Colpaert 1986) on the discriminative effects of prototype opiate compounds. It remains for future work to determine to what extent this theoretical framework can perhaps also account for the full and partial effects that can be observed in the drug discrimination paradigm with nonopiate classes of compounds.

References

Ariëns EJ (eds) (1984) Molecular pharmacology, vol 1. Academic, New York

Colpaert FC (1978) Discriminative stimulus properties of narcotic analgesic drugs. Pharmacol Biochem Behav 9:863–887

Colpaert FC (1986) Drug discrimination: behavioral, pharmacological and molecular mechanisms of discriminative drug effects. In: Goldberg S, Stolerman IP (eds) Behavioral analysis of drug dependence. Academic, New York, pp 161–193

Colpaert FC, Janssen PAJ (1984) Agonist and antagonist effects of prototype opiate drugs in rats discriminating fentanyl from saline: characteristics of partial generalization. J Pharmacol Exp Ther 230:193–199

Colpaert FC, Niemegeers CJE, Janssen PAJ (1982) A drug discrimination analysis of lysergic acid diethylamide (LSD): in vivo agonist and antagonist effects of purported 5-hydroxytryptamine antagonists and of pirenperone a LSD-antagonist. J Pharmacol Exp Ther 221:206–204

Gilbert PE, Martin WR (1976) The effects of morphine- and nalorphine-like drugs in the non-dependent, morphine-dependent and cyclazocine-dependent chronic spinal dog. J Pharmacol Exp Ther 198:66–82

Herling S, Woods JH (1981) Discriminative stimulus effects of narcotics: evidence for multiple receptor-mediated actions. Life Sci 28:1571–1584

Holtzman SG (1982a) Stimulus properties of opioids with mixed agonist and antagonist activity. Fed Proc 41:2328–2332

Holtzman SG (1982b) Discriminative properties of opioids in the rat and squirrel monkey. In: Colpaert FC, Slangen JL (eds) Drug discrimination: Applications in CNS pharmacology. Elsevier, Amsterdam

Iwamoto ET, Martin WR (1981) Multiple opioid receptors. Med Res Rev 1:411–440

Martin WR, Eades CG, Thompson JA, Huppler RE, Gilbert PE (1976) The effects of morphine- and nalorphine-like drugs in the nondependent and morphine-dependent chronic spinal dog. J Pharmacol Exp Ther 197:517–532

Shearman HE, Herz A (1982) Generalization and antagonism studies with rats trained to discriminate an effect of the proposed sigma receptor agonist *N*-allylnormetazocine (SKF 10,047). In: Colpaert FC, Slangen JL (eds) Drug discrimination: applications in CNS pharmacology. Elsevier, Amsterdam

Relationship Between the Discriminative Stimulus Properties and Subjective Effects of Drugs

C. R. Schuster[1,2] and C. E. Johanson[1]

Contents

Abstract

Behavioral pharmacologists have assumed that the properties of drugs that mediate their discriminative stimulus effects are related to aspects of drug actions that result in their subjective effects in humans. The basis of this assumption is examined in this chapter. Evidence to support this assumption includes the formal properties of the learning process involved in acquiring both behaviors. Although the procedures used to train animals to learn a drug discrimination are explicit, an analysis of how humans learn to attach verbal responses to unobservable internal subjective states appears to involve a similar learning paradigm. Additional evidence of the commonality of the two effects is that the results from drug discrimination studies in animals and studies evaluating subjective effects in humans yield similar drug classifications. However, when subjective drug effects are analyzed in more detail, it is clear that the concordance between the two approaches is not always good. On the other hand, when drug discrimination and subjective effects are both measured in humans, an examination of the results generated when individuals respond differently to the same drug indicates that the hypothesis that their discrimination is based upon a profile of subjective effects is supported.

1 Introduction

For many years psychotropic drugs have been characterized and classified using methods designed to measure their subjective effects in humans (Beecher 1959). This approach has two principal purposes: (1) to evaluate the efficacy of a drug in attenuating unwanted subjective states in patients (e.g., pain, anxiety, depression) and (2) to evaluate the dependence potential of new drugs by comparing their profile of subjective effects in experienced drug users to those produced by known drugs of abuse. In regard to the latter, such methods also have been used

[1] Drug Abuse Research Center, Department of Psychiatry, The University of Chicago, 5841 S. Maryland Avenue, Chicago, IL 60637.
[2] *Present address:* National Institute on Drug Abuse, 5600 Fishers Lane, Rm. 10-05, Rockville, MD 20857, USA.

Psychopharmacology Series 4
Transduction Mechanisms of Drug Stimuli
Editors: Colpaert, Balster

to determine whether there are common subjective states produced by all drugs of abuse (e.g., euphoria).

Systematic studies using these methods for the assessment of dependence potential have been conducted for several decades at the Addiction Research Center (ARC) of the National Institute on Drug Abuse. A major activity of the ARC has been to evaluate new analgesic compounds to determine whether they produce a morphine-like spectrum of effects. As part of this evaluation, several studies have shown that morphine and related narcotic analgesics produce a unique spectrum of subjective effects that are different from the subjective effects produced by other psychotropic drugs in experienced narcotic addicts (e.g., Hill et al. 1963; Jasinski 1973). Even within the analgesic class, mixed agonist-antagonists (e.g., cyclazocine) can be readily differentiated from morphine in terms of their subjective effects (Jasinski et al. 1968). Other studies have also shown that these methods can be used to differentiate other classes of abused drugs [e.g., psychomotor stimulants (Martin et al. 1971), sedative-hypnotics (Martin et al. 1974; Jasinski 1977), hallucinogens (Martin 1973)]. Thus, it is possible to determine whether an unclassified drug belongs to the opiate, psychomotor stimulant, sedative-hypnotic, or hallucinogenic drug class on the basis of its subjective effects in experienced drug users. Knowledge of a drug's classification can be useful for making an evaluation of its dependence potential.

Until recently, the measurement of drug-induced changes in subjective effects was only considered possible with humans, since only this species has the necessary verbal skills to describe how a drug makes them "feel." However, behavioral methods have been developed over the past 2 decades which allow animals such as rats and rhesus monkeys to report on discriminations between psychotropic drugs (Schuster and Balster 1977; Colpaert and Slangen 1982). Many researchers in behavioral pharmacology believe that the components of drug action responsible for the discrimination among various classes of psychotropic drugs by animals are the same as those responsible for the differences in the subjective effects of these drugs in humans. Although most investigators involved in drug discrimination research accept this assumption, few attempts have been made to examine the basis of the belief (e.g., Schuster et al. 1981; Woolverton and Schuster 1983). Therefore, the purpose of the present paper is to examine the evidence which can be used to support the contention that drug discrimination and subjective effects are related.

One type of evidence that the two procedures are measuring the same components of drug action is to show that the same learning processes are involved in the two behaviors. The fact that humans apply a topographically unique response (verbal) to drug-induced discriminative stimuli should not mask the fact that learning processes are involved. Another type of evidence is to show that the results generated using the two procedures yield similar predictions. For instance, many studies have demonstrated a striking concordance between drug classes based on similarities in the subjective effects produced in humans and on similarities as discriminative stimuli in animals (e.g., Woolverton and Schuster 1983). However, an even stronger proof of the concordance between the measures is the results from studies where both the discriminative stimulus effects of drugs and their subjective effects are measured simultaneously in the same human subjects (e.g., Chait et al. 1985).

2 Behavior Analysis of Drug-Induced Changes in Self-Reports

An analysis of the processes involved in reporting drug-induced changes in subjective states by humans and discrimination of drugs by animals requires a review of precisely what subjects are asked to do in these experiments. The most common instruments used to measure subjective states in humans are paper and pencil inventories. Some of these instruments are composed of a list of adjectives commonly used to describe mood (e.g., happy, angry) and the subject is asked to rate how he/she feels in relation to that mood. For instance, with one version of the Profile of Mood States (POMS), subjects indicate how they feel at that moment in relation to each of 72 adjectives on a 5-point scale from "not at all" (0) to "extremely" (4). There are eight clusters of adjectives that have been separated using factor analysis (Anxiety, Depression, Anger, Vigor, Fatigue, Confusion, Friendliness, Elation) (e.g., Johanson and Uhlenhuth 1980a, b).

Other instruments for measuring subjective effects consist of statements related to sensations and perceptions about which subjects are asked to indicate their agreement or disagreement. For instance, the Addiction Research Center Inventory (ARCI) developed at the ARC is a true-false questionnaire with empirically derived scales that are sensitive to the effects of a variety of classes of abused drugs (Hill et al. 1963). A short form of the inventory consists of five scales with a total of 49 items (Martin et al. 1971). The five scales are the A (amphetamine), a scale which is a measure of drug-related effects of *d*-amphetamine; the MBG (morphine-benzedrine group), thought to be a measure of "euphoria" common to all drugs of abuse; the BG (benzedrine group), an amphetamine scale consisting of items related to intellectual efficiency and energy; the PCAG (pentobarbital-chlorpromazine-alcohol group), a measure of sedation; and the LSD (lysergic acid diethylamide), a measure of dysphoria and somatic symptoms. Another instrument used to measure subjective effects is the visual analog scale (VAS; e.g., Chait et al. 1986a, b). Such scales are horizontal 100-mm lines, each labeled with an adjective (e.g., "stimulated," "high," "anxious," "sedated," "down," "hungry"). The left ends of the lines are labeled "NOT AT ALL" and the right ends "EXTREMELY" and the subject is instructed to place a mark on the line which represents their subjective state in relation to the adjective.

With all of these instruments, subjects are typically instructed that they should respond in a manner which best reflects how they feel at the moment. The responses before and after drug or under drug and placebo conditions are compared to determine whether the drug has produced a significant change. In using these methods, it is assumed that there is a correspondence between the subject's feelings and the statements or adjectives checked. However, since the individual's feelings are a private internal event, it is difficult for the investigator to determine the degree of accuracy of the subject's verbal report, i.e., how precisely it reflects a feeling state. Furthermore, since the self-reporting response can be viewed as a form of operant behavior which can be controlled by its consequences, such responses are susceptible to change by a variety of influences besides the subjective effects produced by the drug administered. The Greenspoon phenomenon (1955) has amply demonstrated that powerful control can be exerted over the verbal behavior of subjects who were unaware that their behavior was being manipulated by subtle responses of the experimenter. In the Greenspoon study, human subjects were instructed to "say words" during a 50-min experimental session. When the experimenter made a specific verbal response ("mmm-hmm") following every plural noun uttered by the subject, there was a significant increase in the use of plural nouns.

Given the potential malleability of self-reports, one approach to validity is to treat verbal responses as devoid of a referent. Thus, if after being given a sedative drug, subjects say "I feel sleepy," one can record this as a change in verbal behavior induced by the drug, without any inference about changes in subjective state. Thus, a positive answer to that statement would have no value in predicting other behavior of the individual, such as the likelihood of reclining on a bed, or of exhibiting a sleep-appropriate EEG. But if drug-induced changes in verbal responses are treated as devoid of a referent, the meaning or form of the verbal response should be irrelevant. If subjects were properly instructed and trained, they could just as well check off boxes labeled with color names or numbers rather than mood descriptors, and drugs with similar effects would be expected to produce a similar spectrum of responses. But because humans have no previous discriminative training for applying such color names or numbers to internal states, it is unlikely that there would be any consistency in drug effects from one individual to the next unless they received very extensive training. On the other hand, when subjects are allowed to respond using common adjectives or simple descriptive statements of mood states, there is a fair degree of agreement in responses across individuals following administration of psychotropic drugs. This agreement is most likely the result of a common conditioning history in which certain adjectives or mood descriptions have been associated with certain internal states. Stated differently, subjects have already been trained to associate a word with an internal subjective state.

But how do humans learn to apply labels to private events and is it reasonable to assume that most individuals who speak a particular language have similar associations? It is clear how children can be differentially reinforced for correctly labeling colors, sounds, and other publicly observable stimulus events. Internal stimuli represent a special problem for such differential conditioning since the mediator of reinforcement cannot observe the private event of another individual to determine the accuracy of the labeling. Under these conditions, the trainer uses a combination of observing the external environment for significant cues and collateral responses of the speaker as an indication of the appropriateness of the individual's description of their mood state. For example, a trainer would agree with (i.e., reinforce) a child who says she is sad when found sitting hunch-shouldered in her bedroom with tears streaming down her face if it is also observed that her favorite toy had been ruined by the family dog. Conversely, if she comes bounding in the door, whistling and swinging her lunch pail with a smile on her face to show off her good report card, she would be reinforced for saying she is happy. As a child matures, some of the more observable parts of these behavior patterns (e.g., crying, whistling) may diminish, but it is this type of conditioning history which is utilized when subjects are asked to match their internal state with a list of adjectives or statements. It is most fortunate that such conditioning histories are consistent enough across individuals so that drugs induce reliable changes in self-reports of internal states.

3 Drug Discrimination in Animals

Just as humans learn to apply word labels to feeling states, animals can be trained to make an arbitrary response when a particular internal state is present. For instance, it is well established that animals can be trained to discriminate the internal cues associated with food and water deprivation (Hull 1933; Leeper 1935). In a recent experiment (Corwin, unpublished observations), rats were trained to press one lever when deprived of food for 24 h and a second lever when deprived for only 3 hours. After the rats met the criterion for discrimination, intermediate durations of deprivation were tested under conditions where either lever was correct. When tested after 3 h without food, 96% of the responses occurred on the 3-h appropriate lever. In contrast after 24 h less than 1% of the total responses occurred on the 3-h appropriate lever. At the intermediate test values of 6 and 12 h of deprivation the percentage of responses on the 3-h lever, 90% and 62%, respectively, decreased in a systematic manner. Therefore, the rat is correctly identifying an internal state in the same way that a human subject might check the degree of "hunger" under similar food deprivation conditions. In the case of the animal, however, the conditioning history has been totally under the control of the experimenter whereas in the human it is assumed that such discrimination training has already occurred as previously described. As has been amply demonstrated in this volume, animals can also be trained to discriminate between drug states and placebo as well as between various drugs using similar discriminative training procedures. That is, when drug is administered, these animals learn to respond on one manipulandum whereas when administered saline or a different drug, they learn to respond on a different manipulandum. Learning is accomplished by differentially reinforcing the designated response in much the same way that humans are differentially reinforced for attaching certain verbal responses to internal states.

In summary, while it is not possible to actually prove that the type of learning process that enables animals to learn to make a discrimination between a drug and saline is identical to that which enables humans to learn to describe subjective states produced by a drug, the analysis of the contingency relationships does make it reasonable to assume that the two processes are similar.

4 Comparison of the Classification of Drugs in Animals and Humans

In addition to showing that the behaviors involved in reporting subjective effects produced by drugs and drug discrimination are learned in a similar manner, another way of showing the similarity of these two procedures is to determine whether they produce comparable results. There is a vast literature which shows that in drug discrimination studies animals separate drugs into classes which have similar subjective effects in humans. Since these studies have been reviewed previously (e.g., Colpaert and Slangen 1982; Woolverton and Schuster 1983), only a representative example will be given from research conducted at the Drug Abuse Research Center of The University of Chicago. de la Garza and Johanson (1987) trained rhesus monkeys to discriminate between 10 mg/kg pentobarbital

Table 1. Discriminative stimulus properties of drugs in pentobarbital-trained Rhesus monkeys

Drug	Positive/total[a]
Alprazolam	3/3
Diazepam	4/4
Flurazepam	3/3
Lorazepam	3/3
Midazolam	4/4
Triazolam	4/4
CL 218,872	2/4
Methaqualone	3/3
d-Amphetamine	0/4
Fenfluramine	0/3

[a] Number of subjects responding at least 80% on the pentobarbital lever/number of subjects tested.

and placebo delivered intragastrically using a two-lever signaled shock avoidance procedure. Animals readily learned this discrimination with virtually 100% accuracy. Following stabilization of the animals' discrimination performance, a variety of benzodiazepines and other psychoactive substances were evaluated in test sessions where responses on either lever satisfied the avoidance contingency. A test drug was considered to have substituted for pentobarbital if at least 80% of the animal's responses occurred on the pentobarbital-appropriate lever after the administration of at least one dose. As shown in Table 1, all the benzodiazepines tested substituted for pentobarbital. In addition, other sedatives such as methaqualone substituted. In contrast, amphetamine and fenfluramine did not substitute for pentobarbital. These results indicate that intragastric pentobarbital can function as a discriminative stimulus and that its stimulus properties have pharmacological specificity.

It is of interest to compare the profile of subjective effects produced by the same benzodiazepines and pentobarbital to determine whether they share the same spectrum of actions, as suggested by the fact that they share discriminative stimulus properties in rhesus monkeys. Johanson and her colleagues have conducted several studies in which normal human subjects were given various doses of diazepam, flurazepam, lorazepam, and pentobarbital (de Wit et al. 1983, 1984a–c). At appropriate times after the administration of these drugs, the subjects were given a variety of mood questionnaires including the POMS. As can be seen in Table 2, most of these drugs produced a significant decrease in Anxiety, Vigor, and Arousal and an increase in Fatigue and Confusion. In contrast, amphetamine produced opposite effects on all of these scales. It would thus appear on the basis of these studies in both animals and humans that the pharmacological classification of the benzodiazepines and sedative hypnotics would be comparable whether one used the technique of drug discrimination or the measurement

Table 2. Profile of subjective effects across drugs[a]

	DZ 10 mg	FLU 30 mg	LOR 2 mg	PB 160 mg	AMP 10 mg
Anxiety	−		−	−	+
Depression					
Anger					
Vigor	−	−	−		+
Fatigue	+	+	+	+	−
Confusion	+		+	+	−
Friendliness				+	+
Elation				+	+
Arousal	−	−	−	−	+
Positive mood			+	+	

[a] Symbols (+ and −) indicate the overall direction of change relative to placebo for each scale on the Profile of Mood States (POMS) for diazepam (DZ), flurazepam (FLU), lorazepam (LOR), pentobarbital (PB), and *d*-amphetamine (AMP).

of subjective effects. This example, along with similar results with other drug classes, supports the contention that the same drug effects are responsible for their subjective effects and discriminative stimulus effects.

5 Comparison of Discriminative Stimulus and Subjective Effects Measured Simultaneously in Humans

As previously mentioned, many other examples of the concordance between the results generated by the two procedures can be found in the literature. However, despite these many examples, such comparisons do not provide direct evidence of the nature of the relationship. Since it is commonly assumed that the unique pattern of subjective effects produced by different classes of drugs are the basis of their different discriminative stimulus properties, a more direct demonstration of their similarity is to measure both the discriminative stimulus and subjective effects of drugs simultaneously in human subjects. If it is possible to show that drugs which have similar discriminative stimulus properties share common subjective effects, which are different from those produced by drugs with different discriminative stimulus properties, then the assumption that these two measures are comparable is strengthened. Furthermore, if individual differences in response to a particular drug emerge, the degree of covariance of these two measures within subjects can further demonstrate their comparability. For instance, it is of interest to determine whether subjects who do not learn to discriminate drug from placebo differ in their drug-induced subjective effects from those who do learn the discrimination. Finally, there are often individual differences in the extent to which test drugs are generalized from the discriminative stimulus effects of the training drug. Again, whether these differences and the degree of generalization from one drug state to another are correlated with different drug-induced subjective effects is an indication that the discriminative stimulus and subjective effects are measuring similar aspects of the actions of drugs.

Several studies that have measured the discriminative stimulus properties and subjective effects of drugs simultaneously in normal human subjects have been conducted at the Drug Abuse Research Center in Chicago (Chait et al. 1985, 1986a, b). The procedure used in these studies was modeled after animal drug discrimination experiments. In all the experiments to be reviewed, the subjects were normal, healthy adult males and females between the ages of 21 and 35 years. Their task was to learn to discriminate between two different drugs, drug A and drug B, based on the effects produced by each. They were told that they could receive appetite suppressants, sedatives, or placebos. Subjects were further informed that drug A and drug B would be different types but subjects were not told that one was a placebo and one an active drug. Drug A was placebo and drug B was 10 mg *d*-amphetamine for half of the subjects and the assignments were reversed for the other subjects. The drug capsules were all identical.

Subjects reported to the laboratory between 9 and 11 a.m. 3 days/week throughout each study and, upon arrival, they completed subjective effects questionnaires and then ingested a capsule. They were then free to leave for the day, taking additional sets of questionnaires to fill out 1, 3, and 6 h later. The first phase of the study (sampling/training phase) consisted of 2 or 4 sampling days, when subjects were told at the time of ingestion which drug (A or B) they were receiving that day. The purpose of this phase was to familiarize subjects with the two training drugs and to begin initial discrimination learning. The second phase of the study (training/assessment phase) consisted of 5–7 training days designed to determine whether subjects were differentiating between drug and placebo. On these days, subjects were not told at the time of ingestion which capsule they were receiving, but instead indicated on a form at 1, 3, and 6 h which drug (A or B) they believed they had received. After subjects had filled out their 6th-h form, they telephoned the laboratory and reported to the experimenter whether they believed they had received drug A or drug B. They were informed at that time whether or not they were correct. If correct, they received a monetary bonus for the accuracy of their discrimination when they reported to the laboratory the next time. The criteria for learning the discrimination was either a correct 6-h identification 5 days in a row or on 6 of the 7 training days. Only subjects who met at least one of these criteria entered the test phase.

The test phase consisted of 13 additional sessions. Six of these sessions were additional training days in which subjects received either placebo or amphetamine and were differentially reinforced for the accuracy of their judgements. On the other days, however, they were given a test drug, placebo, or amphetamine under conditions in which there was no differential reinforcement. That is, on test days both responses were equally reinforced, and subjects received no feedback as to which drug they had received. Subjects were not told the purpose of test days, nor did they know when test days were scheduled until after they had reported their final drug identification 6 h after ingestion.

Using this procedure, several studies have been completed with a variety of test drugs to determine whether or not they substituted for amphetamine, i.e., exerted similar stimulus control. In the first experiment (Chait et al. 1985), 17 sub-

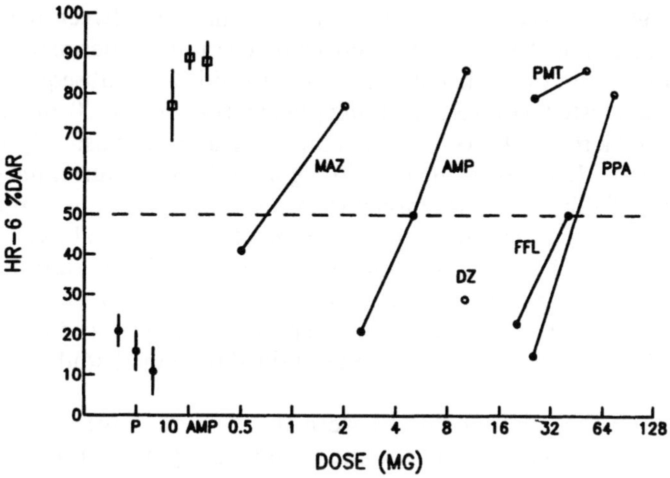

Fig. 1. Group mean hour-6 % drug-appropriate responding for the drugs tested for their ability to substitute for *d*-amphetamine in human subjects trained to discriminate 10 mg *d*-amphetamine from placebo. The *two sets of points to the left* show the data from training days occurring during the test phase of the studies. The drugs tested were mazindol (*MAZ*), *d*-amphetamine (*AMP*), diazepam (*DZ*), fenfluramine (*FFL*), phenmetrazine (*PMT*), and phenylpropanolamine (*PPA*)

jects participated and of these, 7 were able to learn the discrimination between 10 mg *d*-amphetamine and placebo reliably. These seven subjects were subsequently tested with various doses of *d*-amphetamine (2.5, 5, and 10 mg) and diazepam at a dose of 10 mg. A dose of 2.5 mg *d*-amphetamine produced approximately 20% drug identifications, 5 mg produced 50%, and the 10-mg training dose of amphetamine produced almost 90% under test conditions (AMP in Fig. 1). In contrast, when placebo was administered it produced less than 20% drug-appropriate responding. Although 10 mg diazepam was also labeled primarily as placebo (DZ in Fig. 1), this drug did produce measurable subjective effects. Therefore, subjects were not simply learning a drug versus placebo discrimination, but rather a discrimination relating specifically to the unique pattern of effects produced by amphetamine.

In the second experiment (Chait et al. 1986a), of the 20 subjects who underwent discrimination training, 12 reliably learned the amphetamine/placebo discrimination. Subsequently, each discriminator was tested with 25 and 75 mg phenylpropanolamine (PPA) and 0.5 and 2 mg mazindol to determine whether the discriminative stimulus effects of these drugs would substitute for those of amphetamine. Figure 1 shows the percentage of drug identifications for the two different doses of each of these test drugs. Although the low dose of both PPA and mazindol produced primarily placebo-appropriate responding, 75 mg PPA and 2 mg mazindol produced approximately 80% drug-appropriate responding. Therefore, these two drugs have discriminative stimulus properties similar to amphetamine.

In the third experiment in this series (Chait et al. 1986b), the discriminative stimulus and subjective effects of phenmetrazine and fenfluramine were studied.

Again, subjects were trained to discriminate between placebo and 10 mg *d*-amphetamine. Of the 27 subjects who began this experiment, 14 reliably learned the discrimination whereas the other 13 did not. Subsequently, the discriminators were tested with 25 and 50 mg phenmetrazine and 20 and 40 mg fenfluramine. As shown in Fig. 1, both doses of phenmetrazine consistently substituted for amphetamine. In contrast, neither 20 mg nor 40 mg fenfluramine substituted in a majority of subjects for *d*-amphetamine as a discriminative stimulus.

In summary, the three experiments described above indicated that phenmetrazine, PPA, and mazindol all consistently shared discriminative stimulus properties with amphetamine. In contrast, diazepam produced less than 30% drug-appropriate responding whereas fenfluramine produced intermediate levels. Generally similar results have been obtained in animal studies (Schuster and Johanson 1984).

If the commonality between the discriminative stimulus effects of mazindol, PPA, phenmetrazine, and the training drug amphetamine are based upon a similarity in their subjective effects, then a common set of subjective effects would be expected to be produced by this group of drugs. In contrast, diazepam and fenfluramine should produce a different set of subjective effects since they do not substitute for amphetamine as a discriminative stimulus. More specifically, one might expect that the subscales which showed a significant change when subjects are given amphetamine should be the same as those which showed changes after the administration of phenmetrazine, mazindol, and PPA but not diazepam and fenfluramine. As shown in Table 3, there is a good concordance between the changes in subjective effects produced by *d*-amphetamine and phenmetrazine. However, for mazindol and PPA, fewer scales were affected although where there were changes they were similar to those produced by the other two stimulants. Furthermore, the effects of diazepam were usually in the opposite direction whereas fenfluramine did not appear to have any subjective effects. On the other hand, there was some overlap in the subjective effects produced by *d*-amphetamine, phenmetrazine, mazindol, PPA, *and* diazepam since all five of these drugs increased the LSD scale of the ARCI. Therefore, while the concordance between discriminative stimulus and subjective effects was good, an examination of the total pattern of subjective effects produced by these drugs reveals some divergence.

There are several possible reasons why a perfect concordance between the subjective effects of these test drugs and amphetamine and their discriminative stimulus effects was not found. First, the discriminative stimulus effects may have been based upon any one or combination of the changes in mood which these various subscales are purported to measure. Second, since only those subjective effects which were sufficiently changed by drugs to achieve statistical significance are shown, it may very well be that there is a common set of changes in terms of the direction of change produced by all of the drugs but that the magnitude of some of these changes did not reach statistical significance. That is, the relationship of statistical significance on mood measures and the ability of the organism to discriminate these changes is not known. Thus, the measurement of the subjective effects and the measurement of the discriminative stimulus properties of the drug may very well not reflect equal sensitivity. For all these reasons one might not ex-

Table 3. Profile of subjective effects across drugs[a]

	AMP	PMT	MAZ	FFL	PPA	DZ
POMS						
Anxiety	+?	+	+			
Depression						
Anger						
Vigor	+	+			+	
Fatigue	−	−			−	+
Confusion						
Friendliness						
Elation						
Arousal	+	+			+	−
Positive mood						
ARCI						
PCAG	−	−			−	+
BG		+				−
LSD	+	+	+		+	+
MBG	+	+				−
A		+			+	
VAS						
STIM		+			+	
HIGH	+	+		+	+	
ANX		+	+		+	−
SED	−					+
HUNG	−	−	−			

[a] Symbols (+ and −) indicate the overall direction of change relative to placebo for each scale on the Profile of Mood States (POMS), Addiction Research Center Inventory (ARCI) and Visual Analog Scale (VAS) for 10 mg *d*-amphetamine (AMP), 50 mg phenmetrazine (PMT), 2 mg mazindol (MAZ), 40 mg fenfluramine (FFL), 75 mg PPA, and 10 mg diazepam (DZ).

pect to see a perfect concordance between the various measures of subjective effects and the extent to which a drug serves as a discriminative stimulus with this strategy.

Another strategy which can be used to analyze the relationship between subjective effects and the discriminative stimulus properties of drugs takes advantage of individual differences. In the three experiments described above, approximately 50% of the subjects met the training criterion in phase 2 but the other 50% failed. This 50% failure rate may appear at first to be a deficit of the protocol. For the present analysis, however, it is an advantage to have individual differences since it then can be determined whether there is a difference in the subjective effects in those individuals who learn to discriminate the drug and those who do not. The results of this comparison can be summarized as demonstrating that, by and large, discriminators show a greater magnitude of subjective effects than nondiscriminators, suggesting that the basis for the discrimination is a spectrum of discernable subjective effects (Chait et al. 1988). For instance, as shown in Fig. 2, the discriminators in these three studies showed a significantly greater average sensitivity to the effects of *d*-amphetamine on the MBG and A scale of the ARCI, particularly at hour 3, than was observed in the nondiscriminators. Similar differences were found on several scales of the POMS and VAS (Chait et al. 1986c).

To determine the basis of this difference in sensitivity to the subjective effects induced by 10 mg *d*-amphetamine, several additional analyses were performed (Chait et al. 1986c). For instance, a comparison of body weight in the two groups failed to show any significant difference. Thus, it was not a difference in the amount of drug on a milligram per kilogram basis that was responsible for the difference in sensitivity between discriminators and the nondiscriminators. There were also no differences in the history of stimulant use between the discriminator and the nondiscriminator groups. Finally, on a variety of personality inventories such as the Eysenck Personality Inventory (Eysenck and Eysenck 1968), the Sensation Seeking Scale (Zuckerman et al. 1978), Psychopathic State Inventory (Haertzen et al. 1980), Body Consciousness Scale (Miller et al. 1981), Internal-External Locus of Control Scale (Rotter 1966), and a restrained eating scale (Herman 1978), no differences were found between discriminators and nondiscriminators.

The simplest and most likely explanation of the difference between discriminators and nondiscriminators is a quantitative pharmacodynamic difference. In order to determine whether this was the case an additional experiment was conducted with ten subjects who did not meet the criteria for learning the discrimination. These subjects were exposed to two different retraining procedures. One group of five subjects was retrained using the same procedure as used previously. In only one of the five cases was retraining effective, i.e., the subject achieved criterion performance. In contrast, when the other five subjects were retrained using an identical training procedure but using a larger dose of *d*-amphetamine (15 mg), all five met the criteria for discrimination (Chait and Johanson, unpublished observations). It would thus appear that the critical difference between discriminators and nondiscriminators is not one of qualitative drug differences but rather simply a difference in sensitivity to dose. Nevertheless, this difference in sensitivity has given us the opportunity to demonstrate that individuals who could not discriminate 10 mg *d*-amphetamine from placebo also showed significantly lesser subjective effects (Fig. 2). This concordance between the failure to learn the discrimination and the less intense or absence of subjective effects supports the notion that these two phenomena are interrelated. Further, the finding that when the dose of *d*-amphetamine was increased to 15 mg, the subjects now showed significantly greater subjective effects again supports the notion that the individual's ability to tact their subjective effects is the basis of their discrimination.

A second example of how individual differences in drug response can be used to evaluate the relationship of subjective effects to discriminative stimulus effects of drugs involves the results obtained with fenfluramine (Chait et al. 1986b). As previously described, 50% of the subjects who had been trained to discriminate 10 mg *d*-amphetamine from placebo showed drug-appropriate responding when given 40 mg fenfluramine. Thus, 50% of the subjects responded to fenfluramine as if it were amphetamine and 50% responded as if it were placebo. It is therefore of interest to determine whether or not there are differences in the subjective effects produced by fenfluramine in these two groups of subjects. Figure 3 shows the subjective effects of fenfluramine on several subscales of the ARCI, the

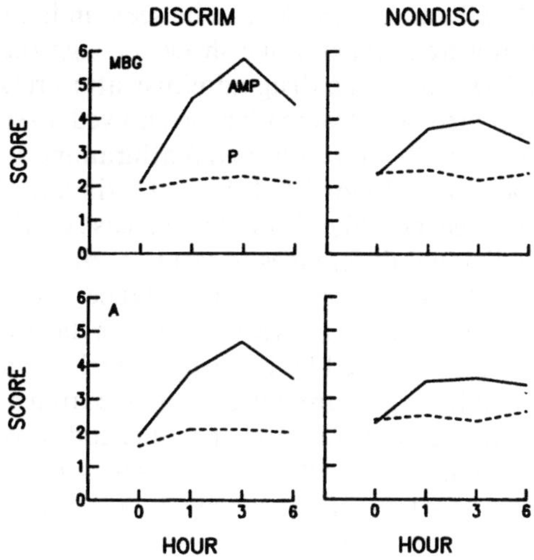

Fig. 2. The effects of 10 mg *d*-amphetamine (*AMP, solid lines*) and placebo (*P, dashed lines*) on the scores for two scales (*MBG* and *A*) of the Addiction Research Center Inventory (*ARCI*) shown separately for subjects that learned the discrimination (*DISCRIM*) and subjects that failed to meet criterion (*NONDISC*). The *abscissa* indicates hours postdrug administration.

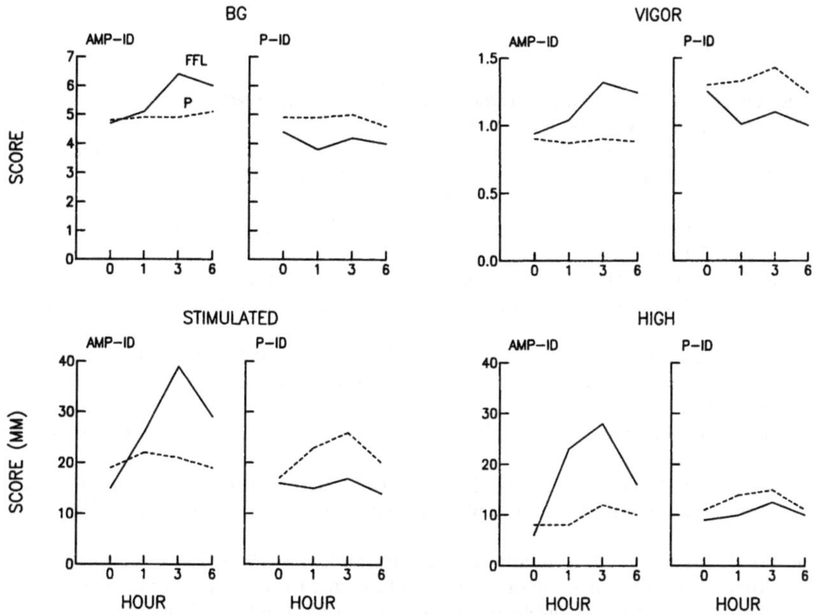

Fig. 3. The effects of 40 mg fenfluramine (*FFL*) and placebo (*P*) on the scores of one scale (*BG*) of the ARCI, one scale (*VIGOR*) of the POMS, and two scales (*STIMULATED* and *HIGH*) of the VAS shown separately for subjects that identified FFL as amphetamine (*AMP-ID*) and subjects that identified FFL as P (*P-ID*). The *abscissa* indicates hours postdrug administration. (Chait et al. 1986b, used with permission)

POMS, and VAS. As can be seen, individuals who responded to fenfluramine as if it were amphetamine showed a significant increase in BG (ARCI) and "high" (VAS) following drug administration relative to placebo, whereas those who responded as if it were placebo showed no significant effect on these scales. Further, individuals who identified fenfluramine as amphetamine-like showed a significant increase in Vigor (POMS) and "stimulated" (VAS), whereas those who identified it as placebo-like showed a decrease on these scales. It would thus appear that the individual differences in terms of drug discrimination and in terms of subjective effects show a striking concordance. This would again support the hypothesis that the basis of drug discrimination is the ability of individuals to tact the subjective effects produced by the drugs which are administered.

Although a great deal more research needs to be done, the power of measuring both the discriminative stimulus and subjective effects of drugs simultaneously in normal human subjects has been amply demonstrated in these studies. It is likely that this type of approach will aid in discerning the basis of drug discrimination which eventually will shed light on the relationship of mood states produced by drugs and their dependence potential.

Acknowledgments. The preparation of this manuscript was supported by grants from the National Institute on Drug Abuse DA 00250 and DA 02812. The authors would like to thank Drs. Larry Chait and Harriet de Wit for their help in assembling the data and Ms. Pamela Goodlow for preparation of the manuscript.

References

Beecher HK (1959) The measurement of subjective responses: quantitative effects of drugs. Oxford University Press, New York

Chait LD, Uhlenhuth EH, Johanson CE (1985) The discriminative stimulus and subjective effects of *d*-amphetamine in humans. Psychopharmacology 86:307–312

Chait LD, Uhlenhuth EH, Johanson CE (1986a) The discriminative stimulus and subjective effects of phenylpropanolamine, mazindol and *d*-amphetamine in humans. Pharmacol Biochem Behav 24:1665–1672

Chait LD, Uhlenhuth EH, Johanson CE (1986b) The discriminative stimulus and subjective effects of *d*-amphetamine, phenmetrazine and fenfluramine in humans. Psychopharmacology 89:301–306

Chait LD, Uhlenhuth EH, Johanson CE (1988) Individual differences in sensitivity to the discriminative stimulus effects of *d*-amphetamine in humans. Psychopharmacology (to be published)

Colpaert FC, Slangen JL (eds) (1982) Drug discrimination: applications in CNS pharmacology. Elsevier, Amsterdam

de la Garza R, Johanson CE (1987) The discriminative stimulus properties of intragastrically administered *d*-amphetamine and pentobarbital in rhesus monkeys. J Pharmacol Exp Ther 243:955–962

deWit H, Johanson CE, Uhlenhuth EH, McCracken S (1983) The effects of two nonpharmacological variables on drug preference in humans. In: Harris LS (ed) Problems of drug dependence, 1982. DHHS publication no. (ADM) 83-1264. Natl Inst Drug Abuse Res Monogr Ser 43:251–257

deWit H, Johanson CE, Uhlenhuth EH (1984a) Reinforcing properties of lorazepam in normal volunteer subjects. Drug Alcohol Depend 13:31–41

deWit H, Uhlenhuth EH, Johanson CE (1984b) Lack of preference for flurazepam in normal volunteers. Pharmacol Biochem Behav 21:865–869

deWit H, Uhlenhuth EH, Pierri J, Johanson CE (1984c) Preference for pentobarbital and diazepam in normal volunteer subjects. Fed Proc 43:931

Eysenck HJ, Eysenck SBG (1968) Manual for the eysenck personality inventory. Educational and industrial testing service, San Diego

Greenspoon J (1955) The reinforcing effect of two spoken sounds on the frequency of two responses. Am J Psychol 68:409–416

Haertzen CA (1974) An overview of addiction research center inventory scales (ARCI): an appendix and manual of scales. US Government Printing Office, Washington DC

Haertzen CA, Martin WR, Ross FE, Neidert GL (1980) Psychopathic state inventory (PSI): development of a short test for measuring psychopathic states. Int J Addict 15:137–146

Herman CP (1978) Restrained eating. In: Stunkard AJ (ed) The psychiatric clinics of North America, vol 1. Saunders, Philadelphia, pp 593–607

Hill HE, Haertzen CA, Wolback AB, Miner EJ (1963) The addiction research center inventory: standardization of scales which evaluate subjective effects of morphine, amphetamine, pentobarbital, alcohol, LSD-25, parahexyl and chlorpromazine. Psychopharmacologia 4:167–183

Hull CJ (1933) Differential habituation to internal stimuli in the albino rat. J Comp Physiol Psychol 16:255–273

Jasinski DR (1973) Assessment of the dependence liability of opiates and sedative hypnotics. In: Goldberg L, Hoffmeister F (eds) Psychic dependence, Bayer-symposium IV. Springer, Berlin Heidelberg New York, pp 160–170

Jasinski DR (1977) Assessment of the abuse potentiality of morphine-like drugs. In: Martin WR (ed) Handbook of experimental pharmacology. Springer, Berlin Heidelberg New York, pp 197–258

Jasinski DR, Martin WR, Sapira JD (1968) Antagonism of the subjective, behavioral, pupillary, and respiratory depressant effects of cyclazocine by naloxone. Clin Pharm Ther 9:217–222

Johanson CE, Uhlenhuth EH (1980 a) Drug preference and mood in humans: *d*-amphetamine. Psychopharmacology 71:275–279

Johanson CE, Uhlenhuth EH (1980 b) Drug preference and mood in humans: diazepam. Psychopharmacology 71:269–273

Leeper R (1935) The role of motivation in learning; a study of the phenomenon of differential motivational control of the utilization of habits. J Genet Psychol 46:3–40

Martin WR (1973) Assessment of the abuse potentiality of amphetamine and LSD-like hallucinogens in man and its relationship to basic animal assessment programs. In: Goldberg L, Hoffmeister F (eds) Psychic dependence, Bayer-symposium IV. Springer, Berlin Heidelberg New York, pp 146–155

Martin WR, Sloan JW, Sapira JD, Jasinski DR (1971) Physiologic, subjective, and behavioral effects of amphetamine, methamphetamine, ephedrine, phenmetrazine, and methylphenidate in man. Clin Pharmacol Ther 12:245–258

Martin WR, Thompson WO, Fraser HF (1974) Comparison of graded single intramuscular doses of morphine and pentobarbital in man. Clin Pharmacol Ther 15:623–630

Miller LC, Murphy R, Buss AH (1981) Consciousness of body: private and public. J Personal Soc Psychol 41:397–406

Rotter JB (1966) Generalized expectancies for internal vs. external control of reinforcement. Psychol Monog 80:609

Schuster CR, Balster RL (1977) The discriminative stimulus properties of drugs. In: Thompson T, Dews PB (eds) Advances in behavioral pharmacology, vol 1. Academic, New York, pp 85–138

Schuster CR, Johanson CE (1984) Efficacy, dependence potential, and neurotoxicity of anorectic drugs. In: Seiden LS, Balster RL (eds) Behavioral pharmacology of psychotropic agents. Liss, New York, pp 263–279

Schuster CR, Fischman MW, Johanson CE (1981) Internal stimulus control and subjective effects of drugs. In: Thompson T, Johanson CE (eds) Behavioral pharmacology of human drug dependence. DHHS publication no. (ADM) 81–1137. Natl Inst Drug Abuse Res Monogr Ser 37:116–129

Woolverton WL, Schuster CR (1983) Behavioral and pharmacological aspects of opioid dependence: mixed agonist-antagonists. Pharmacol Rev 35:33–52

Zuckerman M, Eysenck S, Eysenck HJ (1978) Sensation seeking in England and America: cross-cultural, age and sex comparisons. J Consult Clin Psychol 46:139–149

Similarities and Differences Between Behavioral Control by Drug-Produced Stimuli and by Sensory Stimuli

D. A. OVERTON

Contents

Abstract

Drug-induced state-dependent learning and drug discriminations may be based on sensory stimuli induced by drug actions, or at least on events in the brain that have properties analogous to

---

Departments of Psychiatry and Psychology, Temple University School of Medicine, 3401 North Broad Street, Philadelphia, PA 19140, USA.

Psychopharmacology Series 4
Transduction Mechanisms of Drug Stimuli
Editors: Colpaert, Balster

those of sensory stimuli. A variety of comparisons between behavioral control by drug-induced stimuli and by classically defined interoceptive and exteroceptive stimuli are possible. These allow inferences to be made about the nature of drug stimuli, the properties of behavioral paradigms that we use to investigate them, and the mechanisms by which drugs achieve behavioral control in drug discrimination and state-dependent learning paradigms. Each of these topics is selectively reviewed in this paper.

Drug-induced state-dependent learning (SDL) and drug discriminations (DDs) may be based on sensory stimuli induced by drug actions, or at least on events in the brain that have properties analogous to those of sensory stimuli. This paper will selectively review the similarities and differences between behavioral control by drug-induced stimuli and by other interoceptive and exteroceptive stimuli. Space limitations preclude a comprehensive reference list of the original studies on which this review is based. Hence, at several points the reader will be referred to secondary articles which contain references to the original articles of interest (e.g., Overton 1982b, 1984–1987).

1 Comparisons Along the Quantitative (Intensity) Dimension

1.1 Relative Strength of Control by Drug-Induced and by Sensory Stimuli

In the 1930s, a variety of studies tested the degree to which learned responses were dependent on the environmental stimulus conditions under which they were learned, and most of these studies showed rather small response decrements when stimulus conditions were changed. Interest in this topic then waned, and for several decades there were few published studies of this type. Then, in the late 1970s, interest in contextual control reappeared and several recent reports have shown contextual control to be a stronger effect than the early studies had suggested it to be (Overton 1984).

A priori, one might think that contextual control by drug stimuli would be weaker than control by sensory stimuli since specialized receptors, sensory pathways, and brain mechanisms have evolved to mediate the perception of sensory stimuli, whereas corresponding structures designed to mediate the perception of drug effects are not known to exist. Contrary to this assumption, when experimental investigation of SDL began in the 1930s (Girden and Culler 1937) and again in the 1960s (Overton 1964), very strong SDL effects were observed, and so the investigators who performed these studies concluded that the SDL effects were too strong to be mediated by stimulus effects of the drugs and must be produced by some alternate neurological mechanism (Bliss 1974; Bliss et al. 1971; Overton 1964; Sachs et al. 1966). Later studies showed, however, that although SDL can be a very strong effect under certain conditions, it is more frequently either a weak effect, or unmeasurably small, i.e., obscured by noise or by other drug effects (Eich 1980; Overton 1968a, 1971). If one considers the entire literature of reported studies using designs in which SDL might reasonably be observed, it appears as a significant effect in less than half of these reports. Recognition of the relatively weak response control that is typically produced by contextual drug states (in the absence of DD training) along with the recent demonstrations of relatively strong response control by contextual sensory stimuli has led to a reevaluation of the mechanisms postulated to underlie SDL, and most investigators now believe that drug SDL is an example of drug stimulus control (Overton 1982b, 1985).

Discriminative control of behavior produced by DD training, since it was first reported by Conger (1951), has consistently been considered to be mediated by drug-induced stimuli. The strength of such discriminative control is markedly dependent on the drug and dosage employed. Hence a quantitative comparison of the strength of discriminative control by drug-induced and sensory stimuli is virtually impossible. Nonetheless, a few investigators have attempted this comparison, and depending on the parameters employed have observed that response control by drug states was either stronger or weaker than response control by sensory stimuli (Balster 1970; Kilbey et al. 1971; Overton 1968b, 1971; Spear et al. 1980). Overall, it appears that response control by drug-induced and by sensory stimuli can be considered to be of comparable strength.

1.2 Methods of Measuring the Strength of Drug Stimuli

The preceding paragraph referred to studies which compared the strength of behavioral control by drugs and by sensory stimuli, and this implies that we are able to quantitatively compare the strength (or salience) of drug and sensory stimuli. How might such measurements be made?

In SDL studies, a measure of the amount of retention after a state change is usually obtained. However, we will show later that quantitatively accurate determinations of this type are essentially impossible because of the intermixture of SDL effects with other drug effects.

In DD studies, investigators have generally used either asymptotic accuracy or the speed of acquisition of D vs N (drug versus no drug) discriminations as an indicator of the strength (salience) of drug stimuli. The asymptotic accuracy index is difficult to use, since asymptotic accuracy is strongly influenced not only by the strength of drug stimuli but also by the schedule of reinforcement employed, and by various drug effects which may disrupt discriminative control (Colpaert et al. 1980b; Harris and Balster 1971; Overton and Hayes 1984; Schuster and Balster 1977). Indices of speed of acquisition have produced more orderly data. A linear relationship between sessions to criterion (log transformed) and dosage (log transformed) has been obtained with a variety of drugs (Overton 1982a), suggesting that speed of acquisition may provide a measure of stimulus strength. A detailed analysis of the utility of various indices of discriminability has recently been reported (Overton et al. 1986).

1.3 Width of Generalization Gradients – Quantitative Specificity

After D vs N DD training using a particular drug and dosage, substitution tests can be conducted with various doses of the training drug. Usually a sigmoidally shaped dose-response curve is obtained; the drug response generalizes to doses significantly higher than the training dose and an appreciable percentage of drug responses occur with doses down to about 30% of the training dose (Barry 1978; Colpaert and Slangen 1982; Overton 1966), i.e., rather broad generalization is obtained along the intensity (dosage) continuum. In a few instances, somewhat steeper gradients have been obtained, and the factors that cause such steep gradients have not been determined. After sensory discrimination training (S vs no-S), similarly broad generalization gradients are commonly reported.

In stimulus context and SDL studies, generalization of the response from the training condition during tests with altered stimulus or drug conditions can be regarded as showing very flat generalization gradients. Complete generalization of responding to novel test conditions (lack of SDL) has frequently been reported, and in cases where appreciable SDL does occur, only a few studies have tested the effects of drug dosages intermediate between the training dosage and the no drug condition (Holmgren 1964).

1.4 Size of the JND Along Quantitative Dimensions

How large are the just noticeable differences (JNDs) along the quantitative dimensions of drug and sensory stimuli? When human subjects are employed, a reference stimulus can be presented immediately followed by a test stimulus of equal or different intensity. With such procedures, the JND is relatively small and differs in the various sensory modalities – reported JNDs include brightness, 1/60 of reference stimulus; loudness, 1/10; cutaneous deep pressure, 1/7; lifted weight, 1/50; pain, 1/30; smell, 1/4; taste, 1/3 (Schiffman 1982). In some sensory modalities, analogous procedures have been used in animal preparations to determine JNDs (Blough and Blough 1977).

In DD studies, rats have been trained to discriminate between two doses of a single drug by several investigators. No attempt has been made to find the smallest dosage ratio that could be discriminated. However, the data suggest that the JND varies from drug to drug, and is not too much smaller than 1/2 when determined by this method (Colpaert and Janssen 1982; Overton 1968b, 1977a, 1977b).

Dose vs dose discrimination training does not appear to be directly comparable to procedures used in human subjects because the doses that are to be compared are presented 24 h apart instead of a few seconds apart. This might be expected to increase the JND considerably. Additionally, day-to-day variations in drug absorption and distribution make the actual blood levels achieved during successive training sessions somewhat indeterminate. An altered procedure can be suggested which is more closely analogous to those used with sensory stimuli, and which might yield improved results. Rats could be trained in a two-lever task using sessions comprised of two short minisessions separated by a time out. A large dose could be injected before the beginning of training, and lever 1 reinforced during the first minisession. During the time out either saline or a small additional dose of drug would be injected, and on days when additional drug was administered the reinforced lever would be changed to lever 2 during the second minisession. During early training sessions, a relatively large incremental dose could be employed. Subsequently, the incremental dose could be gradually reduced using a titration procedure (Overton 1979). The smallest incremental dose which could be detected with this paradigm might be considerably smaller than the JND determined by dose vs dose training sessions spaced 24 h apart.

1.5 Comparison of Thresholds for Discriminative and Contextual Control

Several investigators have tried to determine the lowest dosage of a drug that could be discriminated from no drug by using a paradigm in which rats discriminated progressively lower dosages until a dosage was reached at which DDs no longer could be maintained (Colpaert et al. 1980a; Overton 1979; Zenick and Goldsmith 1981). In other studies several groups of rats have been trained, each with a different dose of the training drug, to determine the lowest dose that could acquire control of D vs N DDs (Colpaert et al. 1980b; Greenberg et al. 1975; Overton 1964, 1982a). These experiments have shown that DDs can be maintained at dosages down to about one-tenth of those commonly employed in DD experiments.

In both SDL and sensory-context experiments, thresholds are high. Essentially, the literature contains an enumeration of drug and stimulus changes that are large enough to produce measurable impairment of memory retrieval, and a rather longer list of changes that are not large enough to produce such effects. The threshold dose capable of producing SDL is usually equal to or greater than the highest dose that can be used in D vs N DD experiments.

Theory suggests that moderately large changes in stimulus context will be required to produce memory retrieval failures, whereas after discrimination training animals should be able to discriminate much smaller differences between stimuli (Blough and Blough 1977). Although the literature contains surprisingly few empirical data on this topic, some reports suggest that the threshold for sensory discriminative control is perhaps an order of magnitude lower than the threshold for contextual retrieval effects (Riccio et al. 1966).

Similarly, the literature indicates that the threshold dosages capable of acquiring discriminative control in DD paradigms are considerably lower than the threshold dosages capable of producing measurable SDL effects (Overton 1982b). This author has argued that this datum (when firmly established) could provide a basis for definitively rejecting *all* of the neurological models for SDL and DDs in favor of a sensory model. Since neurological models are outside the scope of this chapter, the reader is referred elsewhere for a detailed presentation of this argument and the relevant data (Overton 1978b, 1982b).

1.6 Dynamic Range of Drug and Sensory Stimuli

The magnitude of the range of stimulus intensities that can be processed by many of the sensory systems is large, with the ratio of the maximum stimulus that can be processed without receptor damage to the threshold stimulus exceeding 1 000 000 in some modalities.

The range of drug dosages that can be discriminated is relatively small. The maximum dosage at which stimulus effects can be evaluated (via techniques based on instrumental behavior) is a dosage just below the one that will prevent instrumental behavior. This varies from task to task, and is higher in the T-maze shock-escape task than in most other tasks. Maximum discriminable doses and their relative discriminabilities have been determined in the T-maze for many drugs (Overton 1982a). The two-lever operant task produces the lowest estimates of threshold dosage obtained thus far, and these have been determined for a few drugs as described previously. For most drugs tested, the ratio of the maximum usable dose to the threshold dose approximates 100. There are a few exceptions. For example, scopolamine is discriminable over a dosage range of 5000 : 1. However, its discriminability is asymptotic over much of that range, suggesting that receptor saturation has occurred and that increasing doses do not produce increasingly strong stimuli (Overton 1977 a). Quite possibly, the relatively narrow dynamic range of drug stimuli directly reflects the properties of their transduction mechanisms, with drug receptors not having developed special mechanisms analogous to those that allow exteroceptive sensory receptors to process the very wide ranges of stimulus intensity that they encounter.

1.7 Occurrence of Peak Shift

After high-dose vs low-dose DD training, peak shift has been reported during test sessions with various doses of the training drug (Akins et al. 1980). However, after D vs N training, peak shift is usually not seen (e.g., White and Appel 1982). This may be an artifact caused by a "ceiling effect" in the vicinity of the training dosage. Additionally, it might be necessary to test with doses as high as twice the training dosage in order to see peak shift, and this is seldom done (but see Emmett-Oglesby et al. 1983). However, also note that in sensory discrimination paradigms, training to discriminate presence vs absence of a discriminative stimulus is not the paradigm that allows observation of peak shift. Hence peak shift in DD paradigms may become more frequently reported if high-dose versus low-dose training is more often employed.

1.8 Stimulus Intensity Dynamism

Stimulus intensity dynamism is commonly seen after presence vs absence discrimination training along an intensity dimension, i.e., rates tend to increase as stimulus intensity increases. This is not generally reported after D vs N DD training, probably for two reasons. First, increasing dosages (greater than the training dosage) usually have intrinsic rate-decreasing effects. Second, the percentage of drug lever responses is already at ceiling (100%) when the training dosage is reached. However, a recent report by Rauch and Stolerman (1986) does indicate stimulus intensity dynamism. These investigators trained rats to discriminate the simultaneous presence of nicotine plus midazolam from the absence of both drugs. Then rats were tested with various doses of each drug individually. During tests with midazolam, the percentage of drug-lever choices continued to increase with test doses considerably higher than the training dosage. In addition to suggesting stimulus intensity dynamism, their results show for the first time that a test drug that only partially recreates the stimulus effects of the training drug(s) may produce relatively complete responding on the drug lever if a sufficiently high test dose is employed. This finding has important ramifications as regards the interpretation of generalization test results.

2 Comparisons Across Qualitative Dimensions and Across Modalities

2.1 Degree of Generalization Between Different Types of Drugs – Qualitative Specificity

After D vs N DD training with a particular drug and dose, tests may be conducted using drugs that differ pharmacologically from the training drug. Such tests reveal very sharp generalization gradients. Typically the drug choice occurs only during tests with drugs that produce pharmacological effects similar to those of the training drug; all other drugs cause N choices. The effect is quite striking, and the results are analogous to those which one might expect if animals were trained using a visual discriminative stimulus and were then tested with the visual stimulus absent and with an auditory stimulus substituted for it. This type of DD experiment has been repeated over and over using different training drugs in successive groups of animals; lack of generalization to novel drugs has been observed irrespective of the training drug employed. The trained animals are said to exhibit high qualitative specificity, and can be used as an assay to detect the presence or absence of the actions of the training drug and its close pharmacological relatives. Indeed, the high qualitative specificity produced by D vs N training is the primary property which has made DD training a useful psychopharmacological assay procedure (Barry 1974; Colpaert and Slangen 1982; Lal 1977).

Since a high degree of qualitative specificity has been observed in a variety of DD training paradigms, it appears to reflect some intrinsic property of drug-induced stimulus control after D vs N training. An analogous effect in sensory discrimination training may be that reported by Jenkins and Harrison, who compared pigeons trained to discriminate presence versus absence of a tone with others that were simply trained to peck in the presence of tone. During subsequent tests, the generalization gradient along the frequency dimension was markedly sharper in the birds that had received tone/no-tone discrimination training than in the control group even though no discrimination training along the frequency dimension had taken place in either group (cited in Rilling 1977).

Recent DD reports suggest that the degree of specificity produced by D vs N training may depend on the type of drug used for training (Overton 1978 a). Three cases can be distinguished. (1) With several types of training drugs, specificity approximates the width of the pharmacological class from which the training drug is taken. This is the case for antimuscarinics, narcotic agonists, tetrahydrocannabinols, and antihistamines, among others. (2) In other cases, specificity appears to be higher, and not all drugs in the training drug's pharmacological class will mimic the training drug. This is true for cholinergics, antipsychotics, and narcotic mixed agonist/antagonists, among others. (3) In still other cases, specificity appears to be lower, as generalization occurs to drugs outside the training drug's pharmacological class. For example, anesthetics, benzodiazepines, muscle relaxants, and some anticonvulsants all tend to generalize to one another, perhaps because they share depressant effects. Other instances of generalization to drugs outside the training drug's class have been reported (see Overton 1984, 1987 for references). Obviously, in the preceding paragraph the term "pharmacological class" is used in its conventional sense to indicate a group of drugs that share a clinically relevant action (e.g., depressants, antimuscarinics, antidepressants).

An alternate interpretation of the same DD data is possible, as it may be the case that a particular D vs N training paradigm produces a fixed degree of specificity with all types of training drugs, and that some pharmacological classes include drugs with more diverse stimulus effects than others, thus allowing cross-generalization between all members in some classes but not in others. At this time we have no way to measure breadth of generalization except by counting the number of drugs to which generalization occurs. Such counting has led to the conclusion that generalization occurs to a greater or smaller portion of the drugs in the training drug's class (and sometimes to drugs in other classes) depending on the type of drug used for training. Whether this result should be interpreted as showing that some classes have more disparate stimulus effects than others, or as showing that specificity is higher after training in some classes than in others, cannot be determined at this time.

Only a few SDL studies have examined the consequences of testing with drugs different than the training drug. Their results showed that if the stimulus effects of a drug were strong enough to produce SDL generalization failures after a D–N state change, then failure of generalization was also observed during tests with drugs dissimilar to the training drug. This suggests that a moderately high degree of qualitative specificity may occur in SDL studies (Sachs et al. 1966).

2.2 Qualitative Specificity in Sensory Discrimination Studies

Having just discussed the degree of generalization among drugs that produce qualitatively dissimilar stimulus effects, the format of this paper indicates that we should next compare those findings to analogous results obtained in sensory discrimination studies. However, it soon becomes apparent that we do not know which variations in classical sensory stimuli are comparable to changes in test drugs. Suppose we train with pentobarbital vs no drug and test with scopolamine. Is this equivalent to training with a red light and testing with a green light? Alternately, is it equivalent to training with a red light and testing with a tone? We know so little about the nature of drug stimuli that we cannot even make an educated guess. Since our ignorance about the qualitative nature of drug stimuli is so complete, it appears that we should review the limited amount of evidence that is available regarding this issue, and then suggest experimental approaches that might provide improved information.

2.3 Reported Tests for Similarities Between Drug and Sensory Stimuli

Identification of specific drug stimuli and sensory stimuli that are equivalent (interchangeable) would provide a substantial enhancement to our knowledge about the qualitative nature of drug stimuli, and several investigators have argued that it might be possible through appropriate manipulations of the internal or external milieu to duplicate the stimulus conditions produced by drugs. There are at least four general types of sensory effects which can apparently be produced by drugs, any or all of which might be responsible for SDL and DDs. (1) Some drugs induce changes in peripheral organs, and these changes produce altered afferent input returning to the brain via the classical afferent pathways. For example, antimuscarinics reduce the flow of saliva and produce sensations of "dry mouth." Sedative drugs such as ethanol produce ataxia which in turn produces altered proprioceptive feedback. (2) Other drugs may modify the processing and perception of interoceptive or exteroceptive stimuli. Blurred or double vision is one example of such a drug-induced modification in sensory processing. Analgesia produced by narcotics provides a second example. (3) Some drugs may directly induce sensory stimuli such as visual hallucinations, tingling sensations, or itches, which may conceivably be referred to different parts of the body depending on what drug is administered. (4) Finally, some drugs may alter the organism's emotions, drive states, or arousal level. Have drug stimuli of any of these four types been identified?

The first type of drug stimulus – one produced by changes in peripheral organs which cause altered sensory feedback – has been tested on several occasions with negative results. For example, antimuscarinic drugs produce altered functioning in a variety of organs enervated by the autonomic nervous system. Tertiary antimuscarinic drugs act at both central and peripheral sites whereas quaternary antimuscarinics cross the blood-brain barrier less easily and thus act more exclusively at peripheral sites. After D vs. N DD training with scopolamine hydrobromide (which produces both central and peripheral actions), animals fail to generalize the D response to quaternary scopolamine compounds which produce only peripheral actions. Hence it appears that the peripheral actions of scopolamine are not responsible for its discriminable actions (Overton 1977 a). Other tests of this type have also been reported, with negative results (Downey 1975; Hazell et al. 1978; Overton 1964). In general, drugs which act only outside the central nervous system have been found to be much less discriminable than centrally acting drugs (Overton 1985, 1987). This has discouraged further attempts to identify specific peripheral sites at which discriminated drug cues might be produced.

The second possibility – that drug-induced alterations in sensory processing might mediate SDL and DDs – has also been tested. For example, Overton (1968 b) hypothesized that blurred vision might mediate the discriminable effects of pentobarbital. To test this possibility, he first blinded rats and then required them to learn a D vs. N discrimination in a T-maze. This discrimination was learned as rapidly by blind as by sighted rats, indicating that drug-induced alterations in visual stimuli were not a prerequisite for the establishment of the discrimination. In another experiment, sighted rats were required to discriminate pentobarbital vs N; these rats were the blinded, and training was continued. Only a transient disruption in discriminative control was noted at the time of blinding, suggesting that even in sighted rats alterations in visual perception do not mediate discriminative control (Overton 1968 b). In a similar vein, Overton hypothesized that pentobarbital vs N discriminations in a shock-escape T-maze task might be mediated by drug-induced analgesia, or at least by a drug-induced insensitivity to some of the consequences of electric shock. However, two pieces of evidence contradicted this hypothesis. First, after D vs. N training with high shock levels, undrugged rats could not be induced to make D choices by the application of low (less painful) shock intensities (Overton 1968 b). Secondly, after D vs. N training with pentobarbital, the D response did not occur during tests with morphine or other narcotic analgesics.

The third possibility – that drugs might directly induce sensations in classical modalities (e.g., itching, auditory sensations, visual hallucinations) has been minimally investigated, perhaps because it appears unlikely on the basis of human subjective reports. For some time, Winter (1975) pursued the hypothesis that hallucinogenic drugs might all have shared stimulus effects (which might in fact be visual hallucinations). However, his studies eventually showed a poor correlation between hallucinogenic activity in man and discriminability in rats. Subsequent studies have shown that several different types of discriminable effects are produced by the several types of drugs which have hallucinogenic actions in humans.

The fourth possibility – that drug stimuli may have effects comparable to those of emotions, drives, or CNS arousal states – raises two questions. First, do such CNS states have contextual or discriminative properties. Second, which if any of the altered CNS states that can be induced by nonpharmacological manipulations produce sensory effects comparable to those induced by drugs.

Regarding the first question, electroconvulsive shock (ECS) produces SDL, and ECS vs no ECS discriminations are robust. Alterations in hunger or thirst are discriminable and produce SDL. Electrical brain stimulation is discriminable. REM sleep deprivation can produce SDL, and learning that occurs during REM sleep is state dependent. These studies show that at least some alterations in CNS state have discriminable or contextual stimulus effects (Overton 1985, 1987).

With regard to the second issue, there are only a few studies that we can cite. Huang (1973) trained rats to press one lever after normal sleep and the second lever after REM sleep deprivation. This discrimination was learned, and subsequent tests showed that amphetamine would partially substitute for REM deprivation in rats that had slept normally and conversely that pentobarbital would cause some responses on the normal-sleep lever in REM-deprived rats. Although the drugs did not completely antagonize (or mimic) the stimulus effects of the sleep manipulations, the results suggested that the discriminated effects of REM deprivation and of normal sleep were to some degree overlapped by the stimulus effects of pentobarbital and/or amphetamine. In a related study, Schechter (1981) trained rats to discriminate pentobarbital vs amphetamine and then tested them with saline at various times of day. During mid-afternoon tests, saline caused responding predominantly on the pentobarbital lever, but during saline tests at 2:00 am the rats showed only 50% responses on that lever. Both of these studies suggest that the drug injections and the time of day (or REM deprivation) manipulations may have moved the animals along some shared sensory dimension (arousal level?) which provided at least part of the basis for discriminative control. In a separate series of experiments, Gardner et al. (1972) reported some similarities between the SDL effects of ECS, physostigmine, and scopolamine. However, Overton et al. (1976) were unable to obtain evidence for comparable effects using a DD paradigm.

Let us review. The problem under consideration is the fact that we do not know the qualitative nature of drug stimuli, and hence do not know what sensory manipulations will produce stimulus effects comparable to those produced by drugs. We reasoned that if drug stimuli had been shown to be interchangeable with any sensory stimuli, this would tell us something about the nature of drug stimuli. However, the results of our literature review have been largely negative. No sensory stimuli have thus far been identified which are directly interchangeable with drug stimuli, and only one or two nonpharmacological manipulations have been found which even interact with drug stimuli. This does not mean that drug stimuli cannot be mimicked by other manipulations. It only means that we have not yet identified instances of interchangeability between drug and sensory cues.

2.4 Methods for Studying the Relationship of Drug Stimuli to Sensory Stimuli

Obviously, our understanding of the relationship between drug stimuli and sensory stimuli would be enhanced if we had more information about the nature of drug stimuli, and specifically if we could identify nonpharmacologically induced stimuli which were similar to drug-induced stimuli. The preceding review suggests that this information may not be easy to obtain. Nonetheless it should be possible to make some progress if we make an effort; four types of experiments come to mind as potentially useful. First, after subjects learn D vs N discriminations, various nonpharmacological manipulations (NMs) could be tested to see if they would directly mimic the effects of the training drug. Second, after D vs. N DD training, test sessions could be performed during which D and NMs were simultaneously present to determine whether any interaction between NM-induced stimuli and drug stimuli could be detected. Third, subjects could be trained to discriminate NM vs N and then tested with various drugs (D1, D2, D3). Fourth, after NM vs. N discrimination training, test sessions could be conducted during which either NM or N were present in combination with one or more drugs. The second and fourth approaches appear to be more sensitive since they will detect even relatively weak interactions (synergism or antagonism) whereas the first and third methods will detect only instances in which the test condition rather completely mimics the training condition.

Nonpharmacological manipulations can interact with drug stimuli either by producing their own stimulus effects which interact with the drug stimuli, or by modifying drug actions so that the drug stimuli are altered. In the present context, only the first type of effect is of interest. This requires that we use NMs which produce discriminable effects. We previously listed several NMs that have been shown to produce SDL-like retrieval decrements or discriminative control, and other NMs with stimulus effects include changes in emotional state, changes in the time of day at which testing is conducted, kindled convulsions produced by low-level brain stimulation, and spreading depression (Overton 1984–1987).

2.5 Size of the Qualitative JND

Previously we described the normal degree of qualitative specificity produced by D vs N training as approximating the width of a pharmacological class. However, that fact is only indirectly related to the present question, which is to identify the smallest qualitative difference betwen the stimulus effects of two drugs (D1, D2) that can be discriminated when animals are explicitly trained to discriminate D1 vs D2 using dosages that produce equally strong discriminable effects. There is no published literature on this topic because there has been no accepted method for selccting equally discriminable dosages of two drugs that produce qualitatively different discriminable effects (but see Overton et al. 1986), and hence it has not been possible to select training drugs for use in such experiments. However, it seems likely that the qualitative JND, when determined, will be considerably smaller than the width of generalization observed during substitution tests conducted after D vs N training.

3 Experimental Paradigms Used To Study Control of Behavior by Sensory and Drug Stimuli

Similar experimental designs have been proposed to evaluate behavioral control by sensory and by drug stimuli. We will describe the advantages and limitations of a few of these methods.

3.1 One-Response 2 × 2 Design

This design normally includes four experimental groups as follows.

Group	Train	Test
1	$N \rightarrow R$	$N \rightarrow R$
2	$N \rightarrow R$	$D \rightarrow R$
3	$D \rightarrow R$	$N \rightarrow R$
4	$D \rightarrow R$	$D \rightarrow R$

A single response (R) is trained in either the drug (D) or no drug (N) condition and subsequently tested for retention in either the same or the alternate drug condition. Latency to respond or rate of responding during the test session is taken to indicate the degree of memory retrieval. The design has frequently been used to study drug-induced SDL. It is also used to study the effects of stimulus context on memory retrieval, in which case unique stimulus conditions (S1 and S2) are substituted for D and N in the design. For the study of drug SDL, this design has several serious limitations (Overton 1974; Swanson and Kinsbourne 1979). Perhaps the most severe is the fact that intrinsic rate-reducing effects of drugs can masquerade as an SDL retrieval deficit in the N–D group 2.

3.2 Two-Response Task with Only One Response Trained

This design uses the same four groups as the 2 × 2 go/no-go design. A behavioral task is used in which two responses are possible (e.g., R1 and R2 in a two-lever compartment) but only one of these responses is trained. During the subsequent tests, response accuracy, R1/(R1 + R2), instead of response occurrence is used as index of memory retrieval. This design has some advantage over the go/no-go 2 × 2 design because the intrinsic effects of many drugs produce a smaller reduction in response accuracy than in response rate. The design also has several obvious problems. In the N–D group, drug-induced disorganization of responding can mimic SDL. In the D–N group, the response must typically be somewhat "overtrained" to achieve criterion in the D condition, and subsequent removal of intrinsic drug effects after the D–N state change may enhance accuracy, making it difficult to see SDL effects.

3.3 Two-Response Task with Reversal Learning

This design differs from the preceding in that the second response is trained after the change in drug state, as follows.

Group	Training	Reversal Training
1	N → R1	N → R2
2	N → R1	D → R2
3	D → R1	N → R2
4	D → R1	D → R2

Slow acquisition of R2 in phase two indicates retrieval of R1 (which interferes with R2), whereas rapid phase 2 acquisition is taken to indicate lack of retrieval (SDL). The advantage of this design lies in the fact that it may be more sensitive than designs in which simple retrieval of R1 is used to measure SDL. For the study of drug SDL, the design also has several problems. In the N–D group, slow relearning may indicate either retrieval of R1 (no SDL) or drug-induced depression of learning. The apparent solution to this problem is to compare phase 2 performance in the N–D group with that in the D–D group. However, in the D–D group, the initial learning of R1 may have required overtraining to overcome the depressant drug effects, or may have been poorly consolidated, or may have induced behavioral tolerance. Any of these effects will contaminate the comparison.

3.4 Two-Response Task with One Response Learned in Each State

This paradigm can be diagramed as follows.

Group	Training	Retraining	Tests
1	N → R1	D → R2	N → ?, D → ?
2	D → R1	N → R2	D → ?, N → ?

The animals learn one response in state one, a second response in state two, and are then tested for state-appropriate retrieval in each state. With this design, retrieval of the state-appropriate response only needs to be substantially stronger than retrieval of the alternate response in order to produce predominantly state-appropriate responding – the amnesia does not have to be complete. Hence this design is almost certainly more sensitive than many of the preceding designs, and able to detect smaller SDL effects. Drug effects on acquisition and/or retrieval can be confounded with SDL effects as in preceding designs, although perhaps less seriously so. Additionally, SDL cannot easily be distinguished from rapidly acquired discriminative control.

3.5 Two-Response Drug Discrimination Paradigm

In this natural extension of the preceding design, repeated D and N training sessions are conducted with one response reinforced during D sessions and the alternate response during N sessions. Typically D and N sessions alternate. When this paradigm is used, rats can learn to discriminate drug stimulus effects that are much too weak to produce measurable SDL effects with any of the preceding designs. This allows utilization of substantially reduced drug dosages, and intrinsic drug effects (typically impairing discriminative control) are correspondingly reduced. Probably because of the reduced strength of intrinsic drug effects, this paradigm has been capable of yielding much more orderly and quantitatively replicable data than any of the preceding designs.

After prolonged training, stable discriminative control is usually obtained and the trained subjects may be tested under a variety of conditions including the following.

- Tests to determine dose response curves and time course
  Training Drug (TD) at various doses
  TD at various time intervals after injection
- Tests to investigate drug interactions
  TD plus putative antagonist
  TD plus transmitter manipulator
  TD plus synergistic drug
  Metabolite of TD
  Proposed precursor(s) of TD
  One or more novel drugs

Such test sessions provide information about the stimulus effects of the test condition (relative to those of the training conditions) and are widely employed to obtain information about the effects of centrally acting drugs.

3.6 Conclusions Regarding Designs

The preceding designs work moderately well for studying the effects of sensory context or sensory discriminative stimuli. At first glance, it might appear that the designs could be simply adopted without change for the study of drug stimuli. However, this has not been the case, and most of the designs have worked rather poorly for studying the effects of drug-induced stimuli. Drug stimuli are more difficult to study for two reasons: (1) They are more difficult to turn on and off from trial to trial and (2) drugs produce a variety of effects in addition to their stimulus effects which can make the results difficult to interpret. These include impairment of acquisition, impairment of long-term consolidation of memories, interference with responding which may (in effect) require overtraining, behavioral tolerance, and drug-induced impairment of retrieval (as distinct from SDL impairment). Interpretation of results obtained with drug stimulus manipulations in most of the preceding designs has posed formidable problems due to the possibilities for confounding the effects of changes in stimulus conditions with other intrinsic drug effects (Overton 1974). Only the DD design has been able to produce consistent, orderly, and replicable results. The conclusion appears warranted that it is extremely difficult to obtain reliable and credible data regarding the contextual stimulus effects of drugs because experimental designs adequate for this purpose have not been developed.

4 Other Comparisons of Drug and Stimulus Control

4.1 Masking of One Stimulus by a Second Concurrently Presented Stimulus

The term "masking" is used here to convey the same meaning that it denotes in sensory psychophysics – a situation in which one stimulus prevents the perception of a second concurrently presented stimulus. There is a large literature on masking effects between sensory stimuli which will not be reviewed except to summarize that cross-modality masking is observed less frequently than is masking between stimuli in the same sensory modality. It appears that if masking occurs in the DD preparation, it should be detectable during tests when a putative masking drug and the training drug are simultaneously administered. After D vs N DD training, rats frequently have been tested while simultaneously drugged with the training drug and with another drug; such tests usually have been performed to identify antagonists of the training drug. In most published data, the *only* occasions when a second drug prevented the subject from detecting the presence of the training drug and performing the D response were those in which the second drug probably pharmacologically antagonized or blocked the actions of the training drug so that its discriminable effects were not present. Thus, the results of these experiments suggest that masking of drug stimuli seldom if ever occurs (but see Young et al., this volume).

However, the conclusion that masking does not occur with drug stimuli must be regarded as only tentatively established, since experimental designs can be envisaged that should be considerably more sensitive to masking than are any of the accidental tests for its occurrence presently in the literature. Apparently, masking will be most likely if a drug with strong sensory effects is used to mask a training drug with weak, slowly discriminated effects. Additionally, if the sensory literature is a useful guide, it may be that drugs producing similar sensory effects will mask one another more effectively than do drugs that produce markedly dissimilar effects. Very few tests using such combinations of drugs have been reported. Finally, there are a few reported data which suggest that masking may have occurred, and possibly these data indicate instances of masking (Overton, 1984). If so, they constitute a first step toward enumerating the conditions under which masking will occur.

4.2 Overshadowing

In sensory conditioning studies, this term is used to describe a situation where two nominal conditioned stimuli (S1 and S2) are simultaneously present during training but where most associative strength becomes attached to one stimulus at the expense of the other. The stimulus which acquires stronger control of behavior (stronger associative loadings) is described as more salient. In the DD paradigm, Järbe et al. (1983a, b) have reported that the discriminative conditions pentobarbital versus N will overshadow the conditions light versus dark.

In the drug stimulus area, a concept closely comparable to salience is that of the strength or degree of discriminability of the stimulus effects of a drug (sometimes referred to as its discriminative efficacy). Consideration of differences between the degree of discriminability of drugs frequently enters into the design of DD experiments, or into the interpretation of their results (Overton et al. 1986). To date, no one has tested the possibility that two drugs which have an equal degree of discriminability (when used individually as training drugs in the D vs N paradigm) may have differing degrees of salience when both are simultaneously injected (thus allowing overshadowing to occur). Hence, for the moment we are spared the complexity of having to consider the possibility that two drugs that are equally discriminable when used independently may exert unequal control over behavior when used conjointly.

4.3 Blocking

The term "blocking" refers to the result diagrammed in group 1 of Table 1 where the stimulus S2 fails to acquire discriminative control (shared with S1) during phase 2 of training. Presumably this occurs because S1 has already acquired control of responding and (by itself) adequately predicts the reinforcement contingencies during phase 2 so that the animal can ignore S2. Such results frequently have been reported in sensory control experiments.

In the DD paradigm, the blocking phenomenon might be observed if the design in groups 2 and 3 in Table 1 was employed. In group 2, the interesting test is that with D2 during which the no drug response (R2) is observed even though D2 has been used as a cue for R1 during phase 2 of training. Group 3 completes

Table 1. Experimental designs for testing the occurrence of blocking, adapted to the drug discrimination paradigm

	Group 1	Group 2	Group 3	Group 4	Group 5
Training (phase 1)	S1 → R1 N → R2	D1 → R1 N → R2	D2 → R1 N → R2	S → R1 N → R2	D → R1 N → R2
Training (phase 2)	S1 + S2 → R1 N → R2	D1 + D2 → R1 N → R2	D1 + D2 → R1 N → R2	S+D → R1 N → R2	S+D → R1 N → R2
Possible test results	S1 → R1 S2 → R2 N → R2	D1 → R1 D2 → R2 N → R2	D1 → R2 D2 → R1 N → R2	S → R1 D → R2 N → R2	S → R2 D → R1 N → R2

the balanced design begun in group 2, and tests whether drugs D1 and D2 were used at dosages sufficiently equated so that D2 could be blocked by D1 and vice versa. Studies using the paradigm described in groups 2 and 3 have not been reported.

Groups 4 and 5 illustrate a design that can be used to test for blocking between a sensory stimulus and a drug stimulus. A similar design has recently been employed by Järbe et al. (1983 b), who showed that initial discriminative training with pentobarbital blocked the subsequent acquisition of discriminative control by the illumination conditions light vs. dark. Conversely, discriminative light vs dark training would partially block the subsequent acquisition of control by pentobarbital vs no drug. The results of this study add another significant entry to the list of paradigms in which drug stimuli and sensory stimuli behave analogously.

4.4 Familiarization or Habituation

In sensory conditioning paradigms simple exposure to a stimulus has been observed to impede the subsequent acquisition of conditioned or discriminative control, presumably by habituating the animal to the to-be-conditioned stimulus. However, this effect has not been observed in studies where animals were exposed to drug on several occasions before the beginning of D vs. N training. Such prior exposure to drug has neither facilitated nor impaired the subsequent acquisition of DDs (Hinderliter 1978; Järbe and Henriksson 1973; Järbe and Holmgren 1977; Kilbey et al. 1971; McKim 1976; Overton 1972).

4.5 Overtraining

In DD and sensory discrimination paradigms, continued training progressively strengthens discriminative control until asymptotic performance is achieved. However, overtraining abolishes SDL retrieval deficits. In other words, a well-learned response will generalize successfully across changes in drug state that will disrupt performance of a response that is not overtrained. Most studies demonstrating this effect have varied the amount of training carried out before a state change occurred (Eich and Birnbaum 1982; Iwahara and Noguchi 1972, 1974; Modrow et al. 1982). It has also been shown that if two responses of unequal difficulty are learned by an animal before a state change occurs, the easier (long since learned) response will generalize whereas the more difficult (recently mastered) response will fail to generalize across the change in drug state (Bliss 1973). In the entire literature reporting SDL effects, all positive findings involve recently mastered responses; there are no reports in which well-established, overly trained habits failed to generalize across a change in drug state. It would be interesting to compare these findings to the literature regarding the influence of overtraining on the effect of changes in environmental sensory stimulus context.

4.6 Interactions Between Exteroceptive Contextual Change and Drug State Change

If a change in drug state and a change in sensory context occur simultaneously, both changes will effect retrieval. In an animal preparation, Duncan (1979) found that the SDL effects of ethanol were insignificant if retrieval testing occurred in the same compartment where the animals had been trained, but were obvious if the animals were tested in a different compartment. Apparently the change in contextual cues acted to unmask an SDL effect that was too weak to be seen in animals tested without a concurrent environmental contextual change (also see Connelly et al. 1973, 1975, 1977). In human subjects, a number of studies have shown that "category cues" could abolish SDL. In these experiments, subjects learned a list of words (red, green, daffodil, chrysanthemum). If a simple test for retrieval was performed after a drug state change, SDL effects were observed. However, if category cues were provided at the time of retrieval ("Name the flowers that you recall") then SDL did not occur (Eich 1977, 1980; Eich et al. 1975; Petersen 1977). Overall, it appears that sensory events and drug state manipulations conjointly determine the efficiency of memory retrieval. Whether the effects of drug and sensory stimuli simply summate, or interact in a more complex fashion, is not known.

4.7 Learning Set Phenomena in DD and SDL Experiments

The large literature on the development of learning sets during sensory discrimination training will not be reviewed here. Similar studies have been carried out in drug stimulus paradigms. In a T-maze DD task, Overton (1971) demonstrated that light vs dark discriminations could be more rapidly learned if rats had previously mastered an ethanol vs N discrimination in the same task. This suggested that a learning set established by DD training could facilitate subsequent acquisition of sensory discriminations. Bliss (1974) required monkeys to learn a series of visual discriminations while drugged. Each individual discrimination was not generalized to the no drug state, but after several had been learned the resulting learning set did generalize to the no drug condition. Modrow et al. (1982) obtained similar results. More studies will obviously be required to obtain a complete understanding of learning set phenomena in drug stimulus paradigms.

4.8 Feature-Positive and Feature-Negative Discrimination Learning

The difference between the ease of learning feature-positive and feature-negative discriminations has recently attracted considerable interest in the sensory discrimination literature, and several theoretical explanations for this effect have been advanced (Hearst 1983; Jenkins and Sainsbury 1970). Feature-positive learning refers to a situation in which a discriminative stimulus is added to a background context, and feature-negative learning to the situation in which the discriminative event is the removal of a stimulus from the context on reinforced trials. Feature-positive learning typically occurs more rapidly than feature-negative learning.

It appears that D vs N training in a go/no-go task will comprise either feature-positive or feature-negative learning, depending on whether the animal is reinforced for responding in the D state or in the N state. A few years ago, Winter reported several such go/no-go DD studies. However, since his interest was primarily in the results of substitution tests, no data regarding the acquisition of these discriminations were reported (Winter 1975, 1977). Harris and Balster (1971) also trained rats in an FR vs extinction D vs N discrimination, but again acquisition data were not presented. Because of the differences between feature-positive and feature-negative learning, it appears that systematic investigation of the applicability of these concepts to the drug discrimination situation might provide a useful enhancement to our knowledge about behavioral control by drug stimuli. Unfortunately most DD studies have used two-lever go/go tasks whereas feature-positive/feature-negative studies use one-response go/no-go tasks. Hence data from the majority of DD studies appear to be irrelevant to this issue.

4.9 Correlation Between DD Results and Human Subjective Experience

Recreational, clinical, and experimental drug use in humans have yielded anecdotal and experimental reports about the strength and nature of the subjective effects of drugs, including hallucinations, changes in affect, and assorted pleasant and unpleasant side effects. One can reasonably ask how well these data about the subjectively reportable stimulus effects of drugs correspond with the results of DD studies. As regards qualitative differences between drugs, the correlation may be good. Although many cases have not been tested, several of the drugs which cross-substitute in animal DD paradigms also produce similar subjective effects in human subjects; for example, all morphine-like drugs produce similar subjective effects in humans and similar discriminable effects in rats (Colpaert 1978). However, quantitative comparisons of the strength of drug stimulus control and the strength of subjective drug effects suggest a lower correlation. For example, chlorpromazine has robust subjective effects in humans (Stitzer et al. 1981) whereas it exerts only weak discriminative control in rats (Overton 1982a). Data such as these suggest that subjective drug effects in humans and discriminable drug effects in animals may not be entirely isomorphic.

5 Conclusion

Fifteen years ago, a lively debate was underway between investigators who believed that state-dependent learning reflected the operation of a sensory mechanism and others who believed it resulted from some other unspecified mechanism not involving sensory control of behavior. The antecedent events which lead to this debate have been reviewed elsewhere (Overton 1978b, 1982b, 1985, 1987). Starting in the 1960s and continuing up to the present, an impressive array of parallelisms between the laws describing control of behavior by sensory and by drug stimuli have been demonstrated. Even though no single experiment has proved that a sensory mechanism underlies DDs and SDL, and some results discordant with this interpretation do exist, nonetheless there is overall an impressively high correlation between the properties of discriminative and contextual control by sensory stimuli and by drug-induced states (Overton 1987). Not surprisingly, scientific opinion has shifted in accord with these facts, and there are now few investigators who would argue against a sensory interpretation of SDL and DDs.

If the issue of mechanism is solved, one might now hope to see a very useful cross-fertilization of ideas between psychologists interested in the mechanisms and properties of discrimination learning and pharmacologists who have provided a whole host of new sensory dimensions to be explored. Unfortunately, this interaction is not happening to a very significant degree. Sensory psychologists have shown little interest in exploring the properties of discriminative control by drug states. Pharmacologists who employ the DD technique have shown little interest in investigating the psychological processes that presumably underlie and determine the properties of the paradigm. In this review several areas have been noted where substantial gaps exist in our knowledge about the nature and properties of drug-induced stimuli. Hopefully, in the not too distant future, an interaction between psychological and pharmacological concepts and methods will lead to research which provides at least part of the missing information.

Acknowledgments. This paper is dedicated to the memory of David K. Bliss, who is recently deceased. David was a careful and thoughtful investigator who had both the wisdom to recognize that neurophysiological investigations of the mechanisms underlying state-dependent learning were important and the stamina and skills which allowed him to carry out such studies. His passing is a loss to all of us.

Preparation of this manuscript was supported in part by NIDA Grant DA02403 and NIMH Grant MH25136.

References

Akins FR, Gouvier WD, Lyons JE (1980) Stimulus control along a drug-dose dimension. Bull Psychonomic Soc 15:33–34

Balster RL (1970) The effectiveness of external and drug produced internal stimuli in the discriminative control of operant behavior. Doctoral dissertation, University of Houston, Houston

Barry H III (1974) Classification of drugs according to their discriminable effects in rats. Fed Proc 33:1814–1824

Barry H III (1978) Stimulus attributes of drugs. In: Anisman H, Bignami G (eds) Psychopharmacology of aversively motivated behavior. Plenum, New York, pp 455–485

Bliss DK (1973) Dissociated learning and state-dependent retention induced by pentobarbital in rhesus monkeys. J Comp Physiol Psychol 84:149–161

Bliss DK (1974) Theoretical explanations of drug-dissociated behaviors. Fed Proc 33:1787–1796

Bliss DK, Sledjeski M, Leiman A (1971) State dependent choice behavior in the rhesus monkey. Neuropsychologia 9:51–59

Blough D, Blough P (1977) Animal psychophysics. In: Honig WK, Staddon JER (eds) Handbook of operant behavior. Prentice Hall, Englewood Cliffs, pp 514–539

Colpaert FC (1978) Discriminative stimulus properties of narcotic analgesic drugs. Pharmacol Biochem Behav 9:863–887

Colpaert FC, Janssen PAJ (1982) Factors regulating drug cue sensitivity: the effects of dose ratio and absolute dose level in the case of fentanyl dose-dose discrimination. Arch Int Pharmacodyn Ther 258:283–299

Colpaert FC, Slangen JL (eds) (1982) Drug discrimination: applications in CNS pharmacology. Elsevier, Amsterdam
Colpaert FC, Niemegeers CJE, Janssen PAJ (1980a) Factors regulating drug cue sensitivity: limits of discriminability and the role of a progressively decreasing training dose in fentanyl-saline discrimination. J Pharmacol Exp Ther 212:474–480
Colpaert FC, Niemegeers CJE, Janssen PAJ (1980b) Factors regulating drug cue sensitivity: the effect of training dose in fentanyl-saline discrimination. Neuropharmacology 19:705–713
Conger JJ (1951) The effects of alcohol on conflict behavior in the albino rat. Q J Stud Alcohol 12:1–29
Connelly JF, Connelly JM, Epps JO (1973) Disruption of dissociated learning in a discrimination paradigm by emotionally-important stimuli. Psychopharmacologia 30:275–282
Connelly JF, Connelly JM, Phifer R (1975) Disruption of state-dependent learning (memory retrieval) by emotionally-important stimuli. Psychopharmacologia 41:139–143
Connelly JF, Connelly JM, Nevitt JR (1977) Effect of foot-shock intensity on amount of memory retrieval in rats by emotionally-important stimuli in a drug-dependent learning escape design. Psychopharmacology 51:153–157
Downey DJ (1975) State-dependent learning with centrally and noncentrally active drugs. Bull Psychonom Soc 5:281–284
Duncan PM (1979) The effect of external stimulus change on ethanol-produced dissociation. Pharmacol Biochem Behav 11:377–381
Eich JE (1977) State-dependent retrieval of information in human episodic memory. In: Birnbaum IM, Parker ES (eds) Alcohol and human memory. Erlbaum, Hillsdale, pp 141–157
Eich JE (1980) The cue-dependent nature of state-dependent retrieval. Memory Cogn 8:157–173
Eich JE, Birnbaum IM (1982) Repetition, cuing, and state-dependent memory. Memory Cogn 10:103–114
Eich JE, Weingartner H, Stillman RC, Gillin JC (1975) State-dependent accessibility of retrieval cues in the retention of a categorized list. J Verbal Learn Verbal Behav 14:408–417
Emmett-Oglesby MW, Wurst M, Lal H (1983) Discriminative stimulus properties of a small dose of cocaine. Neuropharmacology 22:97–101
Gardner EL, Glick SD, Jarvik ME (1972) ECS dissociation of learning and one-way cross-dissociation with physostigmine and scopolamine. Physiol Behav 8:11–15
Girden E, Culler EA (1937) Conditioned responses in curarized striate muscle in dog. J Comp Psychol 23:261–274
Greenberg I, Kuhn DM, Appel JB (1975) Behaviorally induced sensitivity to the discriminable properties of LSD. Psychopharmacologia 43:229–232
Harris RT, Balster RL (1971) An analysis of the function of drugs in the stimulus control of operant behavior. In: Thompson T, Pickens R (eds) Stimulus properties of drugs. Appleton-Century-Croft, New York, pp 111–132
Hazell P, Peterson DW, Laverty R (1978) Inability of hexamethonium to block the discriminative stimulus property of nicotine. Pharmacol Biochem Behav 9:137–140
Hearst E (1983) Stimulus relationships and feature selection in learning and behavior. In: Roitblat HL, Bever TG, Terrace HS (eds) Animal cognition. Erlbaum, Hillsdale, pp 51–88
Hinderliter CF (1978) Hypothermia: amnesic agent, punisher, and conditions sufficient to attenuate amnesia. Physiol Psychol 6:23–28
Holmgren B (1964) Conditional avoidance reflex under pentobarbital. Bol Inst Estud Med Biol 22:21–38
Huang JT (1973) Amphetamine and pentobarbital effects on the discriminative response control by deprivation of rapid eye movement sleep (REMS). Fed Proc 32:786
Iwahara S, Noguchi S (1972) Drug-state dependency as a function of overtraining in rats. Jpn Psychol Res 14:141–144
Iwahara S, Noguchi S (1974) Effects of overtraining upon drug-state dependency in discrimination learning in white rats. Jpn Psychol Res 16:59–64
Järbe TUC, Henriksson BG (1973) Open-field behavior and acquisition of discriminative response control in delta-9-THC tolerant rats. Experientia 29:1251–1253
Järbe TUC, Holmgren B (1977) Discriminative properties of pentobarbital after repeated noncontingent exposure in gerbils. Psychopharmacology 53:39–44

Järbe TUC, Laaksonen T, Svensson R (1983 a) Influence of exteroceptive contextual conditions upon internal drug stimulus control. Psychopharmacology 80:31–34
Järbe TUC, Svensson R, Laaksonen T (1983b) Conditioning of a discriminative drug stimulus: overshadowing and blocking like procedures. Scand J Psychol 24:325–330
Jenkins HM, Sainsbury RS (1970) Discrimination learning with the distinctive feature on positive or negative trials. In: Mostofsky D (ed) Attention: contemporary theory and analysis. Appleton-Century-Crofts, New York, pp 239–275
Kilbey MM, Harris RT, Aigner TG (1971) Establishment of equivalent external and internal stimulus control of an operant behavior and its reversal. Proc Am Psychol Assoc 6:767–768
Lal H (ed) (1977) Discriminative stimulus properties of drugs. Plenum, New York (Advances in behavioral biology, vol 22)
McKim WA (1976) The effects of pre-exposure to scopolamine on subsequent drug state discrimination. Psychopharmacology 47:153–155
Modrow HE, Salm A, Bliss DK (1982) Transfer of a learning set between drug states in monkeys. Psychopharmacology 77:37–42
Overton DA (1964) State-dependent or "dissociated" learning produced with pentobarbital. J Comp Physiol Psychol 57:3–12
Overton DA (1966) State-dependent learning produced by depressant and atropine-like drugs. Psychopharmacologia 10:6–31
Overton DA (1968 a) Dissociated learning in drug states (state-dependent learning). In: Efron DH, Cole JO, Levine J, Wittenborn R (eds) Psychopharmacology, a review of progress, 1957–1967. US Government Printing Office, Washington DC, pp 918–930
Overton DA (1968 b) Visual cues and shock sensitivity in the control of T-maze choice by drug conditions. J Comp Physiol Psychol 66:216–219
Overton DA (1971) Discriminative control of behavior by drug states. In: Thompson T, Pickens R (eds) Stimulus properties of drugs. Appleton-Century-Crofts, New York, pp 87–110
Overton DA (1972) State-dependent learning produced by alcohol and its relevance to alcoholism. In: Kissen B, Begleiter H (eds) The biology of alcoholism II: physiology and behavior. Plenum, New York, pp 193–217
Overton DA (1974) Experimental methods for the study of state dependent learning. Fed Proc 33:1800–1813
Overton DA (1977 a) Discriminable effects of antimuscarinics: dose response and substitution test studies. Pharmacol Biochem Behav 6:659–666
Overton DA (1977 b) Comparison of ethanol, pentobarbital, and phenobarbital using drug vs drug discrimination training. Psychopharmacology 53:195–199
Overton DA (1978 a) Status of research on state dependent learning – 1978. In: Colpaert FC, Rosecrans JA (eds) Stimulus properties of drugs: ten years of progress. Elsevier, Amsterdam, pp 559–562
Overton DA (1978 b) Major theories of state dependent learning. In: Ho BT, Richards DW III, Chute DL (eds) Drug discrimination and state dependent learning. Academic, New York, pp 283–318
Overton DA (1979) Drug discrimination training with progressively lowered doses. Science 205:720–721
Overton DA (1982 a) Comparison of the degree of discriminability of various drugs using the T-maze drug discrimination paradigm. Psychopharmacology 76:385–395
Overton DA (1982 b) Memory retrieval failures produced by changes in drug state. In: Isaacson RL, Spear NE (eds) The expression of knowledge, neurobehavioral transformations of information into action. Plenum, New York, pp 113–139
Overton DA (1984) State dependent learning and drug discriminations. In: Iversen LL, Iversen SD, Snyder SH (eds) Handbook of psychopharmacology, vol 18. Plenum, New York, pp 59–127
Overton DA (1985) Contextual stimulus effects of drugs and internal states. In: Balsam PD, Tomie A (eds) Context and learning. Erlbaum, Hillsdale, pp 357–384
Overton DA (1987) Applications and limitations of the drug discrimination method. In: Bozarth MA (ed) Methods of assessing the reinforcing properties of abused drugs. Springer-Verlag, New York, pp 291–340

Overton DA, Hayes MW (1984) Optimal training parameters in the two-bar fixed-ratio drug discrimination task. Pharmacol Biochem Behav 21:19–25
Overton DA, Ercole MA, Dutta P (1976) Discriminability of the postictal state produced by electroconvulsive shock. Physiol Psychol 4:207–212
Overton DA, Leonard WR, Merkle DA (1986) Methods for quantifying the discriminable effects of drugs. Neurosci Biobehav Rev 10:251–263
Petersen RC (1977) Retrieval failures in alcohol state-dependent learning. Psychopharmacology 55:141–146
Rauch RJ, Stolerman IP (1986) Interactions of midazolam with drugs acting on the GABA system tested through discriminative effects in rats. Paper read at European Behavioral Pharmacology Society workshop on transduction mechanisms for drug stimuli. Beerse, Belgium, July 1986
Riccio DC, Urda M, Thomas DR (1966) Stimulus control in pigeons based on proprioceptive stimuli from floor inclination. Science 153:434–436
Rilling M (1977) Stimulus control and inhibitory processes. In: Honig WK, Staddon JER (eds) Handbook of operant behavior. Prentice-Hall, New Jersey, pp 432–480
Sachs E, Weingarten M, Klein NW Jr (1966) Effects of chlordiazepoxide on the acquisition of avoidance learning and its transfer to the normal state and other drug conditions. Psychopharmacologia 9:17–30
Schechter MD (1981) Rapid acquisition of a two-drug discrimination: time of day effect upon saline state. Pharmacol Biochem Behav 14:269–271
Schiffman HR (1982) Sensation and perception. Wiley, New York
Schuster CR, Balster RL (1977) Discriminative stimulus properties of drugs. In: Thompson T, Dews PB (eds) Advances in behavioral pharmacology, vol 1. Academic, New York, pp 85–138
Spear NE, Smith GJ, Bryan RG, Gordon WC, Timmons R, Chiszar DA (1980) Contextual influences on the interaction between conflicting memories in the rat. Animal Learning Behav 8:273–281
Stitzer ML, Griffiths RR, Bigelow GE, Liebson I (1981) Human social conversation: effects of ethanol, secobarbital and chlorpromazine. Pharmacol Biochem Behav 14:353–360
Swanson JM, Kinsbourne M (1979) State-dependent learning and retrieval: methodological cautions and theoretical considerations. In: Kihlstrom JF, Evans FJ (eds) Functional disorders of memory. Erlbaum, Hillsdale, pp 275–299
White FJ, Appel JB (1982) Training dose as a factor in LSD-saline discrimination. Psychopharmacology 76:20–25
Winter JC (1975) The effects of 2,5-dimethoxy-4-methylamphetamine (DOM), 2,5-dimethoxy-4-ethylamphetamine (DOET), *d*-amphetamine, and cocaine in rats trained with mescaline as a discriminative stimulus. Psychopharmacologia 44:29–32
Winter JC (1977) Morphine and ethanol as discriminative stimuli: absence of antagonism by *p*-chlorophenylalanine methyl ester, cinanserin, or BC-105. Psychopharmacology 53:159–163
Zenick H, Goldsmith M (1981) Drug discrimination learning in lead-exposed rats. Science 212:569–571

Discrimination Strategies in Animal Psychophysics and Their Role in Understanding Sensory Receptor Function

W. C. Stebbins, D. W. Smith, and D. B. Moody

Contents

Abstract

Discrimination is defined in a restricted sense here as a precise and specifiable relation between stimuli and responses. Experimenter control of stimulation is one significant feature of this relation. It is suggested that there may be some important differences between discrimination of drugs and exteroceptive stimulation. However, a discussion of current strategies in animal psychophysics might uncover similarities between these two discrimination formats useful in formulating questions and designing future research. Traditionally animal psychophysics has focused on basic questions of acuity and sensitivity (threshold). Beyond their obvious value in the study of comparative sensory function and evolutionary development, these experiments have provided particularly important new insights into the understanding of peripheral sensory transduction and processing and their relation to perception. Our experiments have been carried out in animals whose hearing has been impaired by drugs and cryolesions, and relations have been sought between the subsequent hearing loss and the histopathological changes occurring in the auditory receptor cells of the inner ear. But threshold sensitivity is only one of many perceptual dimensions and tells us nothing of the discriminations that are made among stimuli which are clearly above minimum detectable levels. For example, discriminations occur along stimulus dimensions such as wavelength of light, sound pressure, acoustic frequency, and so on. Sensory systems also permit the accurate location of the source of stimulation at a distance, the selection of certain frequencies or wavelengths and the rejection of others, and finally the discrimination among very complex but biologically useful signals such as speech. Prototypical experiments are described; the results suggest significant and orderly relations with peripheral sensory processing mechanisms. The power of operant behavior to examine perception from a variety of perspectives is discussed.

Primate Laboratory, Kresge Hearing Research Institute, and Department of Psychology, University of Michigan, Ann Arbor, MI 48109, USA.

Psychopharmacology Series 4
Transduction Mechanisms of Drug Stimuli
Editors: Colpaert, Balster

1 Introduction

We partition the world in an almost infinite variety of ways, many of which depend on how our sensory systems extract information from the external (or internal) environment. The partitioning to which we refer here is largely unlearned, although it is most often an important part of our phylogenetic and ontogenetic histories. The term most often used is discrimination, which is borrowed from the common language and is far too broad and inclusive for our purposes. We use it in a restricted sense here to refer to the normally very precise relation between those measurable energy changes in the environment and measurable changes in a discrete behavioral response. The emphases are on relation, measurable, and behavioral. The relation depends on our ability to demonstrate that the behavior is specifically determined by those energy changes in particular and is not a consequence of other variables. If it is, then they too must be specified and the energy changes quantified. We are interested in a behavioral response from the intact organism, not a physiological response indicative of evoked activity in the afferent pathway.

Of course, learning is involved (discrimination training) when, in order to answer the experimental question, we require that the organism attend to the form and specific dimension of stimulation under analysis. In subjects incapable of the complexities of human language, operant conditioning procedures serve to inform them of those stimulus features to which they must attend so that we, as experimenters, may evaluate the way in which they partition or discriminate between those features. It is the actual discrimination of sensory stimulation, then, that depends, for the most part, on the energy changes or stimuli themselves and on the bottom-up processing mechanisms in the peripheral and central nervous system rather than on learning per se.

Perhaps the simplest or limiting case of a discrimination, in the restricted definition of the term, is in the detection of stimulation. For example, at specifiable sound pressure levels we hear sound, whereas at lower sound levels we hear nothing. The discrimination is between the presence and absence of stimulation. We refer to threshold as the minimum detectable energy level to which we can respond. The stimulus dimension is an intensive one. Above threshold, at detectable levels, there are various stimulus dimensions or continua that we can partition or along which we can discriminate. These include pressure, intensity, frequency, wavelength, and so on. The minimum discriminable difference refers to the smallest change along any of these continua to which we can respond.

It is not yet clear whether the behavioral discrimination of drugs or of certain other interoceptive events can be considered part of the same class of phenomena that we have described here. There are obvious similarities but there are also important differences that may warrant a separate classification for these two forms of discrimination. For example, in the usual manner of administering a drug, certain elements of stimulus control are lacking. Stimulus control is used here in the sense of experimenter control over the various parameters of stimulation. Stimulus duration is perhaps the most obvious example of a parameter not under control; the drug is on – that is, it is in the system – and its duration or stay is a function of variables not directly under the control of the experimenter. It can be difficult then to specify a "tight" relation between drug and behavior in the same way that we specify that relation between stimulus and response in a perceptual experiment where exposure to stimulation is usually under experimenter control.

There are other examples that may lead us to doubt that drug discrimination and sensory discrimination share enough common properties that it would be useful to group them together. Rather than engage that question further, however, it might be more fruitful here in the context of this book to describe some of the current theoretical and methodological issues in, and some of the findings of, animal psychophysics, particularly those related to discrimination, for in so doing it may be possible to uncover features common to both areas of inquiry. If new directions for research are suggested, then the attempt will have been worthwhile. It is significant in animal (or human) psychophysics that almost all of the research has been concerned with auditory and visual perception. Control and specification of stimuli still present a major problem in the study of the chemical senses. There may be instructive parallels to the study of drug discrimination.

2 History of Animal Psychophysics

The history of animal psychophysics is pretty much congruent with that of research on animal behavior (or comparative psychology). It developed in the nineteenth century in the post-Darwinian period from a concern with the origins of human behavior as displayed in the behavior of other animals. The concern even then with experimental rigor and precise specification of stimuli can be seen in an extensive monograph on an elaborate (and presumably standardized) piece of equipment for studying vision in animals developed and built by Yerkes and Watson (1911). The technology of instrumentation, particularly in acoustics, evolved very slowly until after the Second World War. But most importantly it was the development of the behavioral technology that has made the important difference in animal psychophysics. The real power of operant conditioning (Skinner 1938) as an experimental paradigm for questioning animals about their sensory and perceptual experiences was demonstrated by Blough (1966) in an elegant series of experiments on the visual sensitivity of the pigeon.

Blough's early work represented a tour de force in animal psychophysics and showed what was possible with the appropriate methodology. There were several features of this and later experiments that represented a substantial improvement over earlier work and essentially opened a new era in the study of sensory processes in animals other than man. In establishing the reinforcement contingencies for measuring visual threshold, Blough made an important assumption that for the observer or subject "stimulus off" was perceptually equivalent to "stimulus on but below threshold." The risk, therefore, of reinforcing responding to subthreshold stimuli, e.g., guessing, was considerably reduced. Blough's use of a tracking or titration procedure in determining threshold set up an effective feedback loop between stimulus and observer – when the stimulus was correctly detected, its intensity was subsequently reduced, whereas if it went undetected, its intensity was then raised. The tracking procedure was part of the continuous examination in time of the animal's behavior with regard to the stimulus. A record, for example, of the dark adaptation function was a record of the changing state of the animal's visual system in the dark following exposure to a bright light. This was the kind of finding that represented a significant improvement over the less effective procedures that had been used previously in the analysis of sensory function in nonhuman animals.

It was very clear from Blough's research and that of others that the training process was expensive, time-intensive, and quite sophisticated – that is, accomplished only by trained personnel. The procedures were fully automated and thus objective and replicable, and the instrumentation has evolved from switching circuits based on electromagnetic relays to control by computers. It was equally clear that the data were every bit as orderly as those obtained in human psychophysics and could not be obtained with any other available procedures. These statements are valid today as they were then – 30 years ago – although the actual procedures have changed considerably.

3 General Methodological Approaches in Animal Psychophysics

The procedures are not generally free-operant where the behavior is emitted and under continuous observation, but rather they are based on a trial-by-trial format. Responding and/or failure to respond to a change in stimulation are the data of principal interest. Changes in behavior are reflected in the changing proportion of responses as the stimulus is varied along some preselected continuum. For example, as sound level is decreased the number of responses to stimulation decreases as an orderly function. The psychophysical or sensory relationship is characterized by the proportion of responses as a function of (in this example) sound level.

Among the most effective procedures for training an animal to attend and respond accurately to the targeted stimulus dimension are those based on positive reinforcement. Following conditioning, subjects are trained to make an observing response that prepares them for stimulation and, in those experiments where freedom of movement is permitted, places the sensory receptors in a uniform position from one trial to the next to ensure that any variation in stimulation is that which is introduced purposely by the experimenter. A subsequent and correct reporting response is reinforced with food.

3.1 Auditory Thresholds in Monkeys

The following example of a procedure for determining auditory threshold is taken from our laboratory (see also Stebbins et al. 1984). Monkeys are conditioned to grasp and hold a small hollow cylindrical tube in response to a light flashing inside the tube. The light becomes steady when the correct response has been made and remains in that state as long as the subject continues to hold the tube. A brief but varying period later an acoustic stimulus is presented for about 2 s. The response of releasing the tube within that 2-s interval is followed by food. When the tube is released, the light is turned off. Should the tube be released before the stimulus is presented, punishment in the form of a brief time-out is incurred. Failure to release the tube is followed by another stimulus presentation. In order to monitor the subject's guess rate, catch trials (when no stimulus is presented) are randomly given in place of stimulus presentations.

The experiments are conducted in specially constructed sound-proofed rooms. The subjects are generally young adult Old World monkeys (macaque and patas) that have been adapted to restraint chairs for the experimental sessions with earphones tightly coupled to their external ears; in this manner hearing in each ear may be evaluated independently. It is then possible, following experimentally induced hearing loss, to relate the extent of the hearing impairment to the form of the histopathology in the same ear. The sound system is calibrated with great care so that the stimulus parameters may be precisely specified at the entry to the outer ear. The experiments are controlled, and the data collected and analyzed, by computer.

Subjects are considered trained and ready for auditory testing when they are responding reliably and consistently and with few errors to acoustic stimulation. Thresholds for pure tones are determined by a modified tracking procedure and typical results are shown in Fig. 1. As the subject continues to respond correctly, the sound level is reduced in fixed steps. When the subject fails to respond, the level is increased for the next trial. Threshold is calculated by averaging the stimulus levels at which the subject responds with those at which it fails to respond. Thresholds taken in this way at many frequencies are combined to form a sensitivity function, an example of which is shown in Fig. 2.

With a small procedural variation we may examine a subject's differential sensitivity to, for example, frequency or sound level. When the response tube is ini-

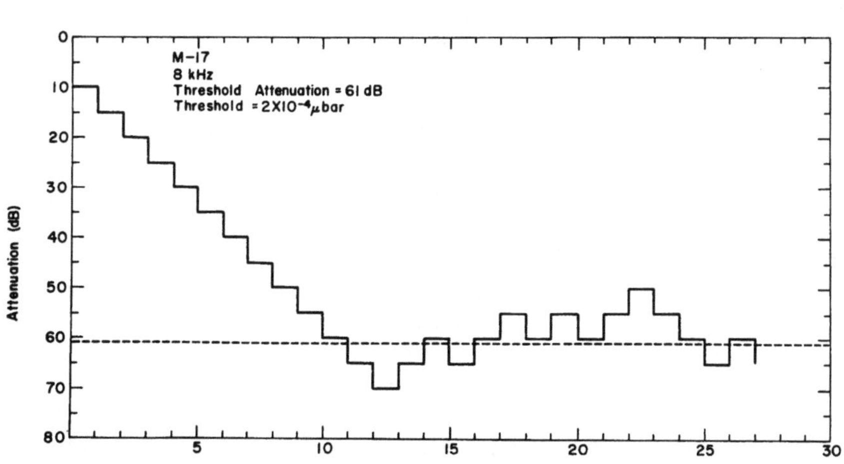

Fig. 1. Use of the tracking method for hearing testing of monkeys. Correct detections cause the tone to be attenuated in 5-dB steps, while failure to hear produces a subsequent 5-dB increase in tone intensity. (Stebbins and Coombs 1975)

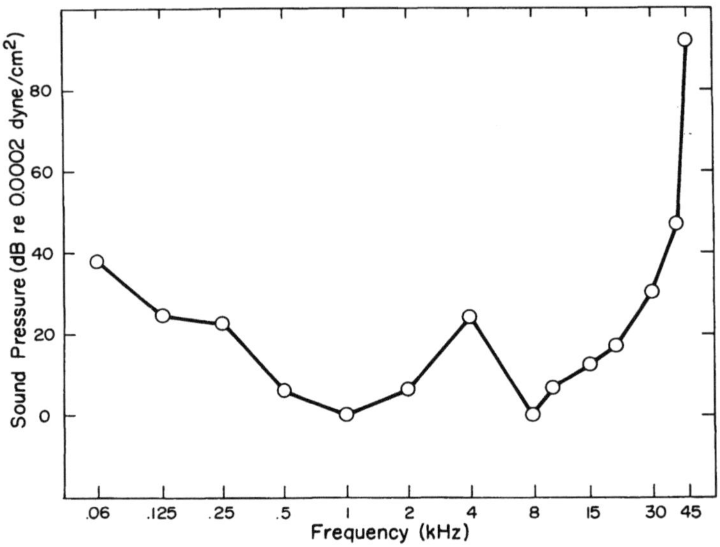

Fig. 2. Hearing threshold function for macaque monkey. (Stebbins 1978)

tially grasped, a very brief and repeating tone signal is introduced. As the subject continues to hold in the observing mode, the frequency, level, or some other characteristic of the signal is changed. The modified signal alternates two or three times with the original or standard signal. If the observer correctly reports the changed signal, reinforcement follows. If the change is not reported, conditions revert to the original repeating standard. Here, too, the tracking procedure is used effectively; correct responses to a stimulus change or difference reduce the extent of that difference on the next trial while failure to respond is followed by an increase in that difference. The difference threshold, like the detection threshold, is based on an average of the stimulus transition values, and it represents the smallest discriminable difference in the selected stimulus parameter to which the subject can respond.

Detection and difference thresholds establish the outer limits on the resolving power of the auditory system for acoustic energy. They specify the frequency and dynamic (intensive) range over which hearing can occur and the discriminable variation in intensity and frequency within the audible range. Within that range the auditory system exhibits a remarkable capability as an acoustic filter in selecting certain frequencies while rejecting others. It also takes advantage of its bilateral and binaural nature to locate sound sources with considerable accuracy, and processes complex signals such as those used in intraspecific communication with facility. These physiological mechanisms have observable and measurable perceptual outcomes. Many animals show behavioral evidence of auditory frequency selectivity, of the ability to locate sound sources at a distance with an angular resolution of less than one degree, and of the capacity to discriminate complex biologically significant signals. The objective of this research is to provide an accurate description of the important stimulus-response relationships with reference ultimately to their significance for the animal in continual interaction with its environment and of their role in understanding the underlying physiological mechanisms involved in transduction, coding, and processing of acoustic information. It is the latter that is the particular concern of this paper.

3.2 Frequency Selectivity

Under natural conditions acoustic signals provide important information to an animal concerning, among other things, the identity, mood, distance, and location of the signaler. Furthermore, the signal's message must be read accurately since survival of the receiver or its kin is often at stake. Discrimination experiments in the laboratory take an analytic perspective in removing and isolating an animal from the natural context and in attempting to determine those features of complex biologically relevant stimuli that are essential to the discrimination.

Frequency selectivity refers to the ability of an animal to select certain acoustic frequencies at the expense of others under conditions of background masking noise – a discrimination required in most natural environments. In the laboratory the experimental paradigm uses a pure-tone masker whose level and frequency may be varied in order to determine the effect on the test tone frequency to which the animal has been trained to respond. For a given series of measurements the test tone is fixed at one level, usually 10 dB above its threshold, and at one frequency. The level of the masker set at one frequency is varied and a threshold determined which is defined as the level of the masker at which the test tone is just audible. A set of such measurements is made at many frequencies for the masker both above and below the test tone frequency: these thresholds are then combined into a single function of the form shown in Fig. 3. This psycho-

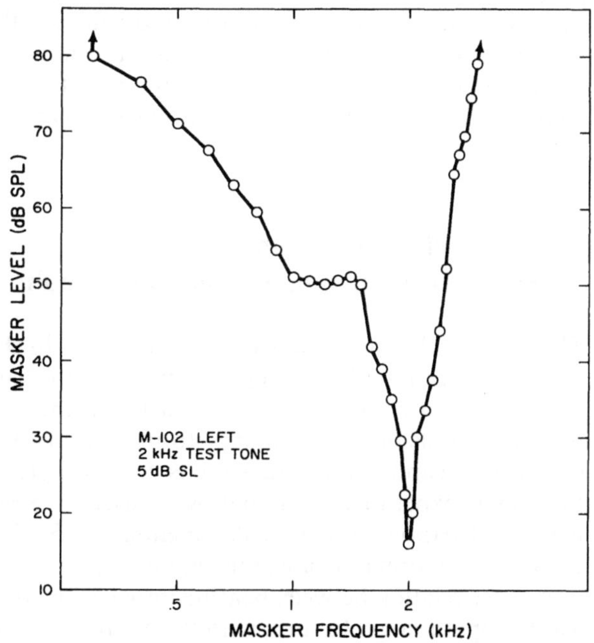

Fig. 3. Psychophysical tuning curve for a macaque monkey. (Stebbins and Moody 1979)

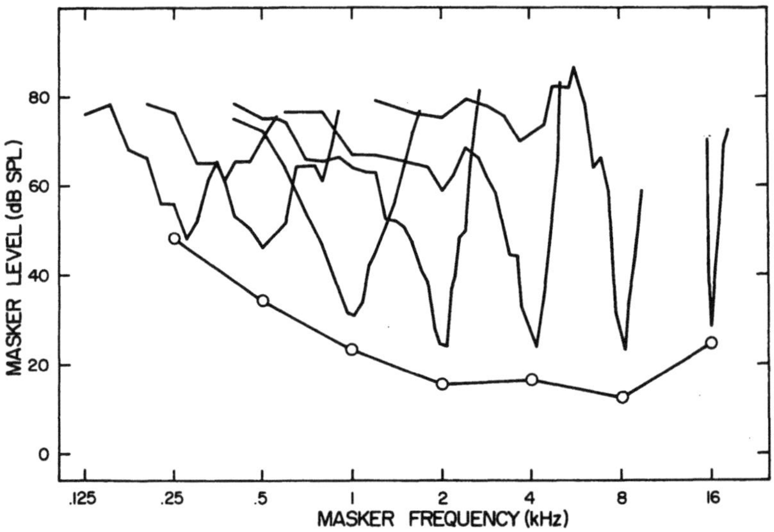

Fig. 4. A family of psychophysical tuning curves shown with underlying pure tone threshold function (*open circles*) for a macaque monkey. (Stebbins 1982)

physical tuning curve resembles a sharply tuned band-pass filter almost identical in shape to the frequency-response function obtained with a microelectrode from a single auditory nerve fiber. This similarity suggests that the underlying physiological mechanisms for this kind of discrimination may be found at the auditory periphery. These behavioral functions or psychophysical tuning curves may be determined at a number of test tone frequencies with the result displayed in Fig. 4. The normal pure-tone threshold function is shown in the same figure. We will examine these functions later in an experimentally hearing-impaired animal and consider their relation to the morphological changes that have occurred in the inner ear.

4 Use of Selective Lesions To Study Transduction of Auditory Stimuli

The concept that acoustic frequency is coded in the peripheral auditory system by spatial location of the receptor cells along the basilar membrane in the cochlea of the inner ear has been with us at least since the time of Helmholtz. But the more intricate details of this frequency-to-place transform have yet to be established. Our own research on this question is based on procedures for evaluating hearing function in experimental animals with discrete receptor cell lesions of varying extent and location on the basilar membrane. Such lesions may be produced by chemical or thermal means; they are always irreversible and permit an examination of the relation between hearing impairment and the exact nature of the cochlear histopathology. In turn, with the appropriate behavioral questions, the analysis of this relationship can provide significant information bearing on the place theory of hearing and of frequency coding in the inner ear.

4.1 Aminoglycoside Antibiotic Lesions

It has been known for some time that the aminoglycoside antibiotics such as kanamycin and streptomycin cause severe sensorineural hearing loss in man and experimental animals by destroying the receptor cells and nerve fibers in the inner ear (Hawkins 1976). More to the point, for our purposes, the destruction proceeds in an orderly manner and direction along the basilar membrane of the cochlea from its base to its apex. The process is very similar to aging in the human ear but considerably accelerated depending, of course, on the dose and length of treatment. The orderly nature of the process is essential to our experiments on nonhuman primates. Unlike aging, the drug treatment can be halted at any time and the hearing loss can be compared with the histopathological changes that have occurred in the inner ear. The highest frequencies in the audible range are presumably processed and initially coded in the extreme cochlear base; the entire range is mapped onto the cochlear spiral from high to low frequency in an apical direction. The aging process and treatment with the aminoglycoside antibiotics result in a progressive hearing loss which starts at the high frequencies.

The effects on the hearing (the detection threshold for pure tones) of a macaque monkey treated for an extended period with the aminoglycoside kanamycin are shown in Fig. 5. By the end of treatment the animal was unable to hear at frequencies of 4 kHz and above but retained normal sensitivity at 2 kHz and below. The final threshold function at the tested frequencies is shown in the lower panel of Fig. 6. The very sharp transition between complete impairment and normal hearing is characteristic of the effects of kanamycin. The severe hearing loss is correlated with complete absence of receptor cells (hair cells) of both types in

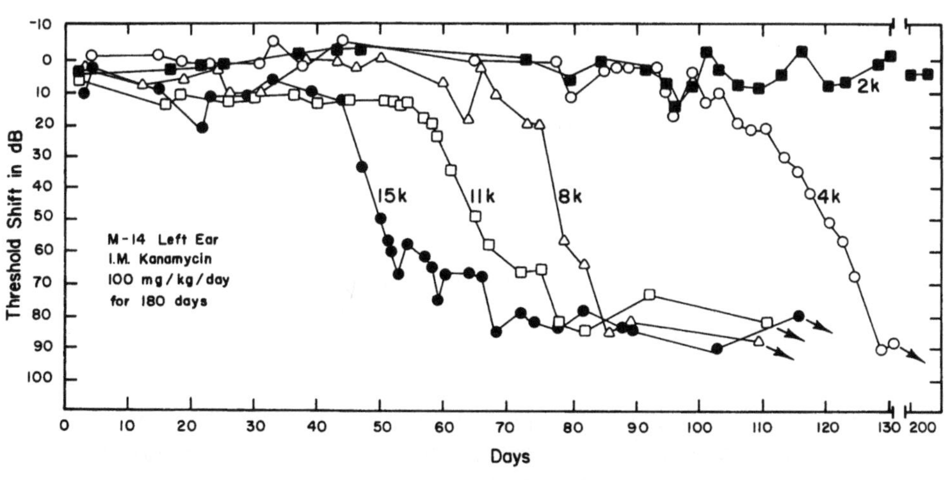

Fig. 5. Progressive changes in threshold for one macaque monkey for different acoustic frequencies during daily kanamycin treatment. The *zero line* represents normal hearing at all frequencies prior to drug treatment. (Stebbins et al. 1969)

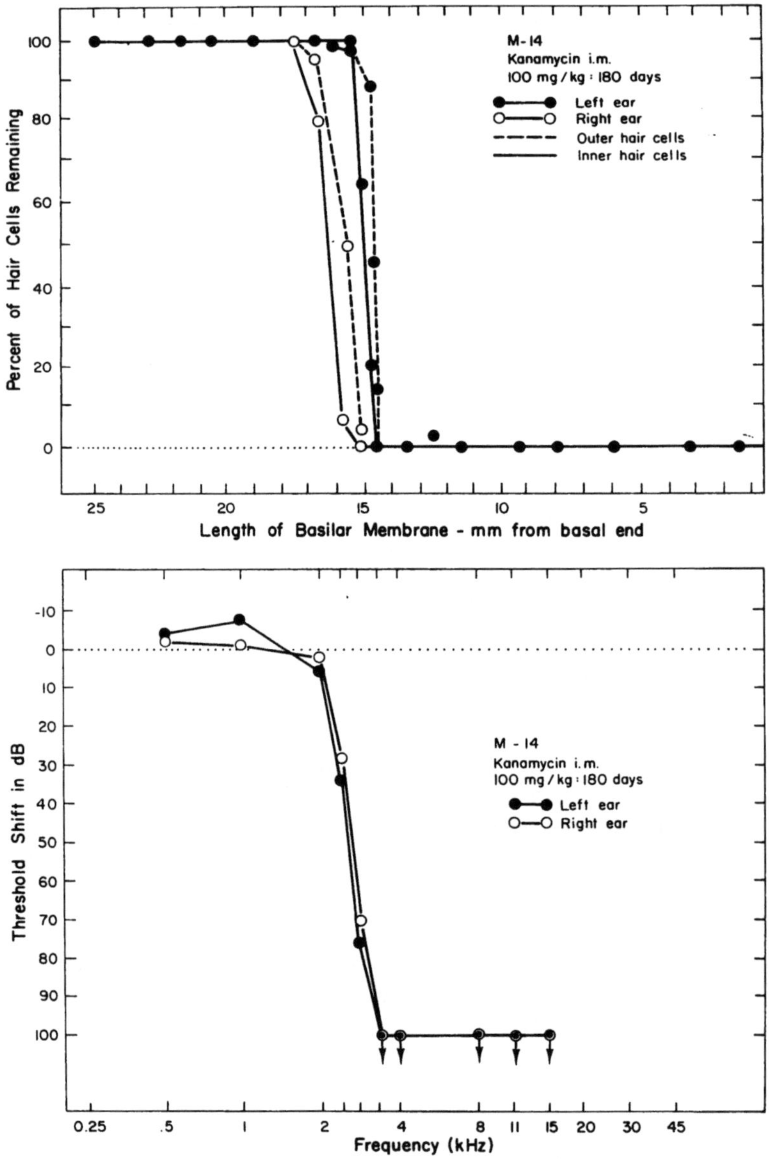

Fig. 6. Auditory threshold shift of the monkey represented in Fig. 5 after 180 days daily kanamycin (*lower panel*); hair cells remaining as a function of position on basilar membrane (*upper panel*). (Stebbins and Coombs 1975)

the basal half of the cochlea as seen in the upper panel of the figure; normal-appearing hair cells remain in the apical half of the cochlea. The relation between hearing and hair cell loss observed in this and many other examples suggests the importance of the hair cells both for normal thresholds and for their role in frequency coding in the auditory periphery (Stebbins et al. 1979).

Our serendipitous discovery that the patas monkey is exquisitely sensitive to the aminoglycoside dihydrostreptomycin (DHSM) complicated the picture considerably by demonstrating what most pharmacologists already know: that a variation in the pharmacological compound and in the subject species can yield a different, although not incompatible, result. While DHSM has no effect on the macaque monkey's hearing, its consequences for the patas' ear are devastating (Hawkins et al. 1977; Stebbins et al. 1987). The progress of the hearing loss from the high to the low frequencies is similar to that for the kanamycin-treated macaque. However, the patas given DHSM invariably retains some hearing at many of the high frequencies together with a considerable number of inner hair cells; most of the outer hair cells are gone except in the extreme apex of the cochlea. The obvious disparity between the number and distribution of inner and outer hair cells is seen in the upper panel of Fig. 7; the threshold detection function appears in the lower panel. A 50- to 60-dB hearing loss at the higher frequencies in the animals' audible range is clearly related to the complete absence of outer hair cells in an extended region of the cochlear base where inner hair cells are present and normal in appearance. The finding has been replicated many times and has

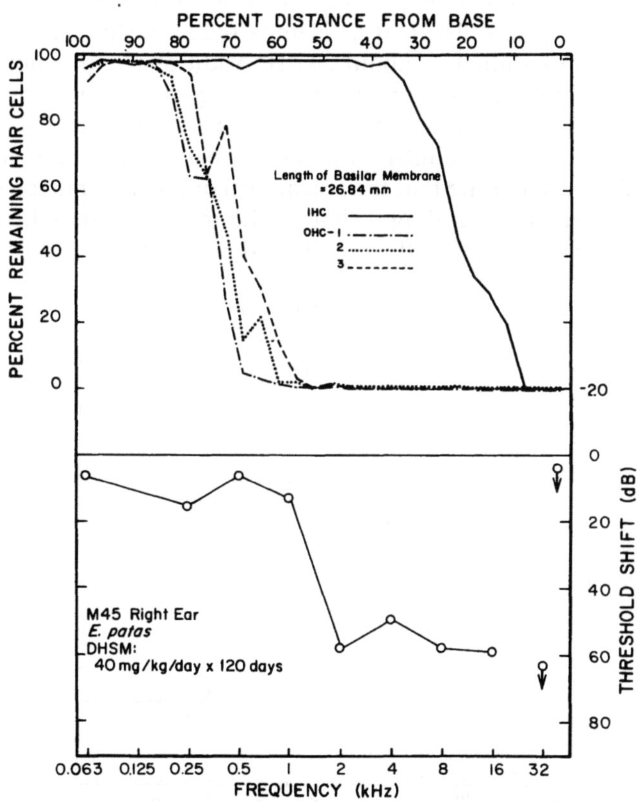

Fig. 7. Auditory threshold shift for a patas monkey after dihydrostreptomycin (DHSM) treatment (*lower panel*); hair cells remaining as a function of position on basilar membrane from the same ear (*upper panel*). (Stebbins et al. 1987)

significant implications for our understanding of differential function between these two morphologically different types of receptor cells (Stebbins et al. 1987).

4.2 Thermal Lesions

It is unfortunate from the perspective of our interest in inner ear function with regard to hearing that there are no drugs that act initially on the apical portion of the cochlea. It has been a legitimate criticism of our work that until recently there has been no way of selectively lesioning the cochlear apex. Our research with intense sound, which does destroy the hair cells and produce a permanent hearing loss, has shown it to be too imprecise and unpredictable as a technique for making discrete inner ear lesions. Recently, Brown and Nuttall (1987) have developed a procedure that uses a small cryoprobe for making apical lesions in the chinchilla cochlea. We have used it successfully in the monauralized, behaviorally trained chinchilla in order to compare hearing threshold changes with loss of receptor cells in the apex (Smith et al. 1987 a). Results from one animal are shown in Fig. 8. The cryoprobe has effectively destroyed both inner and outer hair cells in the apical one-third of the cochlea. Toward the base, hair cells are present and appear viable, and there is an abrupt transition between the two regions of missing and present hair cells. Hearing is impaired at the lowest frequencies but the threshold has only shifted by about 20 dB. Additional experimentation in which high-pass masking noise was applied to the frequencies above 500 Hz suggests that the receptor cells counted as present on the basal side of the lesion were indeed functional and were contributing to the hearing of the lowest frequencies. These recent findings indicate that the apical turn of the mammalian cochlea may be organized for frequency coding – that is, the transform may be frequency-to-place in the same manner as in the middle and basal turns.

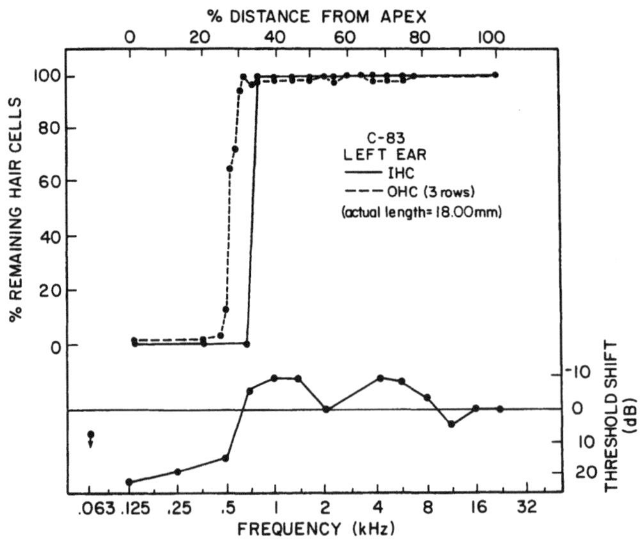

Fig. 8. Auditory threshold shift for chinchilla with cryoprobe lesion (*lower panel*); hair cells remaining as a function of position on basilar membrane (*upper panel*). (Smith et al. 1987 a)

4.3 Dihydrostreptomycin Lesions and Frequency Selectivity

While detection thresholds provide important information regarding the relation between sensory acuity and underlying physiological function, there are additional more specific measures that are capable of revealing other facets of acoustic information processing that in turn may complement the more broadly based threshold functions. One of these, described above, is frequency selectivity measured by psychophysical tuning curves (Figs. 3 and 4). The DHSM-patas monkey model offers a unique opportunity to consider the role that outer hair cells play in frequency selectivity. The characteristic hearing impairment that includes a severe loss at the high frequencies with a more moderate loss in the middle ranges is observed for one DHSM-treated animal in the lower panel of Fig. 9. The cytocochleogram, in the upper panel of the same figure, illustrates the marked disparity between inner and outer hair cell loss in the basal half of the cochlea, similar to that shown in Fig. 7. It is reasonable to suggest on the basis of these findings that normal acoustic sensitivity depends upon the normal functioning of the outer hair cells. Pre-drug thresholds (baseline audiogram) are shown in Fig. 10 for the same animal whose data are shown in Fig. 9. The post-drug audiogram is shown

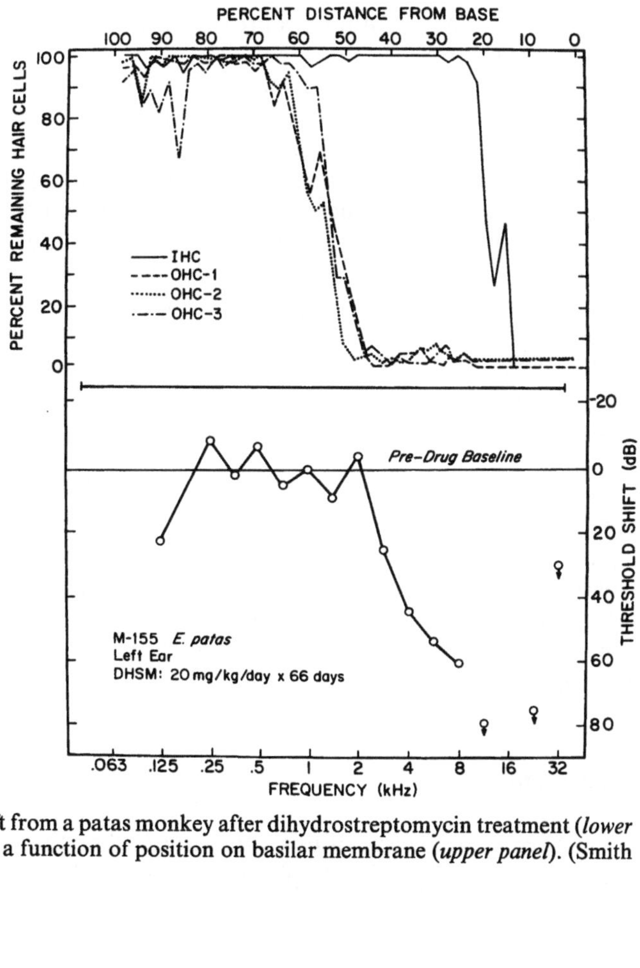

Fig. 9. Auditory threshold shift from a patas monkey after dihydrostreptomycin treatment (*lower panel*); hair cells remaining as a function of position on basilar membrane (*upper panel*). (Smith et al. 1987b)

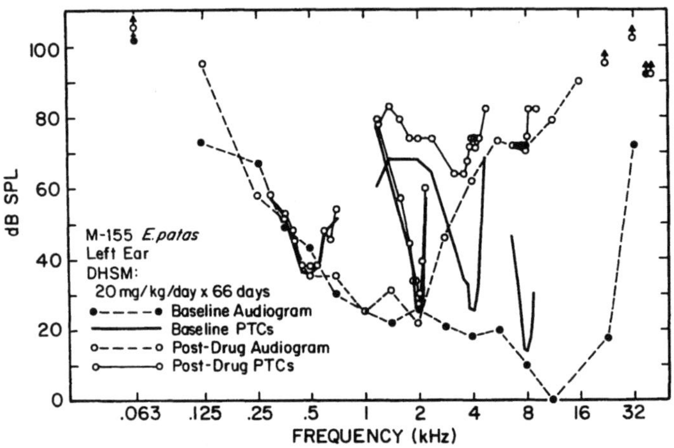

Fig. 10. Direct comparison of baseline (predrug) and postdrug auditory threshold curves and psychophysical tuning curves (PTCs) for patas monkey shown in Fig. 9. (Smith et al. 1987b)

here also. The data, although represented differently, are the same as those presented in Fig. 9.

In Fig. 10, also, pre- and postdrug psychophysical tuning curves are shown at four frequencies (500 Hz, 2, 4, and 8 kHz). At 500 Hz and 2 kHz where threshold sensitivity is normal, presumably because the inner and outer hair cells coding these frequencies are also normal, the postdrug tuning curves lie on top of the predrug curves. At 4 kHz, where the postdrug threshold has increased by more than 50 dB, the tip of the tuning curve has shifted by about 40 dB and some detuning is evident in the increased breadth of the curve; the low frequency tail, which extends below 2 kHz, is only slightly elevated. At 8 kHz the postdrug tuning curve no longer has a well-defined tip; the lowest point on the function is more than 50 dB above the tip of the predrug tuning curve and lies on top of the postdrug audiogram. Complete detuning of the postdrug tuning curve, which resembles a low-pass filter, suggests that frequency selectivity is severely impaired. The absence of outer hair cells on that portion of the basilar membrane corresponding to those frequencies where abnormal tuning is observed suggests their critical importance for frequency selectivity. Further, it suggests that both acoustic sensitivity and frequency selectivity are perceptual functions that are under primary control of processing mechanisms located in the auditory periphery rather than in the central auditory system. These findings, and their implications for auditory function, are more completely documented by Smith (1986) and Smith et al. (1986, 1987b).

5 Summary

It has been the objective of this paper to examine the discrimination paradigm in the context of animal psychophysics with several experimental illustrations. For our purposes sensory discriminations are defined as tight, measurable relations between stimuli and responses. Stimuli, when varied over a wide range of values on a given continuum, yield psychophysical (stimulus-response) functions that precisely characterize the resolving power of an animal's sensory system – that is, its ability to detect and discriminate environmental change. Discrete and localized damage to the sensory system that produces impairment in discrimination can help to reveal the underlying physiological mechanisms involved in stimulus transduction and processing.

Whether there are useful and instructive parallels between the sensory experiments described here and those in the growing field of the behavioral discrimination of drugs is both an important and open question. We offer a caveat that these experiments are difficult to perform, require considerable technical expertise, and are both time and labor intensive. Their apparent simplicity has attracted scientists from other disciplines, who, in their subsequent disillusionment, often reject the approaches described as unfeasible and turn instead to older, quicker, and, unfortunately, cruder methods that are imprecise and incapable of answering questions of the sort that we have posed. The ever-increasing sophistication and resolution of morphological, physiological, and biochemical inquiry in the sensory systems require behavioral procedures of similar resolution and sophistication for studying sensory discrimination and perception of the sort that we have described here. Behavioral inquiry of the form detailed in this paper is particularly important since it is a major objective of biochemical, physiological, and morphological research into the transduction and processing of sensory stimuli to account for the manner in which these stimuli are ultimately perceived.

Acknowledgment. The research was supported by grants N505077 and N505785 from NINCDS.

References

Blough DS (1966) The study of animal sensory processes by operant methods. In: Honig WK (ed) Operant behavior: areas of research and application. Appleton-Century-Crofts, New York, pp 345–379

Brown JN, Nuttall AL (1987) Cryoprobe-induced apical lesions in the chinchilla. I. Effects of lesioning parameters. Hearing Res 26:301–309

Hawkins JE Jr (1976) Drug ototoxicity. In: Keidel WD, Neff WD (eds) Handbook of sensory physiology, vol V, part 3. Springer, Berlin Heidelberg New York, pp 707–748

Hawkins JE Jr, Stebbins WC, Johnsson L-G, Moody DB, Muraski A (1977) The patas monkey as a model for dihydrostreptomycin ototoxicity. Acta Otolaryngol 83:123–129

Skinner BF (1938) The behavior of organisms. Appleton-Century-Crofts, New York

Smith DW (1986) Effects of selective outer hair cell lesions on the frequency selectivity of the patas monkey auditory system. PhD thesis, University of Michigan, Michigan

Smith DW, Moody DB, Stebbins WC, Shaddock LC (1986) The effects of selective outer hair cell loss on psychophysical tuning curves in the patas monkey. Abstracts of the 9th midwinter meeting of the Association for Research in Otolaryngology. Clearwater Beach, February 2–6

Smith DW, Brown JN, Moody DB, Stebbins WC, Nuttall AL (1987 a) Cryoprobe-induced apical lesions in the chinchilla: II. Effects on behavioral auditory thresholds. Hearing Res 26:311–317

Smith DW, Moody DB, Stebbins WC, Norat MA (1987 b) Effects of selective outer hair cell loss on the frequency selectivity of the patas monkey auditory system. Hearing Res 29:125–138

Stebbins WC (1978) Hearing of the primates. In: Chivers DJ, Herbert J (eds) Recent advances in primatology, vol I. Behaviour. Academic, New York, pp 703–720

Stebbins WC (1982) Concerning the need for more sophisticated animal models in sensory behavioral toxicology. Environ Health Perspect 44:77–85

Stebbins WC, Coombs S (1975) Behavioral assessment of ototoxicity in nonhuman primates. In: Weiss B, Laties VG (eds) Behavioral toxicology. Plenum, New York, pp 401–427

Stebbins WC, Moody DB (1979) Comparative behavioral toxicology. Neurobehav Toxicol 1:33–44

Stebbins WC, Miller JM, Johnsson L-G, Hawkins JE Jr (1969) Ototoxic hearing loss and cochlear pathology in the monkey. Trans Am Otol Soc 57:110–128

Stebbins WC, Hawkins JE Jr, Johnsson L-G, Moody DB (1979) Hearing thresholds with outer and inner hair cell loss. Am J Otolaryngol 1:15–27

Stebbins WC, Brown CH, Petersen MR (1984) Sensory processes in animals. In: Darian-Smith I, Brookhart J, Mountcastle VB (eds) Handbook of physiology, sensory processes, vol I. American Physiological Society, Bethesda, pp 123–148

Stebbins WC, Moody DB, Hawkins JE Jr, Johnsson L-G, Norat MA (1987) The species-specific nature of the ototoxicity of dihydrostreptomycin in the patas monkey. Neurotoxicology 8(1):33–44

Yerkes RM, Watson JB (1911) Methods of studying vision in animals. Behav Monogr 2 [2, iv]:90

Molecular Transduction Mechanisms in ACTH-Induced Grooming

W. H. GISPEN [1], D. L. COLBERN [2], and B. M. SPRUIJT [1]

Contents

Abstract

Intraventricular administration of ACTH1-24 induces excessive grooming in the rat. Ethogram analysis shows that the peptide does not alter grooming behavior seen in a novel box, but that it prolongs the duration of the grooming bout. Extensive structure-activity studies have been performed which suggest that the active site lies in a region (5-13) of the ACTH molecule. Interestingly, the (1-24) sequence is fully active, whereas (1-10) and (11-24) alone or in combination are inactive, pointing to a specific stereoconformation necessary to induce grooming. However, despite the fact that there are ACTH- and/or α-MSH-containing peptidergic neurons, no conclusive evidence is available demonstrating stereospecific, saturable binding sites for these peptides in brain.

The analysis of the neural substrate underlying ACTH-induced excessive grooming has been performed by means of electrolytic lesions of specific brain regions and by neuropharmacological manipulations. The data suggest that the periaqueductal gray is the primary target for ACTH and that the activity of neostriatum and accumbens, via a nigro-colliculus-periaqueductal gray pathway, modulates the display of excessive grooming. An important feature of the neural substrate is that it displays single-dose tolerance to the peptide during the first hours after the first peptide injection. It is suggested that the tolerance is a feature of an opioid receptor-containing component of the neural substrate. The molecular mechanism of action of ACTH is complex and may involve different transmembrane signal transduction systems. The peptide decreases the degree of phosphorylation of a neuron-specific, synaptic phosphoprotein B-50 by inhibition of protein kinase C. It is concluded that changes in the degree of phosphorylation of B-50 regulate the

[1] Division of Molecular Neurobiology, Rudolf Magnus Institute for Pharmacology and Institute of Molecular Biology and Medical Biotechnology, University of Utrecht, Padualaan 8, 3584 CH Utrecht, The Netherlands.
[2] Dept. of Physiology and Biophysics, University of Illinois, Chicago, IL 60680, USA.

Psychopharmacology Series 4
Transduction Mechanisms of Drug Stimuli
Editors: Colpaert, Balster

activity of the lipid kinase phosphatidylinositol 4-phosphate kinase. Therefore, the B-50 protein seems to be part of a negative feedback loop in the receptor-activated hydrolysis of phosphatidylinositol 4,5-bis-phosphate (PIP_2).

There is increasing evidence that the molecular mechanism by which ACTH brings about the grooming response involves a change in phosphorylation of B-50. Firstly, the structure-activity relationship of ACTH-induced excessive grooming is nearly identical to that obtained for ACTH-induced inhibition of protein kinase C. Secondly, ACTH can counteract the effects of phorbol esters in hippocampal slices. Phorbol esters are known to strongly activate protein kinase C. Preincubation of hippocampal slices with phorbol esters leads to a reduced hydrolysis of PIP_2 presumably by the postulated B-50 feedback loop. Coincubation with ACTH diminishes the influence of the phorbol ester. Thirdly, phorbol esters that stimulate protein kinase C, but not inactive analogs, are capable of reducing ACTH-induced excessive grooming. Thus, ACTH-induced excessive grooming can be modulated by altering the degree of phosphorylation of protein B-50. As in the rat brain the protein is predominantly localized in presynaptic membranes, it is proposed that the molecular mechanisms underlying ACTH-induced grooming involve a presynaptic modulation of neurotransmission in certain regions.

1 Introduction

Drugs that influence the function of the central nervous system often affect behavior. The behavioral effect is the result of various factors including molecular properties of the drug, the neural substrate that is influenced by the drug, and the behavioral paradigm used in the experiment. Although in general one is aware of the complexity of the mechanism by which drugs influence behavior, oversimplifications are often being used in explaining how or why the particular drug is effective. The measurement of behavior as a function of brain activity is hampered by the fact that the output analysis provides little information on the relevant activity of given brain centers. Furthermore, even if one was to know where in the brain to look for a meaningful correlate of a given behavioral activity, the question remains as to which molecular event is to be studied. For, despite growing insight in the functional wiring of the brain, relatively little is known of the molecular processes that govern the chemical communication in the brain.

It is against this background that we shall review some of the molecular aspects of neuropeptide action on behavior. We shall confine ourselves to the behaviors that rats display in a novel observation cage in response to the intracranial injection of ACTH-like peptides. Simple observational analysis of this behavior has proven to be very useful in the study of the mechanism by which the peptide affects neural activity. Furthermore, it illustrates some of the molecular aspects which play a role in behavioral pharmacology in general.

2 The Neuropeptide ACTH: Multiple Message

The classical biological activity of ACTH is the growth-promoting (trophic) and steroidogenic action on the adrenal cortex. However, the N-terminal part of the ACTH molecule produces extraadrenal effects, including an action on nerve tissue. The study of structure-activity relationships using synthetic peptide fragments has provided insight in the chemical organization of information in the

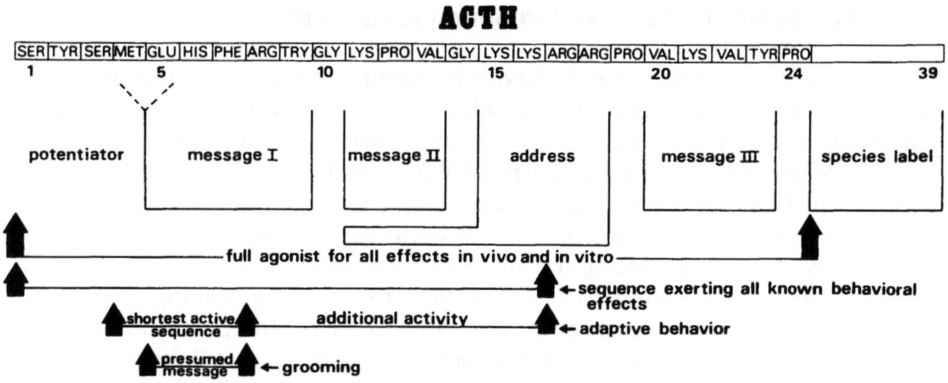

Fig. 1. The multiplicity of information encoded in the primary structure of ACTH

ACTH molecule (Schwyzer 1980). Schwyzer summarized his ideas about the molecular anatomy of ACTH as follows: (1) there are discrete (although sometimes overlapping) sequences of adjacent amino acids, which are responsible for the different components of the total biological action of ACTH; (2) different target cells or different receptors may be stimulated by different portions of the sequence; and (3) partial sequences, obtained by synthesis, can produce effects similar to those elicited by the complete molecule (Schulster and Schwyzer 1980). The ACTH molecule can therefore be divided into different regions with different functions (see Fig. 1). Considering the biological action on adrenal cortex cells, Schwyzer (1980) distinguishes the following sequences: ACTH1-3, potentiator: enhances potency and efficacy; ACTH4-9, message: triggers the receptor; ACTH11-24, address: adds cell-specific affinity; ACTH25-39, species label: antigenicity, transport, protection. Although the ultimate information for the receptor is encoded in the region 4-10, N- and C-terminal extension is necessary to obtain biological activity. It has been shown that some of the extraadrenal effects are brought about by regions different from the message for the adrenal cortex receptor. Thus, a given peptide contains information that is encoded in multiple forms and a given receptor may interact with one part of the molecule, whereas another, different receptor, even on the same cell, may react with another portion of the molecule. As different receptor-mediated responses may differ in effect (quality and quantity) in a heterogeneous system like the nervous tissue, one may expect a multiplicity of responses to a given peptide. The diversity of behavioral, neurophysiological, and neurochemical effects of ACTH is compatible with such a view. This multiplicity of information in neuropeptides makes the study of their neurochemical mechanism of action extremely difficult. Based on their putative neurophysiological and neurochemical actions, neuropeptides can be classified as follows: (1) putative neurotransmitters, (2) neurohormones, and (3) neuromodulators. The value of this classification remains questionable since a given peptide may act via all three mechanisms (Barker and Smith 1979).

3 ACTH-Induced Excessive Grooming in the Rat

"Grooming" or "maintenance" behavior is a common, species-characteristic and fixed action pattern with readily definable components. Both peripheral and central inputs govern the display of this care of the body surface behavior (Fentress 1977; Borchelt 1980; Spruijt and Gispen 1983). Functions other than the maintenance of the body surface are likely to exist. Since grooming in rodents may occur after extremely arousing conditions, it has been suggested that the behavioral response may play a deactivating role in restoring homeostasis as well (Gispen and Isaacson 1981, 1986; Spruijt and Gispen 1983, 1986). The grooming behavior as discussed in this paper consists of the following elements: forepaw vibration, face washing, body grooming, anogenital grooming, scratching, paw licking, head and body shaking and tail preening. The behavior is recorded either by a time sampling technique (Gispen et al. 1975) or by an analysis of grooming patterns (element duration, frequency and sequence; Spruijt and Gispen 1984).

Previously, it was documented that rats, upon placement into a novel observation box, bearing either an intraventricular cannula or not – with or without an intraventricular (i.c.v.) saline injection – initially displayed some grooming behavior but then invariably fell asleep (Colbern et al. 1978; Jolles et al. 1979; Traber et al. 1982). The novelty-related grooming did not differ from that of home cage grooming. Intracranial but not systemic treatment with ACTH and congeners of a variety of animal species such as dogs, cats, rabbits and mice induces a behavioral syndrome which is known as the stretching and yawning syndrome (Ferrari et al. 1963). Especially in rodents the onset of this syndrome is preceded by the display of excessive grooming behavior (Ferrari et al. 1963; Izumi et al. 1973; Gispen et al. 1975). That is, peptide-treated rats spend more time grooming than saline-treated rats. However, if the data are converted to the proportion of time spent per grooming element as a percentage of the total time spent grooming, no significant changes could be detected between the two groups of rats (Gispen and Isaacson 1981). Thus, it would appear that ACTH enhances the display of grooming behavior without changing the composition of the behavioral response relative to saline-injected control rats. The total number of grooming bouts is very similar for saline- and ACTH-treated rats, however, the effect of ACTH is to prolong bout duration (Gispen and Isaacson 1981; Zwiers et al. 1981; Spruijt and Gispen 1984).

Recently, the temporal distribution of the excessive grooming response produced by i.c.v. administration of ACTH has been recognized as an important aspect of the response (Isaacson et al. 1983 a). It is now apparent that the excessive grooming induced by neuropeptide administration produces at least two kinases that are partially independent. The first phase lasts for the first 20–30 min after i.c.v. injection and is relatively independent of most pharmacological interventions. The second kinase is sensitive to systemic administration of neuroleptics and opiate antagonists (see Sect. 5).

The amount of excessive grooming induced by ACTH1-24 does not depend on the physical nature of the observation chambers (Gispen and Isaacson 1986). However, as is important to behavioral pharmacology in general, the behavior is strain dependent (Spruijt et al. 1985) and is sensitive to competing behaviors such as ACTH-related stretching and yawning, foot shock-induced freezing or food or water consumption following deprivation (Gispen and Isaacson 1986; Traber et al. 1987). The advantage of using ACTH-induced excessive grooming as a model in the study of molecular aspects of behavioral pharmacology is the extreme reproducibility – in terms of both quality and quantity – with which it can be elicited across laboratories.

4 Pharmacological Aspects of ACTH-Induced Excessive Grooming

4.1 Dose-Response Relationship

In the study of ACTH action on the nervous system it is not uncommon to obtain an inverted U-shaped dose-response curve (see De Wied and Jolles 1982). This phenomenon has been observed in a number of neurochemical and behavioral studies. The multiplicity of information sites in the peptide and of peptide-sensitive structures in the brain might be responsible for this biphasic dose-response curve. With respect to ACTH-induced grooming, a more common dose-response relationship was observed (Fig. 2; Gispen 1982). As is shown by Isaacson et al. (1983 a) the difference in grooming score as a result of different doses is most pronounced in the second phase of the observation period (55 min starting 15 min following the i.c.v. injection; Gispen et al. 1975). The maximal grooming score is obtained between 1.0 and 10.0 μg ACTH1-24.

4.2 Structure-Activity Relationship

Some of the complexity by which information is encoded in the primary structure of the ACTH molecule can be demonstrated using grooming as the behavioral response to ACTH (Gispen et al. 1975; Wiegant and Gispen 1977). In these studies it was shown that ACTH1-24, but neither its composing sequences (1-10) or (11-24), nor their equimolar combination, is able to elicit the excessive grooming response. A possible explanation could be that a message site in one sequence requires the presence of an address site in the other sequence for expression of its

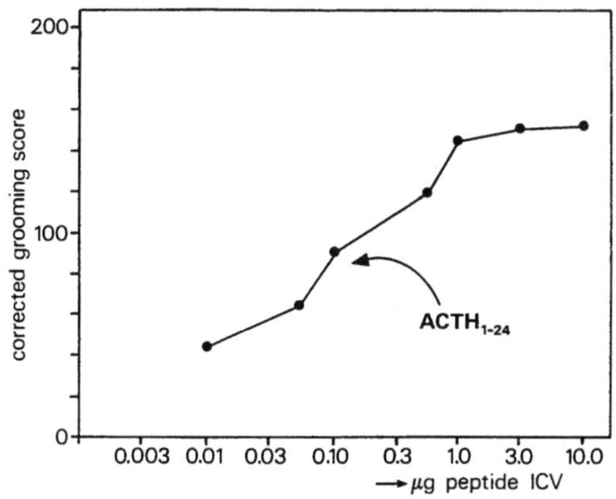

Fig. 2. Dose-response curve of ACTH-induced excessive grooming in the rat

biological activity. In other words, the stereoconformation of the molecule ACTH1-24 at the receptor site favors message-receptor activation.

The grooming message is somewhere in the region (5-13) and may be present in the sequence (5-7). For full expression of the grooming activity, elongation at the C-terminus seems necessary (Gispen et al. 1975). Based on the use of many synthetic ACTH fragments it was concluded that the structural requirements for the peptide to produce grooming differ from several other known nervous system effects of ACTH such as the retardation of extinction of conditioned avoidance behavior and the facilitation of nervous tissue repair (Gispen and De Wied 1984). As stressed in previous reviews, it remains hazardous to make predictions on biological activity of a given peptide based on the features of the primary structure alone (Gispen et al. 1977). Certainly, the stereoconformation at the receptor site is of crucial importance. The induction of grooming activity in the otherwise inactive fragments (4-10) and (1-10) by D-enantiomer substitution of the amino acid phenylalanine, the inactivity of (4-10) and (7-10), and the activity of (4-7) are in support of this view (Wiegant and Gispen 1977). Likewise, the induction of activity in the sequence ACTH1-13-NH_2 by acetylation of the N-terminal serine underscores the importance of peptide stereoconformation (Gispen et al. 1975; O'Donohue et al. 1981). Interestingly, De Graan et al. (1986b) were able to demonstrate the specific potentiating properties of the sequence (1-4) on the biological activity of melanocortins. Using excessive grooming and melanophore dispersion in amphibians as the biological response to i.c.v. peptide treatment, it was found that combined injections of "free" ACTH1-4 and a grooming-inducing peptide such as α-MSH, ACTH1-16, or ACTH5-16 led to a marked enhancement of the response. Thus, although at present no classically defined receptor site for ACTH has been found in the central or peripheral nervous system, clearly the peptide-brain interaction as revealed by ACTH-induced excessive grooming follows certain rules known from well-established peptide-receptor interactions in the periphery.

4.3 Single-Dose Tolerance

Colbern et al. (1978) found that injections of ACTH1-24 given 24 h apart over several days produced comparable amounts of grooming. That is, there was no "carry over" effect from the prior injection(s) under these conditions. Thus, neither tolerance nor sensitization occurred when the injections were separated by this interval. Jolles et al. (1978) confirmed and extended these observations. A detailed analysis of the interval period between two i.c.v. injections of a given dose of ACTH revealed that a substantial suppression of ACTH-induced grooming was seen following the second ACTH injection, provided this injection was given within 8 h following the first injection (Fig. 3). A smaller reduction was seen if the interval ranged from 10 to 18 h, whereas intervals larger than 18 h resulted in equal amounts of grooming after both injections (Jolles et al. 1978). A number of studies were carried out to elucidate the mechanism underlying reduction of grooming after the second injection. The suppression is not the consequence of the release of steroids from the adrenal cortex since the phenomenon was also observed in adrenalectomized rats. The response to the second injection could be

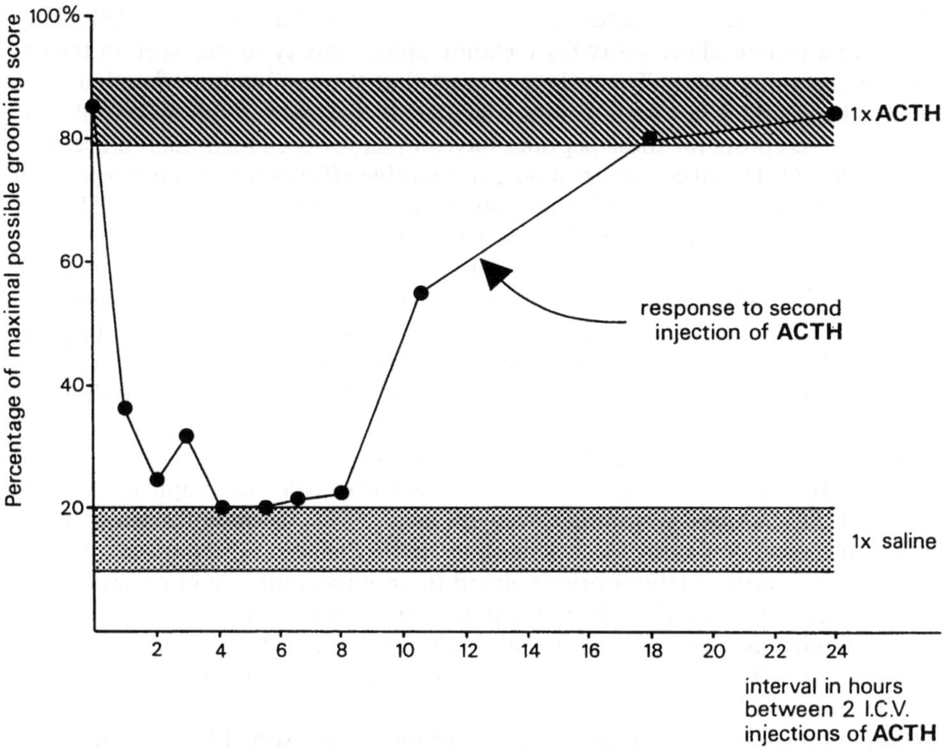

Fig. 3. Single-dose tolerance in ACTH-induced excessive grooming in the rat

normalized if more ACTH was injected at the second injection (Jolles et al. 1978). This phenomenon is therefore reminiscent of single-dose or acute tolerance as seen in temperature responses to morphine and the time course of both processes is remarkably similar (Rosenfeld and Burks 1977). Indeed, cross tolerance could be demonstrated between ACTH1-24, β-endorphin, and morphine (Jolles et al. 1978; Wiegant et al. 1978). This cross tolerance reflects a common element in the neural substrate underlying the behavioral response to ACTH and the opiates (Gispen and Isaacson 1986). The phenomenon of single-dose tolerance has been of great help in the identification of the primary target for ACTH to elicit the excessive grooming behavior (Spruijt et al. 1986b, c).

5 The Neural Substrate Underlying ACTH-Induced Excessive Grooming

Many approaches have been used to unravel the neural pathways that are involved in ACTH-induced excessive grooming. Such studies have involved neuropharmacological manipulations, focal as well as aspiration lesions, and local administration of ACTH and have been summarized recently (Gispen and Isaacson 1986). Here, only some aspects will be discussed in more detail. Spruijt et al.

(1985a) compared the structure-activity relationship of a series of α-MSH fragments in a melanophore assay for melanotrophic activity to that seen in the excessive grooming assay. These data suggest that the mechanism of action of α-MSH-like peptides in the rat brain is similar to that seen in the melanophores. However, receptors for these peptides have not as yet been identified in brain.

While ACTH and congeners have considerable affinity for opiate receptors in the CNS (Terenius 1976) and although opiate antagonists suppress ACTH-induced grooming (Gispen and Wiegant 1976; Dunn et al. 1981), there are data to suggest that the primary effect of the peptides is not via opiate receptor activation. Aloyo et al. (1983a) showed that dynorphin1-13 and dynorphin 2-13 induced excessive grooming of a nature similar to that seen following ACTH1-24, yet dynorphin2-13 lacks the N-terminal amino acid tyrosine known to be crucial for opiate receptor activation (Chavkin and Goldstein 1981). It was thus suggested that opiate receptors are involved in the neural substrate at a site different from the primary target of the grooming-inducing ACTH-like peptides.

After the observation that neuroleptics like haloperidol and fluphenasine suppress ACTH-induced grooming (Wiegant et al. 1977), research focussed on the role that central dopaminergic systems play in this behavioral response to ACTH. Studies from various laboratories pointed to an important role of dopaminergic systems in general and the substantia nigra in particular in ACTH-induced excessive grooming in the rat (Wiegant et al. 1977; Traber et al. 1982; Cools et al. 1978; Drago et al. 1981; Chesher and Jackson 1981; Isaacson et al. 1983b).

Recently we have obtained evidence that suggests that the dopaminergic modulation of ACTH-induced excessive grooming is exerted by a GABAergic, striatonigrocollicular pathway (Spruijt et al. 1986a–c). In the course of these studies, it was found that the local injection of ACTH into the periaqueductal gray induced excessive grooming response (Spruijt et al. 1986c). From a variety of evidence, it was concluded that the periaqueductal gray is the primary site of action of ACTH with respect to the induction of excessive grooming behavior and

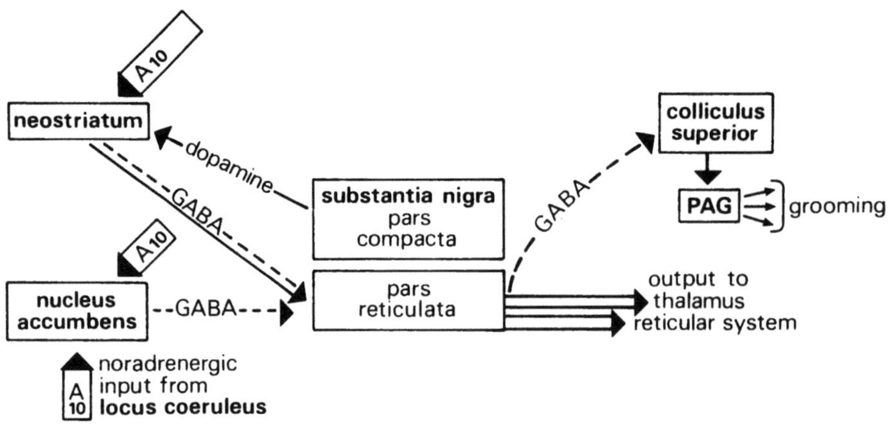

Fig. 4. Scheme of the neural substrate underlying ACTH-induced excessive grooming in the rat. The striato-nigro- and accumbens-nigro-collicular pathways modulate the output of the periaqueductal gray (ACTH-induced excessive grooming). *PAG* periaqueductal gray; *A10* dopaminergic projection from ventral tegmentum

that dopaminergic and GABAergic structures (striatum, colliculus superior) modulate the output of this structure, i.e., the grooming response (Fig. 4).

6 ACTH and Signal Transduction in the Nervous System

As discussed elsewhere (Gispen 1980; Gispen and Zwiers 1985) neuropeptides may act as neurotransmitters, neurohormones, and neuromodulators. In the latter function, they presumably modulate the chemical neurotransmission in certain synapses. Such modulation could take place via several molecular mechanisms involving either pre- or postsynaptic membranes. In the case of ACTH, most research has focussed on the modulation of two transmembrane signal transduction systems, i.e., the receptor-adenylcyclase-cAMP pathway (Wiegant et al. 1986) and the receptor-mediated polyphosphoinositide (PPI) response (Gispen and Zwiers 1985). As the structure-activity relationship for the effects of ACTH on excessive grooming are very similar to that observed for the effects of ACTH on the protein kinase that is part of the PPI response (see below), the significance of this response for ACTH-induced excessive grooming will be considered in detail.

6.1 Receptor-Mediated Polyphosphoinositide Hydrolysis

Activation of various hormone and neurotransmitter receptors results in phosphodiesteratic degradation by a phospholipase C (PLC) of phosphatidylinositol 4,5-bisphosphate (PIP_2) to D-myoinositol 1,4,5-trisphosphate (IP_3) and diacylglycerol (DG). IP_3 can either first be phosphorylated or directly be degraded eventually to D-myoinositol. DG is phosphorylated to phosphatidate and then, through a liponucleotide intermediate, converted to phosphatidylinositol (PI). In two phosphorylation steps, PI is converted via phosphatidylinositol 4-phosphate (PIP) to PIP_2 (Fig. 5; Fisher et al. 1984; Nishizuka 1984a). The relatively low content of PIP_2 and the transient nature of subsequent increases in IP_3, D-myoinositol 1,4-bisphosphate, and D-myoinositol 1-monophosphate and decreases in PIP_2, PIP, and PI suggest that PIP_2 is replenished after its breakdown by PDE. Such replenishment appears to provide a mechanism to control the inositide response (Gispen et al. 1986b).

Both initial PDE products have a second messenger function in the cell (Fig. 5). IP_3 can mobilize calcium from putative endoplasmic reticulum stores (Eichberg and Berti-Mattera 1986). DG can activate protein kinase C (PKC). The activity of PKC is dependent on the presence of an acidic phospholipid such as phosphatidylserine (PS) and is calcium sensitive. DG enhances the calcium sensitivity to PKC to the concentration normally present in an ionized form in the cell (Nishizuka 1984b).

The DG derivative dioctanoylglycerol (DOG) or the phorbol diester phosphoryl 12,13-dibutyrate (PDB; Fig. 5) mimic DG in this respect, but is about 1000-fold more potent and has hence been a useful tool to investigate PKC (Nishizuka 1984b). The original studies with thrombin-stimulated platelets, showing a sequential increase of DG, phosphorylation of a 40-kDa protein (PKC

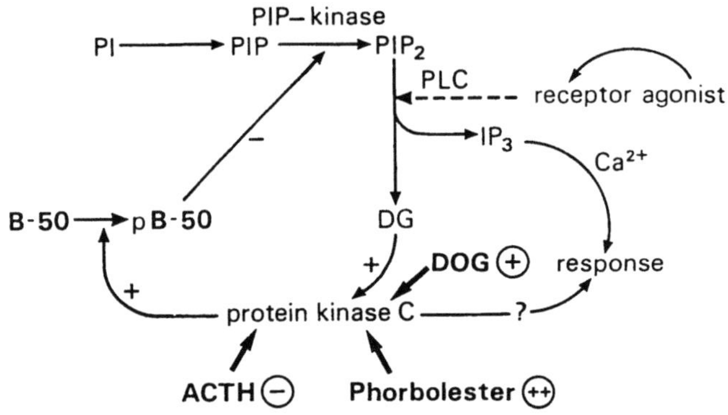

Fig. 5. A working model for the feedback role of phosphoprotein B-50 in receptor-mediated hydrolysis of inositol phospholipids. Activation of the receptor leads to the hydrolysis of PIP_2 by a phospholipase C (PLC) yielding IP_3 (calcium mobilization) and DG. The DG stimulates PKC and this enzyme phosphorylates B-50. Phospho-B-50 inhibits the PIP kinase activity and thus less PIP_2 is available for further receptor-activated hydrolysis. *DOG* dioctanoylglycerol; *PDB* phorbolester

substrate), and secretion of serotonin among others, suggested a coupling of the inositide cycle via protein phosphorylation (PKC) to a cellular response (Nishizuka 1984b). Since then, several studies have provided evidence for the mediating function of PKC in cellular responses. Translocation of PKC from the cytosol to the plasma membrane may be part of its activation mechanism (Kikkawa et al. 1986). Various protein substrates of PKC have been identified, although the function of most of them cannot as yet be defined. One brain phosphoprotein which has been characterized as a substrate of PKC and of which a physiological role has been suggested is B-50.

6.2 Role of the PKC Substrate Protein B-50 in the Polyphosphoinositide Response

The phosphoprotein B-50 (mol. wt., 48 kDa, pI 4.5) is neuron specific and in adult rat brain is predominantly found in presynaptic terminals, presumably associated with the inner side of the plasma membrane and vesicle membranes (Zwiers et al. 1980; Kristjansson et al. 1982; Gispen et al. 1985b). As the endogenous phosphorylation of B-50 is insensitive to cyclic nucleotides but stimulated by calcium, studies were carried out to compare the characteristics of B-50 kinase to the well-known PKC. In view of the similarities in apparent mol. wt., pI, substrate specificity, metal requirements, peptide maps, and sensitivity to modulators, phospholipids, and protease treatment (Aloyo et al. 1982, 1983b), it was concluded that B-50 protein kinase is very similar if not identical to PKC. This is further substantiated by recent findings indicating that DOG and phorbol esters stimulate the phosphorylation of B-50 in synaptic plasma membranes (SPM; De Graan et al. 1986a; Eichberg et al. 1986). These data link B-50 via its B-50 kinase/PKC to the role that enhanced production of DG plays in synaptic membrane function (see Fig. 5).

In a number of chromatographic purification steps, B-50 kinase/B-50 and PIP kinase activity were inseparable, suggesting that they were part of a multifunctional enzyme complex in the presynaptic membrane (Jolles et al. 1980). Under a variety of conditions, it was found that there is a reciprocal relationship between the degree of phosphorylation of B-50 and the activity of PIP kinase (for review see De Graan et al. 1986a). In subsequent experiments, the modulation of PIP kinase activity by the degree of phosphorylation of B-50 was studied in a more direct manner. Partially purified PIP kinase from rat brain cytosol was tested in the presence of added purified B-50 preparations which differed in their degree of phosphorylation. In such a reconstituted system, conditions were found under which phosphorylated B-50 proteins reduced PIP kinase activity while dephosphorylated B-50 did not (Van Dongen et al. 1985). It was suggested, therefore, that the B-50 protein may be an endogenous modulator of PIP kinase activity in rat SPM (Fig. 5).

As discussed above, activation of certain receptors is associated with enhanced hydrolysis of PIP_2. The degraded polyphosphoinositide (PPI) is replenished via PI and PIP, ultimately involving PIP kinase (Downes and Wüstermann 1983). As enhanced phosphorylation of B-50 is accompanied by a decrease in the activity of PIP kinase, it has been proposed that the sequence DG production – PKC activation – B-50 phosphorylation – PIP kinase inhibition may represent a negative feedback control mechanism in the receptor-mediated hydrolysis of inositol phospholipid (Fig. 5; Gispen 1986). In fact, several authors have recently shown that direct activation of PKC by means of phorbol esters reduces the production of inositol phosphates in response to muscarinic receptor activation (Labarca et al. 1984; Vicentini et al. 1985; Orellana et al. 1985).

6.3 ACTH and the Polyphosphoinositide Response

Our interest in the function of B-50 in transmembrane signal transduction processes was raised as the endogenous phosphorylation of B-50 in synaptic plasma membranes was inhibited by ACTH and congeners (Zwiers et al. 1976). Subsequently, it was demonstrated that the inhibition was the result of an inhibition of B-50 kinase (PKC) and not of a stimulation of ATPase or a phosphoprotein phosphatase (Zwiers et al. 1978). Furthermore, the ACTH-induced inhibition of B-50 phosphorylation was, as expected, paralleled by an enhancement of PIP kinase activity (Jolles et al. 1980, 1981; Oestreicher et al. 1983). Thus, it was considered whether part of the behavioral effects of ACTH, i.e., the induction of grooming behavior could be brought about by peptide modulation of signal transduction in certain synapses (Gispen and Zwiers 1985).

Before considering the evidence in favor of this notion, some issues should be considered relating to the question whether the ACTH-induced changes in SPM phosphorylation could be of a physiological nature.

First of all, the doses at which the peptide inhibits PKC in SPM are considerably larger (EC_{50}, 10^{-7} *M*) than those necessary to elicit biological responses in peripheral targets. Secondly, most of our work concerned broken cell preparations rather than intact cells. Furthermore, even if the peptide was operative via PKC inhibition, how is this accomplished across the membrane as PKC most likely is localized at the inner side of cell membrane? These and other issues have led us to propose that the suggested mechanism of action of ACTH on SPM protein phosphorylation would only be of functional significance in brain regions rich in projections of neurons producing melanocortins. For, in such peptidergic synaptic clefts the peptide concentration would be sufficiently high to allow for the train of membrane events as discussed above (Gispen et al. 1985 b). Interestingly, the periaqueductal gray is extremely rich in such peptidergic projections and contains relatively high concentrations of B-50 (Oestreicher et al. 1986). Furthermore, there is evidence that ACTH influences the B-50/phosphatidylinositol cascade in intact cell systems as well (e.g., in hippocampal slices; Schrama et al. 1986; Gispen et al. 1986 a).

There are some considerations which open the possibility of a peptide brain membrane interaction other than classical receptor activation. The N-terminus, known to contain all biological activity of the ACTH molecule (see Sect. 2), has been shown to form an amphipathic helix (Verhallen et al. 1984) which could be inserted in the membrane bilayer upon proper interaction of the positively charged region of the peptide with negatively charged domains in the cell membrane (Gysin and Schwyzer 1984). This insertion of the N-terminus into the bilayer of the membrane itself could conceivably allow for the proposed interaction of ACTH with membrane-bound PKC (Gispen et al. 1985 b). To test whether the proposed molecular mechanism of action was operative in ACTH-induced grooming, we made use of both the behavioral activity of the peptide and the presumed role of B-50 in the feedback control of the inositol-lipid response (Fig. 5).

6.4 Biochemistry of ACTH-Induced Excessive Grooming

First, the proposed feedback role of B-50 in the inositol lipid response was tested in hippocampal slices using the behaviorally active ACTH1-16-NH_2 to inhibit and PDB to stimulate PKC. We confirmed the results of Labarca et al. (1984) and showed that preincubation of hippocampal slices with PDB diminished subsequent receptor-mediated hydrolysis of PIP_2 by incubation with carbamylcholine. More importantly, it was shown that preincubation with ACTH1-16-NH_2, known to inhibit PKC, counteracted the phorbol effect on receptor-mediated PIP_2 hydrolysis. Thus, under appropriate conditions ACTH-like peptides may modulate synaptic transmembrane signal transduction through a reduction of the negative feedback role of the PKC-B-50-PIP kinase loop (Schrama et al. 1986).

There also is a striking similarity in the peptide structure requirements for the induction of excessive grooming in vivo and the inhibition of PKC in SPM in vitro (Fig. 6; Zwiers et al. 1978; Gispen et al. 1979). To our knowledge this is one of the best examples of a correlation between a behavioral and a neurochemical activity of a given peptide. Irrespective of the precise mechanism, the rat brain in situ responds to the information encoded in the ACTH peptide in the same way as the enzyme PKC does when measured in broken cell preparations in vitro.

Finally, there are observations that only phorbol esters known to activate PKC, and not the inactive 4α-phorbol ester, markedly reduce ACTH-induced grooming (Fig. 7; Gispen et al. 1985 a; De Graan et al. 1986 a). Although more and more reports can be expected to show that phorbol esters exert other effects

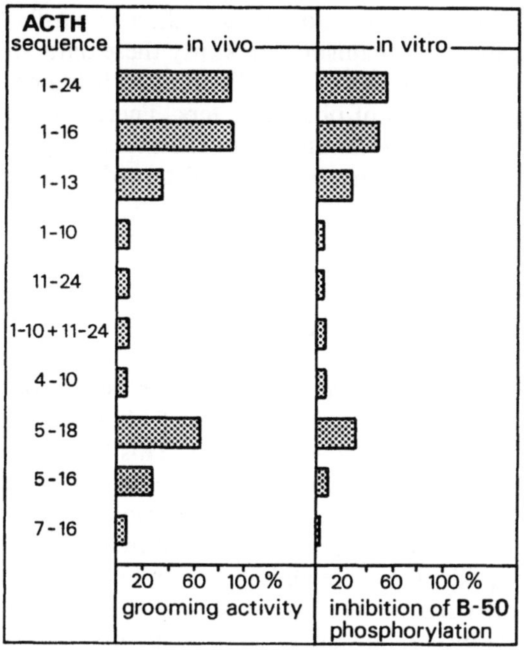

Fig. 6. Structure-activity relations for ACTH-induced excessive grooming and inhibition of PKC

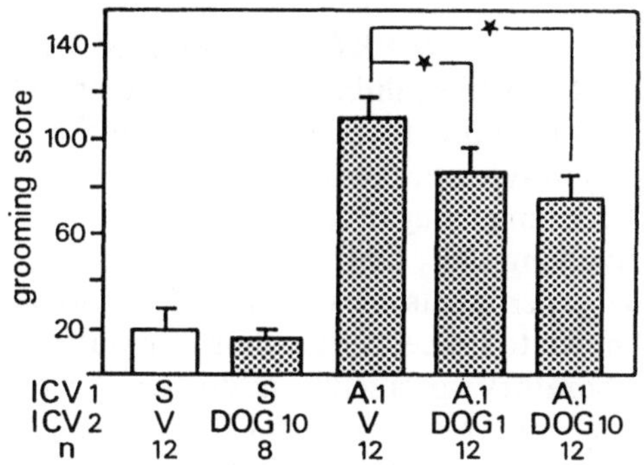

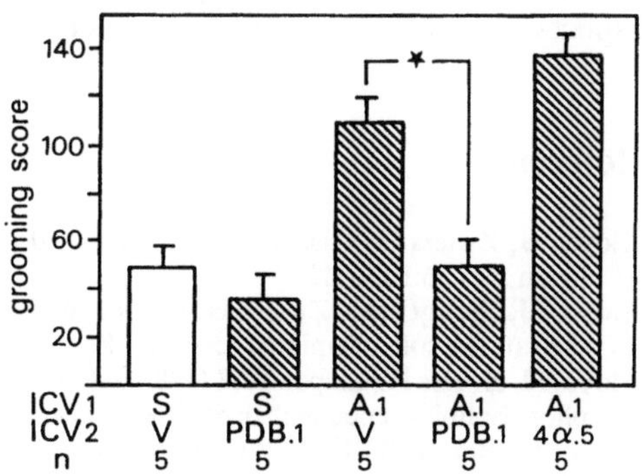

Fig. 7. The effect of DOG and PDB on ACTH-induced excessive grooming in the rat. *ICV*, intraventricular injection; *n*, number of animals; *S*, saline; *V*, vehicle; *A*, 0.1 μg ACTH1-24 i.c.v.; *DOG*, 1 or 10 μg i.c.v.; *PDB*, 0.1 μg i.c.v.; *4α*P, 0.5 μg, 4α-phorbol ester i.c.v.; * denotes significant group differences

than PKC activation alone – there is hardly any drug that acts in vivo through a single mechanism – presently these data are best explained on the basis of the presumed effect of ACTH on PKC being counteracted by the simultaneous administration of phorbol esters. Therefore, it is assumed that the mechanism by which ACTH brings about the excessive grooming response involves a modulation of transmembrane signal transduction in peptidergic synapses in the periaqueductal gray.

7 Concluding Remarks

We have reviewed here molecular aspects of behavioral pharmacology using excessive grooming in the rat as a behavioral assay and the neuropeptide ACTH as the behaviorally active agent. This behavioral response to ACTH can be accurately measured in terms of quality and quantity of the displayed behavior and is useful in studying some of the molecular mechanisms underlying the behavioral response to the peptide. The nature by which information is encoded in an open peptide in itself is ambiguous and undoubtedly the cause of the known multiple and complex interaction of ACTH and congeners with brain function. Nonetheless by carefully studying the structural requirements necessary for the behavioral activity some of the mechanisms by which the brain responds to the peptide became apparent. In the search for a neurochemical correlate of the behavioral activity of the peptide, a neuron-specific protein was identified (B-50), which phosphorylation by PKC was inhibited by ACTH. The peptide-structural requirements for this inhibition were similar to those necessary for the behavioral activity. Therefore, further research on the role that phosphorylation of B-50 would play in synaptic membrane function was undertaken and a model was formulated describing a negative feedback role of B-50 in the transmembrane signal process involving PIP_2 hydrolysis. The predictive value of the model was further tested by experiments in which ACTH-induced grooming was modulated by drugs known to influence the phosphorylation of the B-50 protein. Further work in our laboratory specifically addresses the physiological significance of the peptide-brain interaction in ACTH-induced grooming. Irrespective of the outcome of such studies it appears that the behavioral response has been used successfully as a reliable and relatively simple model system to unravel some of the molecular aspects underlying the behavioral activity of ACTH.

References

Aloyo VJ, Zwiers H, Gispen WH (1982) B-50 protein kinase and kinase C in rat brain. Progr Brain Res 56:303–315

Aloyo VJ, Spruijt BM, Zwiers H, Gispen WH (1983 a) Peptide-induced excessive grooming in the rat: the role of opiate receptors. Peptides 4:833–836

Aloyo VJ, Zwiers H, Gispen WH (1983 b) Phosphorylation of B-50 protein by calcium-activated, phospholipid-dependent protein kinase and B-50 protein kinase. J Neurochem 41:649–653

Barker JL, Smith TG Jr (1979) Three modes of communication in the nervous system. Adv Exp Med Biol 116:3–25

Borchelt PL (1980) Care of the body surface. In: Ray Denny M (ed) Comparative psychology: an evolutionary analysis of animal behavior. Wiley, New York, pp 362–384
Chavkin C, Goldstein A (1981) Specific receptor for the opioid peptide dynorphin: structure-activity relationship. Proc Natl Acad Sci USA 78:6543–6547
Chesher GB, Jackson DM (1981) Swim-induced grooming in mice is mediated by a dopaminergic substrate. J Neurol Transm 50:47–55
Colbern DL, Isaacson RL, Green EJ, Gispen WH (1978) Repeated intraventricular injections of ACTH1-24. The effects of home or novel environments on excessive grooming. Behav Biol 23:381–387
Cools AR, Wiegant VM, Gispen WH (1978) Distinct dopaminergic systems in ACTH-induced grooming. Eur J Pharmacol 50:265–268
De Graan PNE, Oestreicher AB, Schrama LH, Gispen WH (1986a) Phosphoprotein B-50: localization and function. Progr Brain Res 69:37–50
De Graan PNE, Spruijt BM, Eberle AN, Gerard J, Gispen WH (1986b) ACTH1-4 potentiates α-MSH-induced melanophore dispersion excessive grooming. Peptides 7:1–4
De Wied D, Jolles J (1982) Neuropeptides derived from proopiocortin: behavioral physiological and neurochemical effects. Physiol Rev 62:976–1059
Downes CP, Wüsterman M (1983) Breakdown of polyphosphoinositide and phospholipid inositol accounts for muscarinic agonist-stimulated inositol phospholipid metabolism in rat parotid glands. Biochem J 216:633–640
Drago F, Bohus B, Canonico PL, Scapagnini U (1981) Prolactin induces grooming in the rat: possible involvement of nigrostriatal dopaminergic system. Pharmacol Biochem Behav 15:61–63
Dunn AJ, Childers SR, Kramacy NR, Villiger JW (1981) ACTH-induced grooming involves high-affinity opiate receptors. Behav Neural Biol 31:105–109
Eichberg J, Berti-Mattera LV (1986) The role of inositol-phosphate in intracellular calcium mobilization. Progr Brain Res 69:15–28
Eichberg J, De Graan PNE, Schrama LH, Gispen WH (1986) Dioctanoylglycerol and phorbol diesters enhance phosphorylation of phosphoprotein B-50 in native synaptic plasma membranes. Biochem Biophys Res Commun 136:1007–1012
Fentress JC (1977) The tonic hypothesis and the patterning of behavior. Ann NY Acad Sci 290:370–395
Ferrari W, Gessa GL, Vargiu L (1963) Behavioral effects induced by intracisternally injected ACTH and MSH. Ann NY Acad Sci 104:330–345
Fischer SK, Van Rooijen LLA, Agranoff BW (1984) Renewed interest in the polyphosphoinositides. Trends Biochem Sci 9:53–56
Gispen WH (1980) On the neurochemical mechanism of action of ACTH. Progr Brain Res 53:193–206
Gispen WH (1982) Neuropeptides and behaviour: ACTH. Scand J Psychiol [Suppl] 1:16–25
Gispen WH (1986) Phosphoprotein B-50 and phosphoinositides in brain synaptic plasma membranes: a possible feedback relationship. Trans Biochem Soc 14:163–165
Gispen WH, De Wied D (1984) ACTH, structure activity and nervous system effects. In: Hruby VJ, Rich DH (eds) Peptides, structures and function. Proceedings of the 8th American peptide symposium. Pierce, Rockford, pp 399–408
Gispen WH, Isaacson RL (1981) ACTH-induced excessive grooming in the rat. Pharmacol Ther 12:209–246
Gispen WH, Isaacson RL (1986) Excessive grooming in response to ACTH. In: De Wied A, Gispen WH, Van Wimersma Greidanus TjB (eds) Neuropeptides and behavior, vol 1. Pergamon, Oxford, pp 273–312
Gispen WH, Wiegant VM (1976) Opiate antagonists suppress ACTH1-24-induced excessive grooming in the rat. Neurosci Lett 2:159–164
Gispen WH, Zwiers H (1985) Behavioral and neurochemical effects of ACTH. In: Lajtha A (ed) Handbook of neurochemistry, vol 8. Plenum, New York, pp 375–413
Gispen WH, Wiegant VM, Greven HM, De Wied D (1975) The induction of excessive grooming in the rat by intraventricular application of peptides derived from ACTH. Structure-activity studies. Life Sci 17:645–652
Gispen WH, Van Ree JM, De Wied D (1977) Lipotropin and the central nervous system. Int Rev Neurobiol 20:209–250

Gispen WH, Zwiers H, Wiegant VM, Schotman P, Wilson JE (1979) The behaviorally active neuropeptide ACTH as neurohormone and neuromodulator: the role of cyclic nucleotides and membrane phosphoproteins. Adv Exp Med Biol 116:199–224

Gispen WH, Schrama LH, Eichberg J (1985a) Stimulation of protein kinase C reduces ACTH-induced excessive grooming. Eur J Pharmacol 114:399–400

Gispen WH, Leunissen JLM, Oestreicher AB, Verkleij AJ, Zwiers H (1985b) Presynaptic localization of B-50 phosphoprotein: the ACTH-sensitive protein kinase substrate involved in rat brain polyphosphoinositide metabolism. Brain Res 328:381–385

Gispen WH, De Graan PNE, Oestreicher AB, Schotman P, Schrama LH, De Wit M, Spruijt BM (1987) The kinase C substrate protein B-50 and feedback control of synaptic signal transduction. In: Konijn ThM et al. (eds) Molecular mechanisms of desensitization and adaptation to signal molecules. Springer, Berlin Heidelberg New York Tokyo, Vol. 6, pp 197–211

Gispen WH, De Graan PNE, Schrama LH, Van Rooijen LAA (1988) Control mechanism in receptor-mediated polyphosphoinositide hydrolysis. Biochem Pharmacol (to be published)

Gysin B, Schwyzer R (1984) Hydrophobic and electrostatic interactions between adrenocorticotropin-(1-24) tetracosapeptide and lipid vesicles. Amphiphilic primary structures. Biochemistry 23:1811–1818

Isaacson RL, Hannigan JH, Brakkee JH, Gispen WH (1983a) The time course of excessive grooming after stress and after neuropeptide administration. Brain Res Bull 11:289–293

Isaacson RL, Hannigan JH, Springer J, Ryan J, Poplawsky A (1983b) Limbic and neurohormonal influences modulate the basal ganglia and behavior. In: Endröczi E, De Wied D, Angelucci L, Scapagnini U (eds) Integrative neurohumoral mechanisms. Elsevier, Amsterdam, pp 23–34

Izumi K, Donaldson J, Barbeau A (1973) Yawning and stretching in rats induced by intraventricularly administered zinc. Life Sci 12:203–210

Jolles J, Wiegant VM, Gispen WH (1978) Reduced behavioral effectiveness of ACTH(1-24) after a second administration: interaction with opiates. Neurosci Lett 9:261–266

Jolles J, Rompa-Barendregt J, Gispen WH (1979) Novelty and grooming behavior in the rat. Behav Neural Biol 25:563–572

Jolles J, Zwiers H, Van Dongen CJ, Schotman P, Wirtz KWA, Gispen WH (1980) Modulation of brain polyphosphoinositide metabolism by ACTH-sensitive protein phosphorylation. Nature 286:623–625

Jolles J, Zwiers H, Dekker A, Wirtz KWA, Gispen WH (1981) Corticotropin-(1-24)-tetracosapeptide affects protein phosphorylation and polyphosphoinositide metabolism in rat brain. Biochem J 194:283–291

Kikkawa U, Kitano T, Saito N, Fujiwara C, Nishizuka Y (1986) Possible roles of protein kinase C in signal transduction in nervous tissues. Progr Brain Res 69:29–34

Kristjansson GI, Zwiers H, Oestreicher AB, Gispen WH (1982) Evidence that the synaptic phosphoprotein B-50 is localized exclusively in nerve tissue. J Neurochem 39:371–378

Labarca R, Janowsky A, Patel J, Paul SM (1984) Phorbol esters inhibit agonist-induced [^{3}H]-inositol-1-phosphate accumulation in rat hippocampal slices. Biochem Biophys Res Commun 123:703–709

Nishizuka Y (1984a) Turnover of inositol phospholipids and signal transduction. Science 225:1365–1370

Nishizuka Y (1984b) The role of protein kinase C in cell surface signal transduction and tumor promotion. Nature 308:693–697

O'Donohue TL, Handelmann GE, Chaconas T, Miller RL, Jacobowitz DM (1981) Evidence that N-acetylation regulates the behavioral activity of α-MSH in the rat and human central nervous system. Peptides 2:333–344

Oestreicher AB, Van Dongen CJ, Zwiers H, Gispen WH (1983) Affinity-purified anti-B-50 protein antibody: interference with the function of the phosphoprotein B-50 in synaptic plasma membranes. J Neurochem 41:331–340

Oestreicher AB, Dekker LV, Gispen WH (1986) A radioimmunoassay for the phosphoprotein B-50: distribution in rat brain. J Neurochem 46:1366–1369

Orellana SA, Solski PA, Brown JH (1985) Phorbol ester inhibits phosphoinositide hydrolysis and calcium mobilization in cultured astrocytoma cells. J Biol Chem 260:5236–5239

Rosenfeld GC, Burks TR (1977) Single-dose tolerance to morphine hyperthermia in the rat: differentiation of acute from long-term tolerance. J Pharmacol Exp Ther 202:654–659

Schrama LH, De Graan PNE, Eichberg J, Gispen WH (1986) Feedback control of the inositol phospholipid response in rat brain is sensitive to ACTH. Eur J Pharmacol 121:403–404
Schulster D, Schwyzer R (1980) ACTH receptors. In: Schulster D, Levitzki A (eds) Cellular receptors for hormones and neurotransmitters. Wiley, New York, pp 197–217
Schwyzer R (1980) Organization and transduction of peptide information. TIPS 3:327–331
Spruijt B, Gispen WH (1983) ACTH and grooming behaviour in the rat. In: Balthazart J, Pröve E, Gilles R (eds) Hormones and behaviour in higher vertebrates. Springer, Berlin, pp 118–136
Spruijt BM, Gispen WH (1984) Behavioral sequences as an easily quantifiable parameter in experimental studies. Physiol Behav 32:707–710
Spruijt BM, Gispen WH (1984) ACTH and grooming. In: De Wied D, Ferrari W (eds) Central actions of ACTH and related peptides. FIDIA Res Ser Symp in Neurosci, vol 4. Liviana, Padua, pp 179–187
Spruijt BM, De Graan PNE, Eberle AN, Gispen WH (1985) Comparison of structural requirements of α-MSH and ACTH for inducing excessive grooming and pigment dispersion. Peptides 6:1185–1189
Spruijt BM, Cools AR, Ellerbroek BA, Gispen WH (1986a) The colliculus superior modulates ACTH-induced excessive grooming. Life Sci 39:461–470
Spruijt BM, Cools AR, Ellerbroek BA, Gispen WH (1986b) Dopaminergic modulation of ACTH-induced grooming. Eur J Pharmacol 120:249–256
Spruijt BM, Cools AR, Gispen WH (1986c) The periaqueductal gray: a prerequisite for ACTH-induced excessive grooming. Behav Brain Res 20:19–25
Terenius L (1976) Somatostatin and ACTH are peptides with partial antagonist-like selectivity for opiate receptors. Eur J Pharmacol 38:211–213
Traber J, Kleine HR, Gispen WH (1982) Actions of antidepressant and neuroleptic drugs on ACTH and novelty-induced behavior in the rat. Eur J Pharmacol 80:407–414
Traber J, Spencer D, Glaser T, Gispen WH (1987) Activity of psycho-active drugs on ACTH and novelty-induced behavior in the rat. Ann NY Acad Sci USA (to be published)
Van Dongen CJ, Zwiers H, De Graan PNE, Gispen WH (1985) Modulation of the activity of purified phosphatidylinositol 4-phosphate kinase by phosphorylated and dephosphorylated B-50 protein. Biochem Biophys Res Commun 8:1219–1227
Verhallen PJE, Demel RA, Zwiers H, Gispen WH (1984) Adrenocorticotropic hormone (ACTH)-lipid interactions. Implications for involvement of amphipathic helix formation. Biochim Biophys Acta 775:246–254
Vincentini LM, Di Virgilio F, Ambrosini A, Pozzan T, Meldolesi J (1985) Tumor promoter phorbol 12-myristate,13-acetate inhibits phosphoinositide hydrolysis and cytosolic calcium rise induced by the activation of muscarinic receptors in PC12 cells. Biochem Biophys Res Commun 127:310–317
Wiegant VM, Gispen WH (1977) ACTH-induced excessive grooming in the rat: latent activity of ACTH4-10. Behav Biol 19:554–558
Wiegant VM, Cools AR, Gispen WH (1977) ACTH-induced excessive grooming involves brain dopamine. Eur J Pharmacol 41:343–345
Wiegant VM, Jolles J, Gispen WH (1978) β-Endorphin grooming in the rat: single-dose tolerance. Dev Neurosci 4:447–450
Wiegant VM, Zwiers H, Oestreicher AB, Gispen WH (1986) ACTH and brain cAMP and phosphoproteins. In: De Wied D, Gispen WH, Van Wimersma Greidanus TjB (eds) Neuropeptides and behavior, vol 1. Pergamon, Oxford, pp 189–210
Zwiers H, Veldhuis D, Schotman P, Gispen WH (1976) ACTH, cyclic nucleotides and brain protein phosphorylation in vitro. Neurochem Res 1:669–677
Zwiers H, Wiegant VM, Schotman P, Gispen WH (1978) ACTH-induced inhibition of endogenous rat brain protein phosphorylation in vitro: structure activity. Neurochem Res 3:455–463
Zwiers H, Schotman P, Gispen WH (1980) Purification and some characteristics of an ACTH-sensitive protein kinase and its substrate protein in rat brain membranes. J Neurochem 34:1689–1699
Zwiers H, Aloyo VJ, Gispen WH (1981) Behavioral and neurochemical effects of the new opioid peptide dynorphin-(1-13): comparison with other neuropeptides. Life Sci 28:2545–2551

Subject Index

MIX
Papier aus verantwortungsvollen Quellen
Paper from responsible sources
FSC® C105338

If you have any concerns about our products,
you can contact us on
ProductSafety@springernature.com

In case Publisher is established outside the EU,
the EU authorized representative is:
Springer Nature Customer Service Center GmbH
Europaplatz 3, 69115 Heidelberg, Germany

Printed by Libri Plureos GmbH
in Hamburg, Germany